AF598081

Methods in Molecular Biology™

Series Editor
John M. Walker
School of Life Sciences
University of Hertfordshire
Hatfield, Hertfordshire, AL10 9AB, UK

For further volumes:
http://www.springer.com/series/7651

Capillary Electrophoresis of Biomolecules

Methods and Protocols

Edited by

Nicola Volpi and Francesca Maccari

Department of Life Sciences, University of Modena and Reggio Emilia, Modena, Italy

Editors
Nicola Volpi
Department of Life Sciences
University of Modena and Reggio Emilia
Modena, Italy

Francesca Maccari
Department of Life Sciences
University of Modena and Reggio Emilia
Modena, Italy

ISSN 1064-3745 ISSN 1940-6029 (electronic)
ISBN 978-1-62703-295-7 ISBN 978-1-62703-296-4 (eBook)
DOI 10.1007/978-1-62703-296-4
Springer New York Heidelberg Dordrecht London

Library of Congress Control Number: 2013930231

Printed on acid-free paper

Humana Press is a brand of Springer
Springer is part of Springer Science+Business Media (www.springer.com)

Preface

Capillary electrophoresis (CE) is a relatively new separation technique suitable for handling small amounts of sample very important in bioanalytical research and in various clinical, diagnostic, genetic, and forensic applications. CE offers several similarities to high-performance liquid chromatography (HPLC) such as ease of use, high resolution, speed, online detection, and full automation capability. CE encompasses a family of related separation techniques that use narrow-bore fused-silica capillaries to separate a complex array of large and small molecules. High electric field strengths are used to separate molecules with differences in charge, size, and hydrophobic properties. CE may be utilized according to several separation techniques:

1. Capillary zone electrophoresis (CZE) is the simplest form of CE where the separation mechanism is based on differences in the charge-to-mass ratio of the analytes.
2. Capillary gel electrophoresis (CGE) is the adaptation of traditional gel electrophoresis into the capillary by using soluble polymers to create a replaceable molecular sieve allowing size separations.
3. Capillary isoelectric focusing (CIEF) allows amphoteric molecules, proteins, to be separated in a pH gradient generated between the cathode and anode.
4. Isotachophoresis (ITP) is a focusing technique based on the migration of compounds between leading and terminating electrolytes.
5. Micellar electrokinetic capillary chromatography (MECC or MEKC) is a mode of separation in which surfactants are added to the buffer solution at concentrations that form micelles. This technique is useful to resolve both charged and neutral compounds.
6. Microemulsion electrokinetic chromatography (MEEKC) is a technique in which solute partition takes place between moving oil droplets and the aqueous buffer. This allows the separation of both aqueous and water-insoluble compounds.
7. Nonaqueous capillary electrophoresis (NACE) involves the separation of analytes in nonaqueous media that allow additional selectivity options in method development. It is valuable for separations of water-insoluble compounds and for hyphenation with MS detection.
8. Capillary electrochromatography (CEC) is a hybrid separation method that couples the high separation efficiency of CZE with HPLC and uses an electric field rather than hydraulic pressure to propel the mobile phase through a packed bed.

Due to its high resolving power and sensitivity, CE has been applied in the analysis of simple and complex (macro) molecules providing concentration and structural characterization data essential for understanding their biological functions. Although CE technology may be applied to many different types of research, it has gained its reputation from the study of molecules that have traditionally been difficult to separate. In general, CE should be considered first when dealing with highly polar, charged analytes. In fact, CE excels in the analysis of ions when rapid results are desired, and has become the predominant technique for the analysis of both basic and chiral pharmaceuticals. This technology is replacing traditional electrophoresis for the characterization and analysis of macromolecules such as nucleic acids, proteins, and carbohydrates, and promises to be a valuable tool in tackling the characterization challenges posed by proteome-wide analysis and DNA sequencing and genotyping.

This volume on the capillary electrophoresis of biomolecules provides the reader with the latest breakthroughs and improvements in CE and CE techniques applied to several classes of bio(macro)molecules, in particular simple and complex carbohydrates (polysaccharides), amino acids, peptides and proteins, enzymes, and nucleic acids. Along with practical procedures, reviews discussing CE applications related to bio(macro)molecules are also included.

I would like to thank all the contributors for their articles, able to provide a better understanding of the analytical phenomena related to CE and by widening the scope of their possible applications. Acknowledgement is due to Humana Press Editors for their assistance in bringing this issue to publication.

Modena, Italy ***Nicola Volpi***

Contents

Contributors

MITSURU ABO • *Yokohama College of Pharmacy, Yokohama, Japan*
HANS ABROMEIT • *Department of Pharmaceutical Chemistry, School of Pharmacy, Friedrich Schiller University, Jena, Germany*
AMAYA ALBALAT • *BHF Glasgow Cardiovascular Research Centre, Institute of Cardiovascular and Medical Sciences, College of Medical Veterinary and Life Sciences, University of Glasgow, Glasgow, UK*
BENJAMIN J. ALPER • *Department of Biochemistry, St. Jude Children's Research Hospital, Memphis, TN, USA*
NOBUHIRO ANRAKU • *Department of Electric and Electronic Information Engineering, Toyohashi University of Technology, Toyohashi, Japan*
VASILIKI BITSIKA • *Biotechnology Division, Biomedical Research Foundation, Academy of Athens, Athens, Greece*
LUCAS BLANES • *School of Chemistry and Forensic Science, University of Technology, Sydney, NSW, Australia*
M.C. BREADMORE • *Australian Center for Research on Separation Science, School of Chemistry, University of Tasmania, Hobart, TAS, Australia*
CIRIACO CARRU • *Department Biomedical Sciences, University of Sassari, Sassari, Italy*
JULIE M. CHANDLER • *Department of Pathology, McGowan Institute for Regenerative Medicine, University of Pittsburgh School of Medicine, Pittsburgh, PA, USA*
PO-LING CHANG • *Department of Chemistry, Tunghai University, Taitung, Taiwan*
YUQING CHANG • *Department of Chemistry and Chemical Biology, Center for Biotechnology and Interdisciplinary Studies, Rensselaer Polytechnic Institute, Troy, NY, USA*
ALEJANDRO CIFUENTES • *Laboratory of Foodomics, CIAL, CSIC, Madrid, Spain*
ANA S. COROADINHA • *Instituto de Tecnologia Química e Biológica, Universidade Nova de Lisboa, Oeiras, PortugalIBET, Oeiras, Portugal*
FRÉDÉRIC COTTON • *Department of Clinical Chemistry, Erasme Hospital, Université Libre de Bruxelles, Brussels, Belgium*
FRANÇOIS DE L'ESCAILLE • *R&D DIAG, Analis SA, Suarlée, Belgium*
MERCEDES DE FRUTOS • *Institute of Organic Chemistry (IQOG, CSIC), Madrid, Spain*
CLAUDIMIR LUCIO DO LAGO • *Departamento de Química Fundamental, Instituto de Química, Universidade de São Paulo, São Paulo, SP, Brazil*
LUCA DEIANA • *Department of Biomedical Sciences, University of Sassari, Sassari, Italy*
JOSE CARLOS DIEZ-MASA • *Institute of Organic Chemistry (IQOG-CSIC), Madrid, Spain*
JEAN BERNARD FALMAGNE • *R&D DIAG, Analis SA, Suarlée, Belgium*
YI FAN • *Department of Chemistry, University of Illinois at Urbana-Champaign, Urbana, USA*
XUEPING FANG • *Calibrant Biosystems, Rockville, MD, USA*

PAOLO FATTORINI • *Department of Medicine, surgery and Health, University of Trieste, Trieste, Italy*
RAUL GARRIDO-MEDINA • *Institute of Organic Chemistry (IQOG-CSIC), Madrid, Spain*
JOHN F. GILMER • *School of Pharmacy and Pharmaceutical Sciences, Trinity College, Dublin, Ireland*
RAMON GONZALEZ • *Instituto de Ciencias de la Vid y del Vino (CSIC-UR-CAR), Logroño, Spain*
PIERANGELA GRIGNANI • *Department of Legal Medicine, Forensic and Pharmaco-Toxicological Sciences, University of Pavia, Pavia, Italy*
BÉATRICE GULBIS • *Department of Clinical Chemistry, Hôpital Erasme, Université Libre de Bruxelles, Brussels, Belgium*
ANDRAS GUTTMAN • *Horváth Laboratory of Bioseparation Sciences, University of Debrecen, Debrecen, Hungary*
LATANYA HAMMONDS-ODIE • *School of Science & Technology, Georgia Gwinnett College, Lawrenceville, GA, USA*
TOSHIAKI HATTORI • *Department of Electric and Electronic Information Engineering, Toyohashi University of Technology, Toyohashi, Japan*
LI-PING HE • *Department of Medical Statistics and Epidemiology, School of Public Health, Sun Yat-sen University, Guangzhou, P.R. China*
MARTINA HENSE • *Department of Pharmaceutical Chemistry, School of Pharmacy, University of Jena, Jena, Germany*
MING-MU HSIEH • *Department of Chemistry, National Kaohsiung Normal University, Kaohsiung, Taiwan*
JINGWU KANG • *Shanghai Institute of Organic Chemistry, Chinese Academy of Sciences, Shanghai, P.R. China*
VÁCLAV KAŠIČKA • *Institute of Organic Chemistry and Biochemistry, Academy of Sciences of the Czech Republic, Prague, Czech Republic*
RYO KATO • *Cooperative Research Facility Center, Toyohashi University of Technology, Toyohashi, Japan*
TOMOYUKI KATSUBE-TANAKA • *Graduate School of Agriculture, Kyoto University, Kyoto, Japan*
JOANA LAMEGO • *Instituto de Tecnologia Química e Biológica, Universidade Nova de Lisboa, Oeiras, PortugalIBET, Oeiras, Portugal*
CHENG S. LEE • *Department of Chemistry and Biochemistry, University of Maryland, College Park, MD, USA*
ROBERT J. LINHARDT • *Department of Chemistry and Chemical Biology, Center for Biotechnology and Interdisciplinary Studies, Rensselaer Polytechnic Institute, Troy, NY, USA*
CHAO LIU • *Shanghai Institute of Organic Chemistry, Chinese Academy of Sciences, Shanghai, P.R. China*
GIORGIO MARRUBINI • *Department of Pharmaceutical Chemistry, University of Pavia, Pavia, Italy*
KIARACH MESBAH • *Laboratory of Proteins and Nanotechnologies in Separation Sciences, Faculté de Pharmacie, University of Paris-Sud, UMR-CNRS 8612, Châtenay-Malabry, France*

Mehdi Moini • *Museum Conservation Institute, Smithsonian Institution, Suitland, MD, USA*
William Mullen • *BHF Glasgow Cardiovascular Research Centre, Glasgow, UK*
Simon M. Mwongela • *School of Science and Technology, Georgia Gwinnett College, Lawrenceville, GA, USA*
Reine Nehmé • *Institut de Chimie Organique et Analytique (ICOA), Université d'Orléans, UMR CNRS 7311, Orléans, France*
Thiago Nogueira • *Departamento de Quimica Fundamental, Instituto de Quimica, Universidade de São Paulo, São Paulo, SP, Brazil*
Akira Okubo • *Yokohama College of Pharmacy, Yokohama, Japan*
Cristina Pelaez-Lorenzo • *Institute of Organic Chemistry (IQOG-CSIC), Madrid, SpainEuropean University of Madrid, Madrid, Spain*
Catherine Perrin • *Institut des Biomolécules Max Mousseron (IBMM), UMR 5247, Université Montpellier 1-CNRS, Montpellier, France*
Carlo Previderé • *Department of Legal Medicine, Forensic and Pharmaco-Toxicological Sciences, University of Pavia, Pavia, Italy*
Angel Puerta • *Institute of Organic Chemistry (IQOG-CSIC), Madrid, Spain*
Zuly Rivera-Monroy • *Horváth Laboratory of Bioseparation Sciences, University of Debrecen, Debrecen, HungaryNational University of Colombia, Bogota, Colombia*
Renata Mayumi Saito • *Departamento de Química Fundamental, Instituto de Química, Universidade de São Paulo, São Paulo, SP, Brazil*
Kae Sato • *Yokohama College of Pharmacy, Yokohama, Japan*
Walter K. Schmidt • *Department of Biochemistry and Molecular Biology, University of Georgia, Athens, GA, USA*
Gerhard K.E. Scriba • *Department of Pharmaceutical Chemistry, School of Pharmacy, Friedrich Schiller University, Jena, Germany*
Carolina Simó • *Laboratory of Foodomics, CIAL, CSIC, Madrid, Spain*
Ana L. Simplício • *Instituto de Tecnologia Química e Biológica, Universidade Nova de Lisboa, Oeiras, PortugalIBET, Oeiras, Portugal*
Justyna Siwy • *Mosaiques diagnostics and therapeutics, Hannover, Germany*
Solange Sorçaburu-Cigliero • *Department of Medicine, Surgery and Health, University of Trieste, Trieste, Italy*
Salvatore Sotgia • *Department of Biomedical Sciences, University of Sassari, Sassari, Italy*
Myriam Taverna • *Laboratory of Proteins and Nanotechnologies in Separation Sciences, Faculté de Pharmacie, University of Paris-Sud, UMR-CNRS 8612, Châtenay-Malabry, France*
Heather R. Trenary • *Kremers Urban Pharmaceuticals Inc., Seymour, IN, USA*
Maria Teresa Veledo • *Institute of Organic Chemistry (IQOG, CSIC), Madrid, Spain PharmaMar, Madrid, Spain*
Romain Verpillot • *Laboratory of Proteins and Nanotechnologies in Separation Sciences, Faculté de Pharmacie, University of Paris-Sud, UMR-CNRS 8612, Châtenay-Malabry, France*
Gary R. Walker • *Center for Applied Chemical Biology, Youngstown State University, One University Plaza, Youngstown*

CHENCHEN WANG • *Department of Chemistry and Biochemistry, University of Maryland, College Park, MD, USA*
AMANDA WEYERS • *Department of Chemistry and Chemical Biology, Center for Biotechnology and Interdisciplinary Studies, Rensselaer Polytechnic Institute, Troy, NY, USA*
FLEUR WOLFF • *Laboratory of Clinical Chemistry, CHU Brugmann, Brussels, Belgium*
BO YANG • *Key Laboratory of Urbanization and Ecological Restratio, Department of Environmental Science, East China Normal University, Shanghai, China*
PHILIP M. YANGYUORU • *Department of Chemistry, Kent State University, Kent, OH, USA*
JOSEPH ZAIA • *Department of Biochemistry, Center for Biomedical Mass Spectrometry, Boston University Medial Campus, Boston, MA, USA*
QIANQIAN ZHANG • *Shanghai Institute of Organic Chemistry, Chinese Academy of Sciences, Shanghai, P.R. China*
ANGELO ZINELLU • *Department Biomedical Sciences, University of Sassari, Sassari, Italy*
PETRA ZURBIG • *Mosaiques diagnostics and therapeutics, Hannover, Germany*

Chapter 1

Recent Advances in Capillary Electrophoresis-Based Proteomic Techniques for Biomarker Discovery

Chenchen Wang, Xueping Fang, and Cheng S. Lee

Abstract

Due to the inherent disadvantage of biomarker dilution in complex biological fluids such as serum/plasma, urine, and saliva, investigative studies directed at tissues obtained from the primary site of pathology probably afford the best opportunity for the discovery of disease biomarkers. Still, the large variation of protein relative abundances with clinical specimens often exceeds the dynamic range of currently available proteomic techniques. Furthermore, since the sizes of human tissue biopsies are becoming significantly smaller due to the advent of minimally invasive methods and early detection and treatment of lesions, a more effective discovery-based proteomic technology is critically needed to enable comprehensive and comparative studies of protein profiles that will have diagnostic and therapeutic relevance.

This review therefore focuses on the most recent advances in capillary electrophoresis-based single and multidimensional separations coupled with mass spectrometry for performing comprehensive proteomic analysis of clinical specimens. In addition to protein identification, monitoring quantitative changes in protein expression is essential for the discovery of disease-associated biomarkers. Comparative proteomics involving measurements in changes of biological pathways or functional processes are further expected to provide relevant markers and networks, molecular relationships among different stages of disease, and molecular mechanisms that drive the progression of disease.

Key words: Biomarker discovery, Capillary electrophoresis, Mass spectrometry, Reversed-phase liquid chromatography, Tissue proteomics

1. Single-Dimension Capillary Electrophoretic Coupled with Mass Spectrometry

Capillary zone electrophoresis (CZE) resolves proteins and peptides based on their differences in electrophoretic mobility which is a function of the charge-to-size ratio. Due to its high throughput and excellent resolving power, the coupling of CZE with electrospray ionization-mass spectrometry (ESI-MS) has been employed for the analysis of low-molecular-weight proteins (below 20 kDa)

Nicola Volpi and Francesca Maccari (eds.), *Capillary Electrophoresis of Biomolecules: Methods and Protocols*, Methods in Molecular Biology, vol. 984, DOI 10.1007/978-1-62703-296-4_1, © Springer Science+Business Media, LLC 2013

and peptides for the discovery of biomarkers in human urine. Samples were investigated from patients suffering from a variety of diseases including ureteropelvic junction obstruction (1), cancer (2–4), vasculitis (5), coronary artery diseases (6–8), kidney diseases (9–11), lithium-induced nephropathy (12), graft-versus-host disease (13), and diabetes (8).

The CZE–ESI-MS data were presented by plotting the measured molecular masses against their migration times and compared among healthy and diseased patients. Known and potential new urine biomarkers have been identified using subsequent MS/MS experiments (14, 15). Although a variety of different proteins/peptides were discovered, most of these putative markers are derived from the most abundant proteins in the body such as collagen—mainly, types I, II, and III, albumin, β-2-macroglobulin, and uromodulin (16).

In addition to human urine, single-dimension CZE separation was also employed for the proteomic analysis of other body fluids such as human plasma (17) and ventricular cerebrospinal fluid (18). Potential biomarkers of vascular disease in plasma from patients with chronic kidney disease were discovered by CZE–ESI--MS (17). In contrast to the use of ESI-MS or ESI-MS/MS for the detection and identification of protein and peptide markers, off-line matrix-assisted laser desorption/ionization-time of flight/time of flight (MALDI-TOF/TOF) MS coupled with iTRAQ labeling (19) was employed for multiplexed quantification of proteins in human ventricular cerebrospinal fluid samples collected from a patient with traumatic brain injury during patient recovery (18).

2. Capillary Electrophoresis-Based Multidimensional Separation Coupled with Mass Spectrometry

2.1. Capillary Isoelectric Focusing

Combined capillary isoelectric focusing (CIEF)/nano-reversed phase liquid chromatography (nano-RPLC) separations have been developed and demonstrated to achieve comprehensive and ultrasensitive analysis of minute protein digests extracted from microdissected tissue specimens (20, 21). In addition to protein identification, the capabilities of the CIEF-based proteomic platform coupled with the spectral counting approach (22, 23) to confidently and reproducibly quantify proteins and changes in protein expression levels among samples were evaluated by the measurements of coefficient of variation (CV) and the Pearson correlation coefficient (24). Analytical reproducibility of relative protein abundance was determined to exhibit a Pearson R^2 value greater than 0.99 and a CV of 14.1%. The platform was capable of measuring changes in protein expression as low as 1.5-fold with confidence as determined by *t*-test followed by Benjamini–Hochberg multiple testing adjustments.

The protein expression profiles from two distinct ovarian endometrioid tumor-derived cell lines have been compared using CIEF-based multidimensional separations coupled with ESI-MS/MS (25). Differentially expressed proteins were further investigated by ingenuity pathway analysis to reveal their association with important biological functions and signaling pathways such as the P13K/AKT pathway. The results illustrated the utility of high-throughput proteomic profiling combined with bioinformatic tools to provide insights into the mechanisms of deregulation in neoplastic cells.

In addition to CIEF, microscale in solution IEF was employed as the first separation dimension for the fractionation of intact proteins, followed by tryptic digestion and subsequent CZE peptide separation coupled off-line to MALDI-TOF/TOF MS (26). The platform was used for the analysis of human follicular fluid with clinical implication. A total of 73 unique proteins were identified, including mostly acute phase proteins and proteins that are known to be extensively involved in follicular development.

2.2. Transient Capillary Isotachophoresis/ Capillary Zone Electrophoresis

Besides proteome complexity, the greatest bioanalytical challenge facing comprehensive proteomic analysis, particularly in the identification of low-abundance proteins, is related to the large variation of protein relative abundances. For example, the protein concentration dynamics range from 10^6-fold in cells to 10^{12}-fold in blood (27, 28). In contrast to universally enriching all analytes by a similar degree, the result of the capillary isotachophoresis (CITP) stacking process is that major components may be diluted, but trace compounds are concentrated (29). Furthermore, CITP offers the benefits of speed and straightforward manipulation/switching between the stacking and separation modes in transient CITP/CZE. Transient CITP/CZE further provides seamless combination with nano-RPLC (Fig. 1) as two highly resolving and completely orthogonal separation techniques critically needed for analyzing complex proteomes (30, 31).

While CITP has been widely used for analyte preconcentration prior to electrophoretic separation, the application of CITP to selectively enrich trace amounts of proteins/peptides in complex proteome mixtures represents a completely untapped avenue in proteomic technology development. As illustrated by Fang and coworkers for the analysis of human saliva and mouse brain mitochondrial proteomes (30, 31), the application of CITP to selectively enrich trace amounts of proteins in targeted cells significantly augments the ability to perform ultrasensitive and global proteomic analysis, particularly toward the identification of low-abundance proteins.

2.2.1. Proteomic Analysis of Glioblastoma Multiforme-Derived Cancer Stem Cells

The ultrahigh resolving power of transient CITP/CZE as the first separation dimension has been demonstrated by significantly low peptide fraction overlapping for the analysis of protein expression within glioblastoma multiforme-derived cancer stem cells (32).

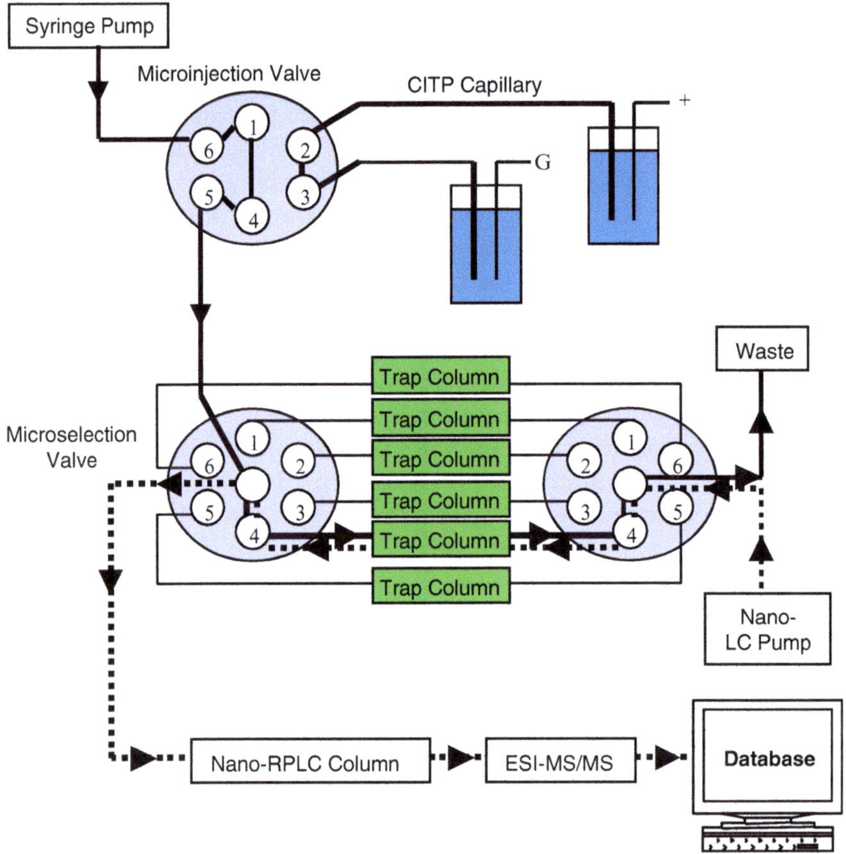

Fig. 1. Schematic of on-line integration of CITP with nano-RPLC for achieving selective analyte enrichment and multidimensional proteome separation. Solid and dashed lines represent the flow paths for the loading of CITP fractions and the injection of fractions into a nano-RPLC column, respectively.

Approximately 89% of distinct peptides were identified in only a single CITP fraction. In contrast, a high degree of peptide overlapping in strong cation-exchange (SCX) chromatography, as the first separation dimension of the multidimensional protein identification technique (MuDPIT) (33), was observed with at least 40% of carry-over peptides that were identified in previous salt gradients. A high degree of peptide overlap in SCX unnecessarily burdens the subsequent nano-RPLC separation and greatly reduces the overall peak capacity in a multidimensional separation system. The presence of high-abundance peptides in multiple SCX fractions further negatively impacts the selection of low-abundance peptides for tandem MS identification.

For the evaluation of overall cancer stem cells' proteome performance (32), the CITP proteomic platform demonstrated significant enhancements in total peptide, distinct peptide, and distinct protein identifications over a corresponding MuDPIT run by 119, 192, and 79%, respectively. The CITP proteomic technology,

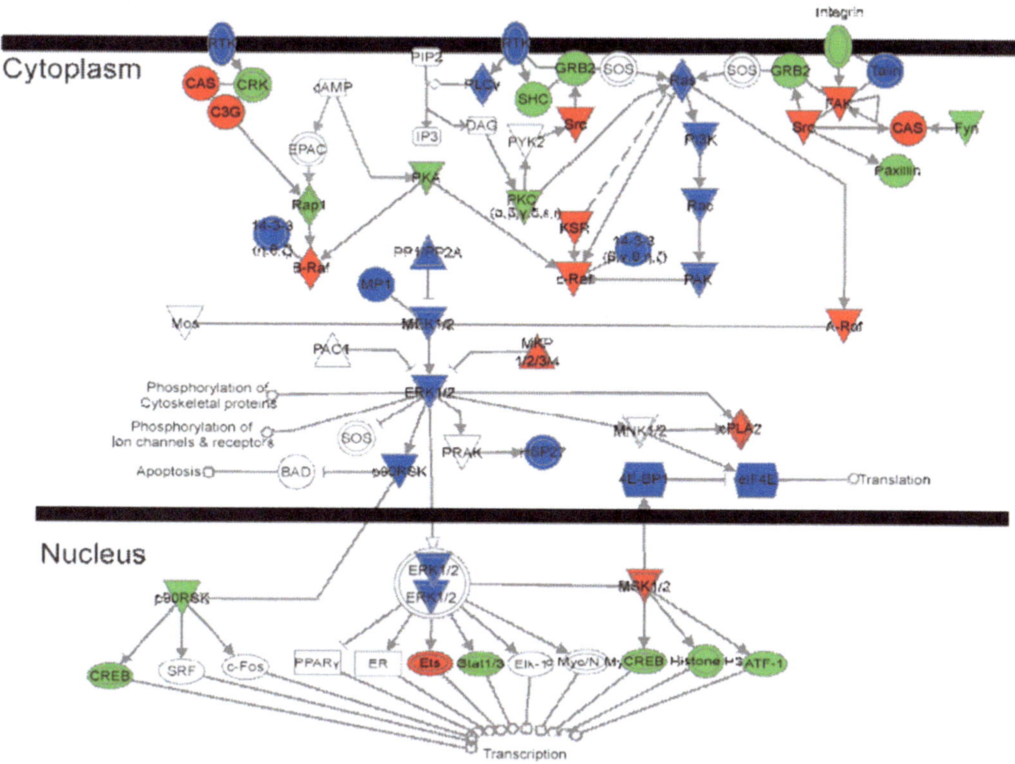

Fig. 2. Comparison of the coverage in the ERK/MAPK pathway achieved by the MuDPIT and CITP proteomic platform. *Red*: proteins only identified by CITP; *blue*: proteins identified by CITP with higher confidence (larger numbers of spectral counts and distinct peptide identifications per protein) than that achieved by MuDPIT; *green*: proteins identified by both CITP and MuDPIT with approximately equal confidence (32).

equipped with selective analyte enrichment and ultrahigh resolving power, further accomplished superior coverage in key pathways than that of the MuDPIT. For example, many biologically relevant proteins, including MKP, the Raf family, and Src in the ERK/MAPK pathway, were only identified by the CITP technology (Fig. 2).

2.2.2. Tissue Proteomic Studies of Primary and Recurrent Ovarian Tumors

Ovarian cancer is the most lethal malignant gynecologic neoplasm and serous carcinoma represents the most common type of ovarian cancer. Most patients with ovarian serous carcinoma are diagnosed at advanced stages and the patients routinely undergo debulking surgery followed by a chemotherapy regimen. Although chemotherapeutic agents have improved patients' 5-year survival rate, the overall mortality of ovarian cancer has remained largely unchanged because most patients eventually develop recurrent carcinomas after chemotherapy and succumb to their disease. Thus, it is widely accepted that recurrent tumors represent the true "killer" as the primary tumors are always removed by surgery or local irradiation.

Despite considerable efforts aimed at elucidating the molecular mechanisms of chemoresistance in ovarian carcinoma, its pathogenesis is still unknown due to the lack of technologies to globally identify the changes in protein expression secondary to chemoresistance. Jinawath and coworkers (34) have applied the CITP proteomic technology to perform comparative proteomic analysis of paired primary and recurrent post-chemotherapy ovarian high-grade serous carcinomas from nine ovarian cancer patients. A total of 30,884 distinct peptides were identified using a 0.1% false discovery rate for total peptide identifications, leading to the identification of 5,063 nonredundant proteins from the Swiss-Prot human database. Of the proteins identified from primary ovarian tumor cells, a number of them, including merlin, ezrin, moesin (35), selenium-binding protein 1 (36), glutathione S-transferase, epidermal growth factor receptor, E-cadherin, α- and β-catenins (37), β-tubulin, ubiquitin carboxyl-terminal hydrolase, glyoxalase 1, F-actin α-subunit, and cofilin (38), have been reported to be present in ovarian epithelial cells.

Furthermore, the increase in ovarian cancer proteome coverage, attributed to CITP-based selective analyte enrichment, allowed the application of protein network and pathway analysis toward the discovery of ovarian carcinoma biomarkers. For example, low-abundance proteins such as cytokine IL-6 and signal transducer and activator of transcription 3 (STAT3), as well as many other proteins known to participate in the IL-6 signaling pathway, have been identified and compared for their expression levels within primary and recurrent ovarian tumors (Fig. 3). Both STAT3 and IL-6 were found to be over-expressed in the recurrent tumor cells. As supported by recent in vitro studies of ovarian (39) and non-small cell lung (40) cancer cell lines, the STAT3 pathway plays a significant role in the development of high-grade ovarian cancer and the drug-resistant ovarian cancer. STAT3 can be activated by various protein tyrosine kinases including epidermal growth factor receptor, interleukin, and IFN ligands. Inhibition of the STAT3 pathway has been shown to enhance paclitaxel-induced apoptosis. In addition, IL-6 expression was found to be associated with the generation of drug resistance.

The comparative proteomic results have further identified RELA which is the p65 subunit of the NF-κB complex (Fig. 3). p65 was over-expressed more than three folds in recurrent tumors as compared to the primary tumors. The NF-κB/RELA family of transcription factors is one of the most important and well-characterized signaling pathways in both normal and pathological conditions. It controls a variety of cellular functions including inflammatory and immune responses, cell growth and survival, and drug resistance to several chemotherapeutic agents (41).

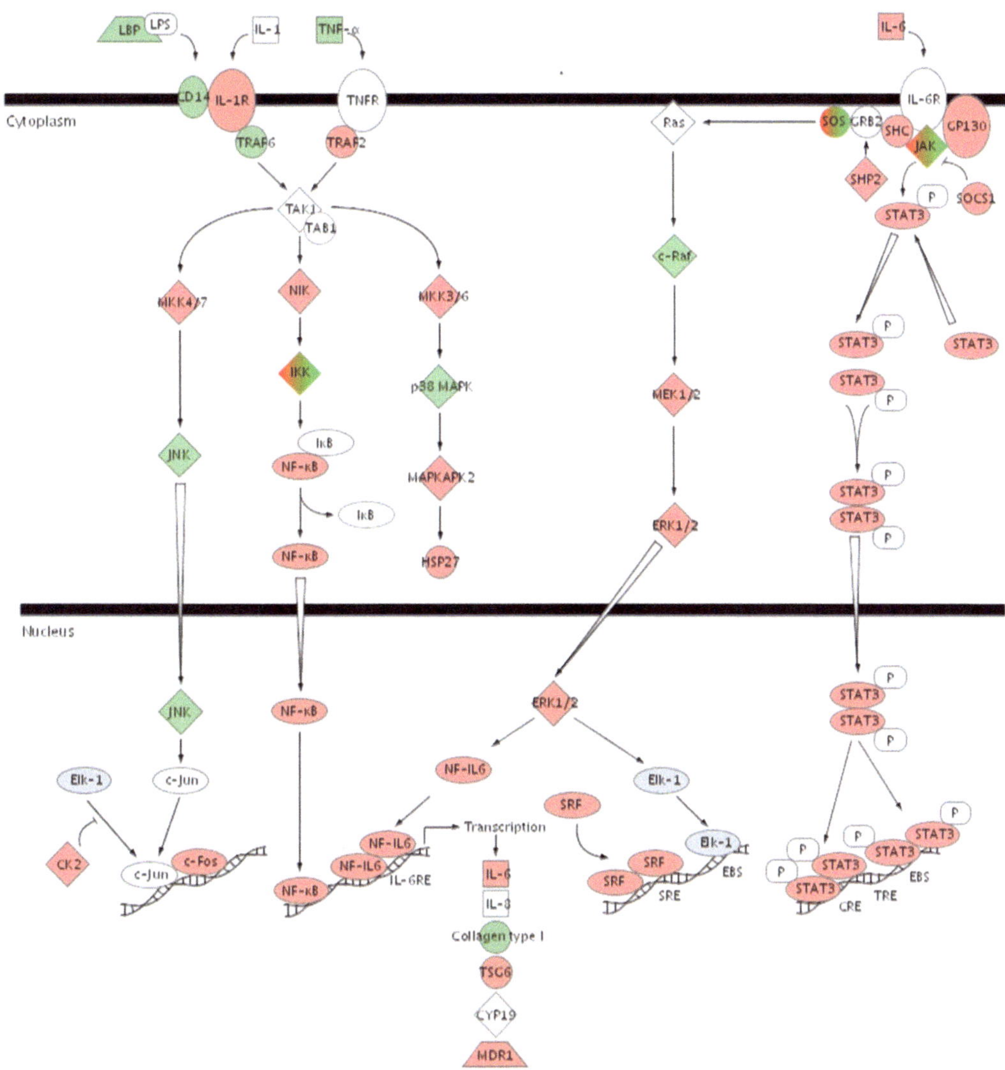

Fig. 3. Differentially expressed proteins among primary and recurrent ovarian tumors in the IL-6 signaling canonical pathway using the Ingenuity System™ (34).

2.2.3. Effects of Archival Time on Formalin-Fixed and Paraffin-Embedded Tissue Proteomes

In human disease research, where knowledge of disease outcome is critical for the evaluation of the significance of phenotypic or genotypic profiles, as well as response to therapy and outcome, it may take 5, 10, or more years to gain a relatively complete picture of the pathophysiology of a disease. Because of the long history of the use of formalin as the standard fixative for tissue processing in histopathology, there are a large number of archival formalin-fixed and paraffin-embedded (FFPE) tissue banks worldwide. These FFPE tissue collections, with attached clinical and outcome

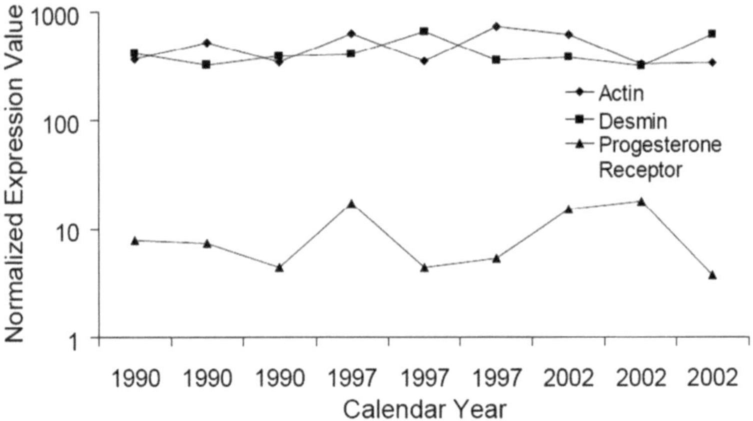

Fig. 4. Distribution of protein expression of three leiomyoma markers, including actin, desmin, and progesterone receptor, over the archival years from 1990 to 2002 (43).

information, present invaluable resources for conducting retrospective biomarker investigations.

From a practical point of view, the possible protein degradation as a result of prolonged archival duration represents one of the important considerations in comparing protein expression among FFPE tissues. By coupling an effective antigen retrieval approach (42) with the CITP proteomic platform, Balgley and coworkers (43) have therefore performed proteomic analysis over ten archived mesenchymal tumor tissue blocks, including nine uterine leiomyomas dating from 1990 to 2002 and a single 1980 vaginal alveolar soft part sarcoma (ASPS) case to investigate potential variability in protein composition and retrieval during different storage periods.

Although the 1990 group of leiomyomas demonstrated a slightly worse proteome performance in terms of total peptide, distinct peptide, and distinct protein identifications compared to leiomyomas cataloged in 1997 and 2002, excellent correlation among leiomyomas in quantitative proteomics was still achieved with a Pearson R^2 value of greater than 0.97. The expression values of three commonly used leiomyoma markers, including actin, desmin, and progesterone receptor, were remarkably consistent over 12 years of archival time from 1990 to 2002 (Fig. 4) and validated using immunohistochemistry measurements.

Furthermore, high confidence and comparative proteomic analysis between the uterine leiomyomas and the vaginal sarcoma (ASPS) was achieved using the sarcoma tissue block dating back as many as 28 years ago. Despite sharing over 1,800 common proteins in a core set, a total of 80 proteins were uniquely identified in the sarcoma tissues. The single 1980 sarcoma case was well distinguished from the nine leiomyomas by an unsupervised hierarchical cluster analysis even though the analyzers had no knowledge that one of the samples they had received was a substantially different tumor type. The results indicated the possibility and potential of using archived specimens as an extensive resource of normal and diseased tissue for conducting retrospective and prospective screening of disease biomarkers.

3. Conclusion

While proteomic measurements have been heralded to provide a wealth of information complementary to the transcriptomic data, biologically relevant proteome data can only be generated if the tissue samples investigated consist of homogeneous cell populations, in which no unwanted cells of different types and/or development stages obscure the results. Also, it has been well accepted that molecular profiling in tumor lesions is fundamental to understand the molecular etiology in tumor development and to provide the biomarkers for early detection and prevention. Furthermore, the need to detect small but biologically important changes in protein expression profiles remains, as cancer researchers explore the initial steps in biological-signaling cascades and compensatory processes. However, in the absence of PCR-like protein amplification, comprehensive analysis of protein expression within small populations of tumor cells microdissected from limited tissue samples represents a significantly challenging task which necessitates highly sensitive analytical approaches, exceeding the dynamic range of currently available proteomic techniques.

Besides sample amount constraints, the greatest bioanalytical challenge facing comprehensive proteomic analysis of microdissected tumor specimens is related to the large variation of protein relative abundances, particularly in the identification of low-abundance proteins. Developments in capillary electrophoresis-based multidimensional separations coupled with tandem MS, capable of achieving ultrasensitive and comprehensive analysis of minute protein amounts extracted from targeted cells in tissue specimens, are particularly highlighted for their roles within the broader context of a state-of-the-art clinical proteomic efforts. The coupling of tissue microdissection for diseased cell enrichment with CITP-based selective analyte concentration not only presents a synergistic strategy toward the detection and characterization of low-abundance proteins but also offers a novel biomarker discovery paradigm toward the identification of tumor-associated markers, exploration of molecular relationships among different tumor states and phenotypes, and a deeper understanding of molecular mechanisms that drive cancer progression.

Acknowledgment

We thank the National Cancer Institute (CA143177), the National Center for Research Resources (RR032333), and the National Institute of General Medical Sciences (GM103536) for supporting portions of our research activities reviewed in this chapter.

References

1. Drube J, Zurbig P, Schiffer E, Lau E, Ure B, Gluer S, Kirschstein M, Pape L, Decramer S, Bascands J-L, Schanstra JP, Mischak H, Ehrich JHH (2010) Urinary proteome analysis identifies infants but not older children requiring pyeloplasty. Pediatr Nephrol 25:1673–1678
2. Theodorescu D, Schiffer E, Bauer HW, Douwes F, Eichhorn F, Polley R, Schmidt T, Schofer W, Zurbig P, Good DM, Coon JJ, Mischak H (2008) Discovery and validation of urinary biomarkers for prostate cancer. Proteomics Clin Appl 2:556–570
3. Schiffer E, Vlahou A, Petrolekas A, Stravodimos K, Tauber R, Geschwend JE, Neuhaus J, Stolzenburg J-U, Conaway MR, Mischak H, Theodorescu D (2009) Prediction of muscle-invasive bladder cancer using urinary proteomics. Clin Cancer Res 15:4935–4943
4. Schiffer E, Bick C, Grizelj B, Pietzker S, Schofer W (2012) Urinary proteome analysis for prostate cancer diagnosis: cost-effective application in routine clinical practice in Germany. Int J Urol 19:118–125
5. Haubitz M, Good DM, Woywodt A, Haller H, Rupprecht H, Theodorescu D, Dakna M, Coon JJ, Mischak H (2009) Identification and validation of urinary biomarkers for differential diagnosis and evaluation of therapeutic intervention in anti-neutrophil cytoplasmic antibody-associated vasculitis. Mol Cell Proteomics 8:2296–2307
6. Zimmerli LU, Schiffer E, Zurbi P, Good DM, Kellmann M, Mouls L, Pitt AR, Coon JJ, Schmieder RE, Peter KH, Mischak H, Kolch W, Delles C, Dominiczak AF (2008) Urinary proteomic biomarkers in coronary artery disease. Mol Cell Proteomics 7:290–298
7. von Zur Muhlen C, Schiffer E, Zuerbig P, Kellmann M, Brasse M, Meert N, Vanholder RC, Dominiczak AF, Chen YC, Mischak H, Bode C, Peter K (2009) Evaluation of urine proteome pattern analysis for its potential to reflect coronary artery atherosclerosis in symptomatic patients. J Proteome Res 8:335–345
8. Snell-Bergeon JK, Maahs DM, Ogden LG, Kinney GL, Hokanson JE, Schiffer E, Rewers M, Mischak H (2009) Evaluation of urinary biomarkers for coronary artery disease, diabetes, and diabetic kidney disease. Diabetes Technol Ther 11:1–9
9. Jantos-Siwy J, Schiffer E, Brand K, Schumann G, Rossing K, Delles C, Mischak H, Metzger J (2009) Quantitative urinary proteome analysis for biomarker evaluation in chronic kidney disease. J Proteome Res 8:268–281
10. Kistler AD, Mischak H, Poster D, Dakna M, Wuthrich RP, Serra AL (2009) Identification of a unique urinary biomarker profile in patients with autosomal dominant polycystic kidney disease. Kidney Int 76:89–96
11. Good DM, Zurbig P, Argiles A, Bauer HW, Behrens G, Coon JJ, Dakna M, Decramer S, Delles C, Dominiczak AF, Ehrich JHH, Eitner F, Fliser D, Frommberger M, Ganser A, Girolami MA, Golovko I, Gwinner W, Haubitz M, Herget-Rosenthal S, Jankowski J, Jahn H, Jerums G, Julian BA, Kellmann M, Kliem V, Kolch W, Krolewski AS, Luppi M, Massy Z, Melter M, Neususs C, Novak J, Peter K, Rossing K, Rupprecht H, Schanstra JP, Schiffer E, Stolzenburg J-U, Tarnow L, Theodorescu D, Thongboonkerd V, Vanholder R, Weissinger EM, Mischak H, Schmitt-Kopplin P (2010) Naturally occurring human urinary peptides for use in diagnosis of chronic kidney disease. Mol Cell Proteomics 9:2424–2437
12. Raedler TJ, Wittke S, Jahn H, Koessler A, Mischak H, Wiedemann K (2008) Capillary electrophoresis mass spectrometry as a potential tool to detect lithium-induced nephropathy: preliminary results. Prog Neuropsychopharmacol Biol Psychiatry 32:673–678
13. Weissinger EM, Schiffer E, Hertenstein B, Ferrara JL, Holler E, Stadler M, Kolb H-J, Zander A, Zurbig P, Kellmann M, Ganser A (2007) Proteomic patterns predict acute graft-versus-host disease after allogeneic hematopoietic stem cell transplantation. Blood 109: 5511–5519
14. Haselberg R, de Jong GJ, Somsen GW (2011) Capillary electrophoresis-mass spectrometry

for the analysis of intact proteins 2007–2010. Electrophoresis 32:66–82

15. Albalat A, Mischak H, Mullen W (2011) Clinical application of urinary proteomics/peptidomics. Expert Rev Proteomics 8:615–629
16. Mischak H, Schanstra JP (2011) CE-MS in biomarker discovery, validation, and clinical application. Proteomics Clin Appl 5:9–23
17. Metzger J, Chatzikyrkou C, Broecker V, Schiffer E, Jaensch L, Iphoefer A, Mengel M, Mullen W, Mischak H, Haller H, Gwinner W (2011) Diagnosis of subclinical and clinical acute T-cell-mediated rejection in renal transplant patients by urinary proteome analysis. Proteomics Clin Appl 5:322–333
18. Zuberovic A, Wetterhall M, Hanrieder J, Bergquist J (2009) CE MALDI-TOF/TOF MS for multiplexed quantification of proteins in human ventricular cerebrospinal fluid. Electrophoresis 30:1836–1843
19. Wiese S, Reidegeld KA, Meyer HE, Warscheid B (2007) Protein labeling by iTRAQ: a new tool for quantitative mass spectrometry in proteome research. Proteomics 7:340–350
20. Wang W, Guo T, Rudnick PA, Song T, Li J, Zhuang Z, Zheng W, Devoe DL, Le CS, Balgley BM (2007) Membrane proteome analysis of microdissected ovarian tumor tissues using capillary isoelectric focusing/reversed-phase liquid chromatography-tandem MS. Anal Chem 79:1002–1009
21. Guo T, Wang W, Rudnick PA, Song T, Li J, Zhuang Z, Weil RJ, DeVoe DL, Lee CS, Balgley BM (2007) Proteome analysis of microdissected formalin-fixed and Paraffin-embedded tissue specimens. J Histochem Cytochem 55:763–772
22. Liu H, Sadygov RG, Yates JR (2004) A model for random sampling and estimation of relative protein abundance in shotgun proteomics. Anal Chem 76:4193–4201
23. Ishihama Y, Oda Y, Tabata T, Sato T, Nagasu T, Rappsilber J, Mann M (2005) Exponentially modified protein abundance index (emPAI) for estimation of absolute protein amount in proteomics by the number of sequenced peptides per protein. Mol Cell Proteomics 4:1265–1272
24. Balgley BM, Wang W, Song T, Fang X, Yang L, Lee CS (2008) Evaluation of confidence and reproducibility in quantitative proteomics performed by a capillary isoelectric focusing-based proteomic platform coupled with a spectral counting approach. Electrophoresis 29:3047–3054
25. Dai L, Li C, Shedden KA, Misek DE, Lubman DM (2009) Comparative proteomic study of two closely related ovarian endometrioid adenocarcinoma cell lines using cIEF fractionation and pathway analysis. Electrophoresis 30:1119–1131
26. Hanrieder J, Zuberovic A, Bergquist J (2009) Surface modified capillary electrophoresis combined with in solution isoelectric focusing and MALDI-TOF/TOF MS: a gel-free multidimensional electrophoresis approach for proteomic profiling–exemplified on human follicular fluid. J Chromatogr A 1216:3621–3628
27. Hood L (2003) Systems biology: integrating technology, biology, and computation. Mech Ageing Dev 124:9–16
28. Aebersold R, Cravatt BF (2002) Proteomics–advances, applications and the challenges that remain. Trends Biotechnol 20:1–2
29. An Y, Cooper JW, Balgley BM, Lee CS (2006) Selective enrichment and ultrasensitive identification of trace peptides in proteome analysis using transient capillary isotachophoresis/zone electrophoresis coupled with nano-ESI-MS. Electrophoresis 27:3599–3608
30. Fang X, Yang L, Wang W, Song T, Lee C, Devoe D, Balgley B (2007) Comparison of Electrokinetics-Based Multidimensional Separations Coupled with Electrospray Ionization-Tandem Mass Spectrometry for Characterization of Human Salivary Proteins. Anal Chem 79:5785–5792
31. Fang X, Wang W, Yang L, Chandrasekaran K, Kristian T, Balgley BM, Lee CS (2008) Application of capillary isotachophoresis-based multidimensional separations coupled with electrospray ionization-tandem mass spectrometry for characterization of mouse brain mitochondrial proteome. Electrophoresis 29:2215–2223
32. Fang X, Balgley BM, Wang W, Park DM, Lee CS (2009) Comparison of multidimensional shotgun technologies targeting tissue proteomics. Electrophoresis 30:4063–4070
33. Wolters DA, Washburn MP, Yates JR (2001) An automated multidimensional protein identification technology for shotgun proteomics. Anal Chem 73:5683–5690
34. Jinawath N, Vasoontara C, Jinawath A, Fang X, Zhao K, Yap K-L, Guo T, Lee CS, Wang W, Balgley BM, Davidson B, Wang T-L, Shih I-M (2010) Oncoproteomic analysis reveals co-upregulation of RELA and STAT5 in carboplatin resistant ovarian carcinoma. PLoS One 5:e11198
35. Chen Z, Fadiel A, Xia Y (2006) Functional duality of merlin: a conundrum of proteome complexity. Med Hypotheses 67:1095–1098
36. Huang K-C, Park DC, Ng S-K, Lee JY, Ni X, Ng W-C, Bandera CA, Welch WR, Berkowitz RS, Mok SC, Ng S-W (2006) Selenium binding protein 1 in ovarian cancer. Int J Cancer 118:2433–2440

37. Chekhun VF, Lukyanova NY, Urchenko OV, Kulik GI (2005) The role of expression of the components of proteome in the formation of molecular profile of human ovarian carcinoma A2780 cells sensitive and resistant to cisplatin. Exp Oncol 27:191–195
38. Smith-Beckerman DM, Fung KW, Williams KE, Auersperg N, Godwin AK, Burlingame AL (2005) Proteome changes in ovarian epithelial cells derived from women with BRCA1 mutations and family histories of cancer. Mol Cell Proteomics 4:156–168
39. Duan Z, Foster R, Bell DA, Mahoney J, Wolak K, Vaidya A, Hampel C, Lee H, Seiden MV (2006) Signal transducers and activators of transcription 3 pathway activation in drug-resistant ovarian cancer. Clin Cancer Res 12:5055–5063
40. Haura EB, Zheng Z, Song L, Cantor A, Bepler G (2005) Activated epidermal growth factor receptor-Stat-3 signaling promotes tumor survival in vivo in non-small cell lung cancer. Clin Cancer Res 11:8288–8294
41. Chen L-F, Greene WC (2004) Shaping the nuclear action of NF-kappaB. Nat Rev Mol Cell Biol 5:392–401
42. Xu H, Yang L, Wang W, Shi S-R, Liu C, Liu Y, Fang X, Taylor CR, Lee CS, Balgley BM (2008) Antigen retrieval for proteomic characterization of formalin-fixed and paraffin-embedded tissues. J Proteome Res 7:1098–1108
43. Balgley BM, Guo T, Zhao K, Fang X, Tavassoli FA, Lee CS (2009) Evaluation of Archival Time on Shotgun Proteomics of Formalin-Fixed and Paraffin-Embedded Tissues. J Proteome Res 8:917–925

Chapter 2

Capillary Electrophoresis–Mass Spectrometry of Carbohydrates

Joseph Zaia

Abstract

The development of methods for capillary electrophoresis (CE) with on-line mass spectrometric detection (CE/MS) is driven by the need for accurate, robust, and sensitive glycomics analysis for basic biomedicine, biomarker discovery, and analysis of recombinant protein therapeutics. One important capability is to profile glycan mixtures with respect to the patterns of substituents including sialic acids, acetate, sulfate, phosphate, and other groups. There is additional need for an MS-compatible separation system capable of resolving carbohydrate isomers. This chapter summarizes applications of CS/MS to analysis of carbohydrates, glycoproteins, and glycopeptides that have appeared since 2008. Readers are referred to recent comprehensive reviews covering earlier publications.

Key words: Capillary electrophoresis, Mass spectrometry

1. Introduction

Protein glycosylation occurs through a series of biosynthetic events in the endoplasmic reticulum and Golgi apparatus that are under complex regulation. Structures of glycoproteins depend on numerous factors including glycosyltransferase enzyme and nucleotide donor concentrations and rates of passage through these compartments. As a result, glycosylation at a given protein site typically reflects a mixture of glycoforms that elaborate a core structure. With regard to function, glycoprotein glycans elaborate the physicochemical properties of proteins and enable binding interactions with carbohydrate-binding protein domains. Key functions during biosynthesis include protein folding quality control and protein sorting. Cell surface glycans contain an array of antigenic epitopes

Nicola Volpi and Francesca Maccari (eds.), *Capillary Electrophoresis of Biomolecules: Methods and Protocols*, Methods in Molecular Biology, vol. 984, DOI 10.1007/978-1-62703-296-4_2, © Springer Science+Business Media, LLC 2013

(1) including those that form the ABO blood groups. These epitopes bind to carbohydrate-binding domains and enable the glycoprotein to bind lectin domains including the galectins, C-type, P-type, and I-type lectins. Expression of carbohydrate epitopes is regulated according to cell phenotype during development and as expressed in adult physiology. Such interactions are key to cell–cell and cell–pathogen interactions as well as to development of aberrant cell growth including cancers.

As described in recent reviews, mass spectrometry is an enabling technology in glycomics (2–5). Precise measurement of a glycan mass combined with assumptions regarding the biosynthetic reactions determines the monosaccharide compositions of glycans present in the sample and their approximate relative abundances. Thus, mass spectral analysis of glycans released from a glycoconjugate defines the types of glycans present. Such analysis of glycopeptides produced by proteolytic digestion serves to map the glycosylated peptides. Because the mixtures are typically complex, there is a need to combine separations with mass spectrometry analysis in order to produce a comprehensive map of glycoconjugate glycans.

In order to meet the emerging needs for glycomics analysis, it is desirable to have separations methods that are (1) high in peak capacity, (2) high in dynamic range, and (3) robust. High peak capacity is necessary to separate glycan structural isomers. Sensitivity is necessary to enable detection of low-abundance glycans of biological interest in the presence of high-abundance glycans. Robustness is key to dissemination of any emerging mass spectrometric technology; it must be possible to use the technology on a routine basis. As summarized in recent reviews (6–10), on-line separations combined with mass spectrometry have become extremely useful for profiling of released glycans, but do not fill all three needs. Liquid chromatography-based methods are based on hydrophilic interaction chromatography, reversed-phase chromatography, reversed-phase ion pairing chromatography, size exclusion chromatography, or porous graphitized carbon chromatography. At the present time, the extent of isomeric separation of complex glycans available using any of these methods is quite limited.

Capillary electrophoresis (CE) separates analytes based on charge, size, and shape. The fact glycoconjugate glycans contain positional isomers makes CE attractive as a separations method. The topic of CE of glycans has been reviewed recently (11). While CE separations complement liquid chromatography profiling of glycans and glycoconjugates using a mass spectrometric detector, not all methods are appropriate for interfacing with a mass spectrometer. Capillary electrophoresis/mass spectrometry (CE/MS) for analysis of glycans, glycoproteins, and other glycoconjugates has been the topic of recent comprehensive reviews (12–14). The intention here is to give an update on applications of CE/MS for glycans, glycopeptides, and glycoproteins that have appeared since the most recent previous review (2009).

2. General Considerations for CE/MS of Carbohydrates, Glycopeptides, and Glycoproteins

Analysis of glycans using CE is carried out typically using reducing end derivatives that facilitate optical detection. Commercial kits are available for glycan reductive amination. Using these methods, it is necessary to match observed migration times against those of standard compounds. As a result, peaks for which standards are not available cannot be identified. The use of MS as a detector for CE has the advantage that the mass dimension defines the glycan composition with respect to types of monosaccharides and substituents. This extra dimension of information provides a mass profile for the glycan migrating through the column; however, the use of an MS detector dictates that the electrolyte composition be compatible, specifically that all components be volatile (12, 13). This requirement eliminates many CE methods because of the presence of nonvolatile electrolyte components. CE/MS is generally performed in free solution, rather than in gel-filled capillaries because of compatibility issues. In order to maintain a sufficient flow of solution into the MS source, a coaxial sheath flow of a few μL/min is often used at the distal end of the capillary (e.g., see ref. 15). The problem is that this sheath flow necessarily dilutes the analytes migrating from the capillary. Another interface option employs a junction at the capillary tip to add sufficient flow of solution for the MS source (for recent examples see refs. 16, 17). Typically, the flow for such a makeup flow is 0.2–0.4 nL/min, considerably lower than that used by the coaxial sheath flow.

3. Applications

3.1. CE/MS of Lipopolysaccharides

Lipopolysaccharide (LPS) is found in the outer membrane of Gram-negative bacteria and acts as a strong stimulator of the innate immune system in host organisms. LPS from *E. coli* exists as a set of nonstoichiometric substituents on an inner core structure. The substituents include phosphate, ethanolamine, and acidic KDO monosaccharide residues. Because the presence of these substituents influences the LPS charge, CE is a useful means of profiling the variants. CE/MS using a sheath flow with ion trap MS detector has been used for this purpose.

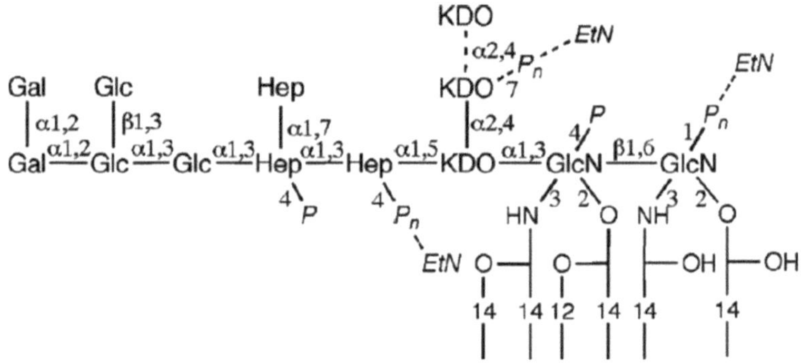

Fig. 1. Chemical structure of LPS of *E. coli* C strain. Gal, d-galactose; Glc, d-glucose; Hep, l-glycero-d-manno-heptose; KDO, 3-deoxy-d-manno-octulosonic acid; GlcN, 2-amino-2-deoxy-d-glucose; EtN, ethanolamine; P, phosphate (Copyright 2009, Elsevier, used with permission).

Figure 1 shows the structure of *E. coli* LPS with positions of possible substituents labeled (18). Liquid chromatographic separation of deacylated LPS has proven challenging because the compounds do not bind reversed-phase columns. Successful analysis of the deacylated LPS required optimization of the sheath flow composition, and 50% methanol without additives was selected as the best choice. The running electrolyte for the CE separations was pH 9.0 ammonium acetate and the applied potential was 30 kV. The concentration of ammonium acetate for fully deacylated LPS was 10 mM and that for partially deacylated LPS was 50 mM. The latter electrolyte required use of lower separation potentials to minimize problems with capillary heating. The sheath flow solution was either methanol/water or methanol/water/formic acid depending on the analyte. Using this approach, it was possible to separate LPS variants based on the number of KDO and phosphate groups. The CE conditions depended primarily on the presence or absence of acyl chains on the LPS.

3.2. CE/MS of N-Glycans

The fact that many protein therapeutics are glycosylated drives the need for effective methods for rapid and precise testing of glycosylation in recombinant protein batches. Release of *N*-glycans using peptide *N*-glycosidase F (PNGase F) is often combined with reductive amination to add desired optical properties (19). Reductive amination using aminopyrene trisulfonate (APTS) has been used for CE/MS with on-line laser-induced fluorescence detection of *N*-glycans released from recombinant antibody preparations (15). Reductive amination using APTS adds a group with fluorescent properties appropriate for commercial laser-induced fluorescence detectors with excitation at 488 nm and emission at 520 nm. The sulfonate groups ensure that all glycans are negatively charged in the electrolyte, enabling use of coated neutral capillaries. The sulfonate groups also facilitate detection using on-line negative ion MS.

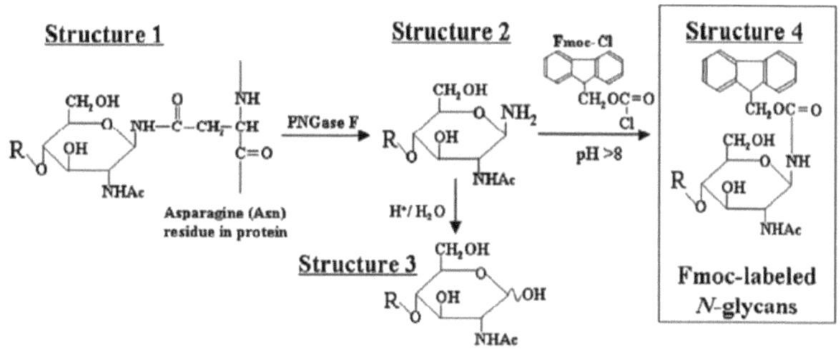

Fig. 2. Flowchart for the preparation of Fmoc-labeled *N*-glycans after the release of *N*-glycans from protein. The *N*-glycan (structure 1) linked to the Asn residue of the core protein/peptide is released with PNGase F as *N*-glycosylamine (structure 2). Structure 2 is hydrolyzed to yield free *N*-glycan (structure 3) under acidic conditions. However, at above pH 8, *N*-glycosylamine (structure 2) can be stabilized; therefore, it can be directly reacted to the amino groups with Fmoc-Cl. The Fmoc-labeled *N*-glycan (structure 4) was subjected to analysis by CE–ESI MS (Copyright 2008, Oxford University press, used with permission).

The combination of the fluorescence detector and MS provides clear value in that it provides absolute quantification of glycans using fluorescent peak areas; the MS dimension determines the composition of the eluting APTS-glycans. The running electrolyte consisted of 40 mM aminocaproic acid, pH 4.5 and a sheath liquid of 50% isopropanol, 0.2% ammonia was used. This enabled migration of the APTS-labeled glycans with an applied potential of −30 kV and without the need to condition the capillary. The basic sheath liquid was appropriate for efficient negative-ion electrospray ionization. This approach was used to assess the extent of sialylation of the recombinant antibody molecules. Sialylation affects clearance of antibody molecules in the blood stream and is thus important to measure. The authors' use of a time-of-flight mass spectrometer to produce accurate masses provided confident determination of glycan compositions in the CE/MS data sets.

As shown in Fig. 2, PNGase F releases *N*-glycans from such glycoproteins as glycosylamines (20). The glycosylamine is rapidly hydrolyzed at slightly acidic pH but is more stable under basic conditions. Glycosylamines may be derivatized with amine-reactive reagents to stabilize the glycan reducing ends and add a chromophore or fluorophore (21). The advantage to this approach is that the stereochemistry of the reducing end is preserved as a single anomer. Recently, fluorenylmethyloxycarbonyl chloride (Fmoc-Cl) has been used to derivatize *N*-glycans released using PNGase F (20). The authors analyzed purified glycoproteins including fetuin, α1 acid glycoprotein, immunoglobulin G, and transferrin to validate the method. The CE/MS mass electropherograms were acquired using a bare fused silica capillary, an electrolyte pH of 6.8, and an applied potential of 30 kV. A sheath liquid of water/methanol/formic acid was used with positive-ion electrospray MS detection. The data showed FMOC-derivatized

N-glycans migrating according to increasing degree of sialylation due to the electroosmotic flow (EOF). This enabled profiling of the *N*-glycans based on degree of sialylation. The method was applied to the analysis of Fmoc-labeled *N*-glycans from the therapeutic monoclonal antibody trastuzumab (Herceptin), rituximab (Rituxan), and palivizumab (Synagis) for which 10 μg aliquots were analyzed. The MS dimension provided clear value by facilitating detection of fucosylated *N*-glycans that are not easily resolved using CE alone. The authors also subjected 10 μg quantities of the glycoproteins to SDS–PAGE followed by PNGase F digestion of the excised bands and Fmoc derivatization. The *N*-glycans migrated in a diffuse pattern due to the presence of acrylamide polymer in the solution. Nonetheless, it was possible to extract mass spectra from the data showing the presence of the *N*-glycans from the glycoprotein samples.

3.3. CE/MS of Glucose Ladders

It is desirable to use the lowest flow rate of solution after the CE capillary that suffices to produce a stable electrospray. Given that reductive amination with APTS is a popular means of introducing both fluorophore and charge to glycans, researchers have optimized a junction interface that introduces a makeup flow of 0.3–0.4 μL/min of flow to supply the electrospray source (16). The makeup flow was composed of isopropanol/methanol/formic acid. The electrolyte solution was methanol/water/formic acid and the separation potential was −16 kV. These APTS-glycans were negatively charged in this acidic electrolyte and migrated against the electroosmotic flow toward the LIF and MS detectors. The makeup flow also backfilled the capillary as electrolyte flowed toward the cathode due to the EOF. As shown in Fig. 3, it was possible to analyze a mixture of glucose oligomers containing from 1 to 24 monosaccharides. Baseline resolution was obtained up to G15 (an oligomer of 15 monosaccharide residues). These results showed the potential for use of such APTS-labeled glucose ladders to calibrate CE/MS data acquired for glycans released from glycoproteins. Such glucose ladders have been used with great success for high-performance liquid chromatography-based studies of the human *N*-glycome (22, 23). Such population-based studies require a rapid, reproducible, and stable analytical platform. CE/MS using glucose ladders may enable increased information through the MS dimension to be produced for such studies.

3.4. CE/MS Determination of Sialic Acids

Sialic acid residues cap many nonreducing end termini in glycoconjugate glycans and are present in many glycan epitopes (1). Their presence helps determine the carbohydrate–protein-binding interactions that occur among mammalian cells and in host–pathogen interactions. Sialylation strongly influences the biological lifetime and antigenicity of proteins; the asialoglycoprotein receptor recognizes un-sialylated galactose residues and facilitates the removal of

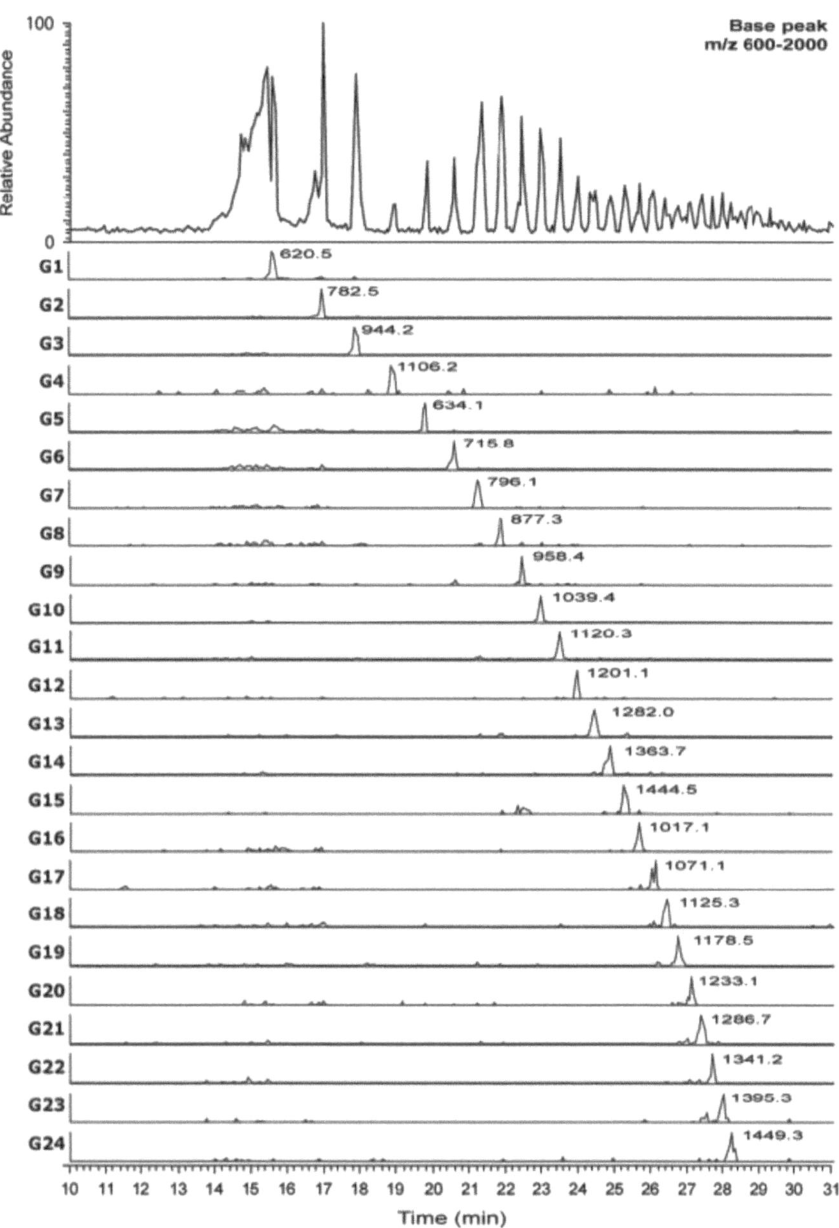

Fig. 3. Base peak and extracted ion electropherograms for APTS-labeled glucose ladder standard (G1–G24, referring to the number of glucose monosaccharides in each oligosaccharide) using optimized conditions (BGE 2.0% formic acid, 30% methanol; separation potential −16 kV) (Copyright 2011 WILEY-VCH Verlag GmbH & Co. KGaA, Weinheim).

asialoglycoproteins from circulation. Humans lack the functional enzyme that converts Neu5Ac to Neu5Gc and in this respect differ from all other mammalian species including the closest primate relatives. As a result, the presence of Neu5Gc in a recombinant protein therapeutic has the potential to cause undesirable immune reactions in humans (24). Most recombinant glycoproteins are expressed in animal cells and/or using animal-derived serum or serum factors. Thus there is some level of incorporation of Neu5Gc into glycoproteins. As a result, there are strict standards on how much Neu5GFc can be present in a lot of a therapeutic glycoprotein. This drives the need for accurate and precise methods for analysis of sialic acids. The use of CE/MS has been demonstrated for this purpose (25). The authors released sialic acids from glycoconjugates using mild acid hydrolysis. Uncoated fused silica capillaries were used with an electrolyte solution of ammonium acetate pH 9.5. The sheath liquid was water/methanol/ammonium acetate and the separation potential was 30 kV. Peaks corresponding to Neu5Ac and Neu5Gc were well separated. The results showed that it was possible to detect Neu5Ac released from human serum in a 9 min CE/MS run.

3.5. CE/MS of Glycopeptides

Analysis of glycopeptides provides information on the glycan composition and the peptide to which glycans are attached. There is an associated challenge, however, because glycopeptide physicochemical properties reflect those of both the peptide and glycan portions of the molecule. As a result, glycopeptides elute as broad peaks using standard reversed-phase columns in proteomics LC/MS workflows. It is therefore not surprising that use of CE/MS for profiling glycopeptides has been investigated in recent years. This application also drives the use of high-resolution mass analyzers so as to define the peptide and glycan compositions with confidence.

CE/MS with a time-of-flight mass analyzer has been used to map *N*- and *O*-linked glycopeptides from recombinant human erythropoietin (EPO) (26). The investigators used uncoated fused silica capillaries, electrolyte solution of acetic acid/formic acid pH 2.2, separation potential of 18 kV, and a sheath flow of 50% isopropanol modified with formic acid. The mass spectrometer was operated in the positive ion mode. The use of uncoated capillaries eliminates potential problems with bleeding of capillary coatings into the MS source. Glycopeptides may interact with the walls of uncoated capillaries, however, necessitating a washing procedure between each CE/MS sample injection. Glycopeptides were separated based on number of sialic acid groups and it was possible to identify variants in the number of acetyl modifications and the presence of *N*-glycolylneuraminic acid.

Glycoprotein glycans undergo sulfonylation in the Golgi apparatus mediated by specific sulfotransferases. Such modifications change the physico-chemical properties of the glycan and thereby create new protein binding activities. Examples include the sulfated Lewis glycan epitopes. Other complex glycan-sulfated motifs include 4-sulfonylated-LDN and the HNK antigen (1). The sulfonate group increases the acidity of the glycoprotein; at the same time it poses particular analytical challenges. In particular, the sulfonylation may occur to only a small percentage of the population of glycoprotein molecules. Also, the MS ionization of sulfonylated molecules is likely to be suppressed in the positive ion mode. At the same time, ionization in the negative mode is likely to be enhanced by the increased acidity. For CE separations, it is not necessary to add charge to a sulfonylated glycan. Thus, CE methods can be designed to favor the migration of sulfonylated glycans over those of less-acidic glycans. Investigators have employed an acidic electrolyte of ammonium formate pH 3.3, a sheath liquid of methanol/water/acetic acid, and an applied potential of −30 kV for this purpose (27). They applied their method to analysis of a proteinase K digest of thyroid-stimulating hormone (TSH). This glycoprotein contains several sulfonylated *N*-glycans including the 4-sulfonylated-LDn type that contains sulfonylated GalNAc in the antennae. The CE/MS data, acquired using negative-ion electrospray, showed abundant ions, demonstrating effective enrichment, from sulfonylated glycopeptides. The authors also showed that the glycopeptides could be analyzed using positive-ion electrospray MS by including a basic peptide of sequence KKK in the sheath liquid.

3.6. CE/MS of Intact Glycoproteins

Glycoproteins exist as a population of molecules that are heterogeneous with respect to extent of glycosylation. It is desirable to analyze glycoproteins intact in order to assess the extent to which such populations are likely to display differences in function based on glycan structure. Such an analysis would require high resolution and mass accuracy in order to define the composition of the observed peaks accurately. It is also important that the method be reproducible so as to enable comparison of recombinant glycoprotein lots.

α1 Acid glycoprotein (AGP) levels are altered in blood plasma during inflammation. Its glycosylation microheterogeneity has been observed to vary according to disease state. As a result, it is a good candidate for development of methods for intact glycoprotein profiling. The 42 kDa protein consists of approximately 45% carbohydrate by mass and is highly heterogeneous (28). For CE/MS analysis, bare fused silica capillaries were coated with acrylamide–pyrrolidine methacrylate copolymer (29, 30). The coating was repeated every two CE runs. The separation electrolyte was 6-aminocaproic acid/ammonia/methanol, the sheath liquid was isopropanol/water, and the separation potential was 28 kV. Analysis of deglycosylated AGP required use of a different capillary coating method. The CE/MS analysis was performed using a time-of-flight analyzer with moderate–high resolution. The method was used to compare AGP profiles from 16 serum samples. AGP was detected between 6 and 10 min as a broad peak envelope. Plots of the EIEs show an extremely complex series of peaks ranging from m/z 2,000 to 3,000. Despite the complexity, it was possible to deconvolute the data and determine the neutral masses of AGP glycol

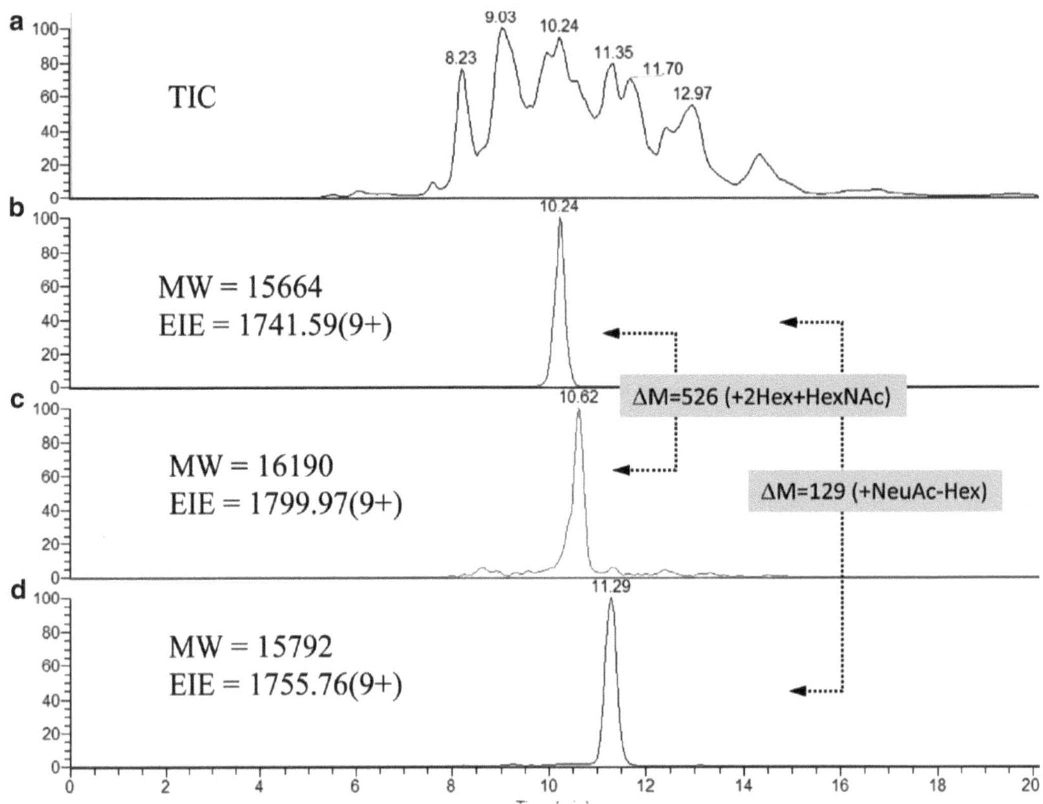

Fig. 4. Demonstration of the high-resolution separation of intact r-RhCG derived from a murine cell line using CE/MS. (**a**) Total ion current and (**b**–**d**) selected extracted ion electropherograms (EIEs). The dotted arrows indicate mass shifts between pairs of masses (Copyright 2010, American Chemical Society, used with permission).

variants. The data were used to make qualitative comparisons of the AGP molecular weights present among the sample set. A statistical analysis of the data was not presented.

A Fourier transform MS system has been used for analysis of intact recombinant human chorionic gonadotropin (rHCG) (17). For this purpose, a polyvinyl alcohol-coated capillary was used with an electrolyte of 2% acetic acid (pH 2.5) and a separation potential of 8 kV. A pressurized electrolyte reservoir system was used to provide a post-capillary makeup flow of ~200 nL/min into the MS source. Figure 4 shows the total ion electropherogram for rHCG (A) and extracted ion electropherograms (EIEs) (B-D). The data were acquired using a limited *m*/*z* range (1,400–2,000) to maximize the scan speed of the MS instrument. The EIEs show peak widths of ~12 s, demonstrating impressive ability to resolve glycoforms. The data serve as a means for comparison of ion abundances among different process batches. It was not appropriate to compare the abundances of different rHCG ions to each other due to potential differences in ionization efficiencies. Coefficients of variation for replicate analyses on the same day were less than 10%

for the 20 most abundant glycoforms of rHCG. This example demonstrates the potential power for high-resolution MS for CE/MS of intact glycoproteins.

CE/MS using non-covalently coated capillaries was used to determine changes in recombinant glycoprotein structure caused by a variety of storage conditions (31). The glycoproteins analyzed were recombinant human growth hormone (HGH) and oxytocin. The non-covalently bound capillary coating was Polybrene–poly(vinyl sulfonic acid) or Polybrene–dextran sulfate–Polybrene. The coating method, sheath flow composition, electrolyte, and applied potential differed according to the glycoprotein being analyzed. The method was useful for profiling recombinant glycoforms based on the extracted mass spectra. It was also possible to observe oxidations, sulfonate formation, and deamidation occurring in heat-exposed glycoprotein preparations.

4. Conclusions

Glycans may be analyzed using CE/MS as underivatized or reducing end derivatized forms. The analysis of underivatized glycans has the advantage of simplicity. For such separations, CE separations using forward (positive potential) polarity are often used so that the EOF sweeps all glycans toward the detector. The migration time increases with the number of acidic groups on the glycan. Thus, the most acidic glycans display long migration times and broad peak shapes. *N*-Glycans released using PNGase F exist as glycosylamines that may be derivatized using amino-reactive reagents. This approach preserves the stereochemistry of the reducing end as a single anomer and provides a chromophore/fluorophore for optical detection. Forward-polarity CE allows separation of the derivatized glycans based on number of acidic groups. Sialic acids released from glycoproteins are readily analyzed using forward-polarity CE/MS. Reductive amination of glycans using APTS has the advantage that it renders all glycans very acidic; they may therefore be separated using reversed-polarity CE (negative applied potential). This enabled separation of glucose oligomers that are likely to be useful for standardizing CE/MS profiles of glycoconjugate glycans.

CE/MS has been used for mapping glycopeptides. For this purpose, the need for high-accuracy mass analysis increases. This is because there is need to define the compositions of both the glycan and peptide portions of the molecule. The number of possible compositions for an observed glycopeptide m/z is the multiple of the possible peptide and glycan variants. Thus, the higher the mass accuracy, the greater the confidence in the interpretation of glycopeptide mass spectra. One approach for glycopeptide separation is to use an acidic electrolyte and forward polarity with uncoated capillaries and positive-ion MS detection. This approach has been shown for analysis of *N*-glycopeptides, some of which are sialylated. Use of reversed polarity and acidic electrolyte enriches for sulfonylated glycopeptides, the MS detection of which is facilitated with negative-ion electrospray.

Methods for CE/MS of intact glycoproteins have advanced in the period covered by this review. There is a clear need for the highest possible mass spectrometric resolution and mass accuracy so as to assign the glycoprotein composition with the greatest confidence possible. In addition, there is a need to minimize interactions between the glycoprotein analytes and the capillary wall. One approach is to use non-covalently coated capillaries and forward-polarity CE. Another approach is to use covalently coated neutral capillaries at acidic pH and forward-polarity CE.

With regard to CE/MS interfaces, the majority of articles published in the period of this review used a coaxial sheath to provide 2–5 μL/min flow to the MS source. The use of junction-type interfaces that provide 0.2–0.5 μL/min makeup flow dilutes the CE analytes to a lesser degree but is not as commonly reported for glycan or glycoconjugate CE/MS in the published literature.

Acknowledgment

The author's effort is supported by US National Institute of Health grants P41RR10888 and R01HL098950.

References

1. Cummings RD (2009) The repertoire of glycan determinants in the human glycome. Mol Biosyst 5:1087–1104
2. Zaia J (2004) Mass spectrometry of oligosaccharides. Mass Spectrom Reviews 23:161–227
3. Zaia J (2010) Mass spectrometry and glycomics. OMICS 14:401–418
4. Bielik AM, Zaia J (2010) Historical overview of glycoanalysis. Methods Mol Biol 600:9–30
5. Zaia J (2008) Mass spectrometry and the emerging field of glycomics. Chem Biol 15:881–892
6. Zaia J (2009) On-line separations combined with MS for analysis of glycosaminoglycans. Mass Spectrom Rev 28:254–272
7. Wuhrer M, de Boer AR, Deelder AM (2009) Structural glycomics using hydrophilic interaction chromatography (HILIC) with mass spectrometry. Mass Spectrom Rev 28:192–206
8. Mechref Y, Novotny MV (2009) Editorial: glycomics through hyphenated techniques. Mass Spectrom Rev 28:191–191
9. Mechref Y, Novotny MV (2006) Miniaturized separation techniques in glycomic investigations. J Chromatogr B Analyt Technol Biomed Life Sci 841:65–78
10. Wuhrer M, Deelder AM, Hokke CH (2005) Protein glycosylation analysis by liquid chromatography-mass spectrometry. J Chromatogr B Analyt Technol Biomed Life Sci 825:124–133
11. Volpi N, Maccari F, Linhardt RJ (2008) Capillary electrophoresis of complex natural polysaccharides. Electrophoresis 29:3095–3106
12. Mechref Y, Novotny MV (2009) Glycomic analysis by capillary electrophoresis-mass spectrometry. Mass Spectrom Rev 28:207–222

13. Campa C, Coslovi A, Flamigni A, Rossi M (2006) Overview on advances in capillary electrophoresis-mass spectrometry of carbohydrates: a tabulated review. Electrophoresis 27:2027–2050
14. Amon S, Zamfir AD, Rizzi A (2008) Glycosylation analysis of glycoproteins and proteoglycans using capillary electrophoresis-mass spectrometry strategies. Electrophoresis 29:2485–2507
15. Gennaro LA, Salas-Solano O (2008) On-line CE-LIF-MS technology for the direct characterization of N-linked glycans from therapeutic antibodies. Anal Chem 80:3838–3845
16. Maxwell EJ, Ratnayake C, Jayo R, Zhong X, Chen DD (2011) A promising capillary electrophoresis-electrospray ionization-mass spectrometry method for carbohydrate analysis. Electrophoresis 32:2161–2166
17. Thakur D, Rejtar T, Karger BL, Washburn NJ, Bosques CJ, Gunay NS, Shriver Z, Venkataraman G (2009) Profiling the glycoforms of the intact alpha subunit of recombinant human chorionic gonadotropin by high-resolution capillary electrophoresis-mass spectrometry. Anal Chem 81:8900–8907
18. Kojima H, Inagaki M, Tomita T, Watanabe T, Uchida S (2009) Separation and characterization of lipopolysaccharide related compounds by HPLC/post-column fluorescence derivatization (HPLC/FLD) and capillary zone electrophoresis/mass spectrometry (CZE/MS). J Chromatogr B Analyt Technol Biomed Life Sci 877:1537–1542
19. Anumula KR (2006) Advances in fluorescence derivatization methods for high-performance liquid chromatographic analysis of glycoprotein carbohydrates. Anal Biochem 350:1–23
20. Nakano M, Higo D, Arai E, Nakagawa T, Kakehi K, Taniguchi N, Kondo A (2009) Capillary electrophoresis-electrospray ionization mass spectrometry for rapid and sensitive N-glycan analysis of glycoproteins as 9-fluorenylmethyl derivatives. Glycobiology 19:135–143
21. Hase S (1994) High-performance liquid chromatography of pyridylaminated saccharides. Methods Enzymol 230:225–237
22. Lauc G, Essafi A, Huffman JE, Hayward C, Knezevic A, Kattla JJ, Polasek O, Gornik O, Vitart V, Abrahams JL, Pucic M, Novokmet M, Redzic I, Campbell S, Wild SH, Borovecki F, Wang W, Kolcic I, Zgaga L, Gyllensten U, Wilson JF, Wright AF, Hastie ND, Campbell H, Rudd PM, Rudan I (2010) Genomics meets glycomics-the first GWAS study of human N-Glycome identifies HNF1alpha as a master regulator of plasma protein fucosylation. PLoS Genet 6:e1001256
23. Knezevic A, Polasek O, Gornik O, Rudan I, Campbell H, Hayward C, Wright A, Kolcic I, O'Donoghue N, Bones J, Rudd PM, Lauc G (2009) Variability, heritability and environmental determinants of human plasma N-glycome. J Proteome Res 8:694–701
24. Ghaderi D, Taylor RE, Padler-Karavani V, Diaz S, Varki A (2010) Implications of the presence of N-glycolylneuraminic acid in recombinant therapeutic glycoproteins. Nat Biotech 28:863–867
25. Ortner K, Buchberger W (2008) Determination of sialic acids released from glycoproteins using capillary zone electrophoresis/electrospray ionization mass spectrometry. Electrophoresis 29:2233–2237
26. Gimenez E, Ramos-Hernan R, Benavente F, Barbosa J, Sanz-Nebot V (2011) Capillary electrophoresis time-of-flight mass spectrometry for a confident elucidation of a glycopeptide map of recombinant human erythropoietin. Rapid Commun Mass Spectrom 25:2307–2316
27. Imami K, Ishihama Y, Terabe S (2008) On-line selective enrichment and ion-pair reaction for structural determination of sulfated glycopeptides by capillary electrophoresis-mass spectrometry. J Chromatogr A 1194:237–242
28. Fournier T, Medjoubi NN, Porquet D (2000) Alpha-1-acid glycoprotein. Biochim Biophys Acta 1482:157–171
29. Ongay S, Neususs C (2010) Isoform differentiation of intact AGP from human serum by capillary electrophoresis-mass spectrometry. Anal Bioanal Chem 398:845–855
30. Neususs C, Pelzing M (2009) Capillary zone electrophoresis-mass spectrometry for the characterization of isoforms of intact glycoproteins. Methods Mol Biol 492:201–213
31. Haselberg R, Brinks V, Hawe A, de Jong GJ, Somsen GW (2011) Capillary electrophoresis-mass spectrometry using noncovalently coated capillaries for the analysis of biopharmaceuticals. Anal Bioanal Chem 400:295–303

Chapter 3

Approaches to Enhancing the Sensitivity of Carbohydrate Separations in Capillary Electrophoresis

M.C. Breadmore

Abstract

Electrophoresis in both capillaries (CE) and microchips (ME) is an extremely powerful liquid phase-separation technique that is indispensable for the separation of carbohydrates. It is capable of separating both small mono- and disaccharides, through to more complex oligo- and polysaccharides, with high resolution, but as with all CE and ME separations, the detection limits are often inferior to those that can be achieved with liquid chromatographic methods. One avenue to address this is to use an on-line concentration strategy. Various approaches have been developed over the past 20 years, and this chapter will highlight their application to improve the sensitivity of carbohydrate separations in both CE and ME.

Key words: Capillary electrophoresis, Microchip

1. Introduction

Capillary electrophoresis (CE) is one of the most powerful liquid phase-separation techniques known and has found widespread application for the separation of biomolecules, such as proteins, amino acids, and carboxylic acids, and was one of the main technologies used to sequence the human genome. It has also found application for the separation of the entire range of carbohydrates, from small mono- and disaccharides right through to more complex oligo- and polysaccharides. Their separation by CE can be achieved with and without prior derivatization. Derivatization is frequently used to attach a chromophore or fluorophore to facilitate detection, and while there is a wide range of reagents that can be used, 1-aminopyrene-3,6,8-trisulfonic acid (APTS) is by far the most common (1, 2). This reagent contains 3 sulfonic acid groups which introduce multiple charges over a wide pH allowing very

Nicola Volpi and Francesca Maccari (eds.), *Capillary Electrophoresis of Biomolecules: Methods and Protocols*, Methods in Molecular Biology, vol. 984, DOI 10.1007/978-1-62703-296-4_3, © Springer Science+Business Media, LLC 2013

high-resolution separations of long-chain oligosaccharides to be obtained. When used with an argon ion laser, excellent sensitivity can be obtained, with LODs in the range of 10^{-12} M being reported (3, 4). When separating underivatized sugars, they must be either natively charged, as is the case with those released from glycosaminoglycans, or made charged through either deprotonation at high pH (12–13) or complexation with borate which imparts a partial negative charge to compounds containing vicinal diols through formation of a boronate ester. Detection can be through indirect absorbance/fluorescence (5–8), directly through electrochemical methods, primarily amperometric detection (9–13), and with a mass spectrometer (14–19). While the separation of underivatized carbohydrates is simpler in that it does not require derivatization, detection limits are typically 3–6 orders of magnitude higher than what can be achieved with laser-induced fluorescence. This is sufficient for many applications, particularly food and food products, but not for many biological and medical applications.

Irrespective whether or not the carbohydrates are derivatized and the type of detector used, it is possible to improve detection limits by employing one of a number of pre-concentration approaches in-line, i.e., within the separation capillary (20–23). Originally, samples were prepared in the separation electrolyte for continuity; however, it was quickly discovered that preparing the sample in a less-conductive solution improved the separation performance through "stacking" as analyte ions from the sample became concentrated on the sample/electrolyte interface. Since then, there have been a number of methods developed, which are either based on the exploitation of electrophoretic phenomena, broadly categorized as "stacking," or chemical partitioning onto or into a heterogeneous phase, which can be considered as "extraction." Extraction methods are traditionally performed off-line, i.e., separate to the electrophoretic separation, and while there are a number of ways to perform extraction in-line, they are almost exclusively limited to solid-phase extraction which requires special customized capillaries. Stacking methods rely on chemical changes in composition within the capillary to achieve concentration and are thus practically much simpler to implement and have been used for carbohydrates to a much higher extent than extraction methods.

2. Materials

See refs. 20–23.

3. Methods

3.1. Stacking

The basic premise of stacking is that analytes injected in a large volume of sample travel with a higher velocity in the sample than they do in the separation electrolyte, thus causing them to "stack" or concentrate at the sample/electrolyte interface (depicted in Fig. 1a). For example, if an analyte has a velocity in the sample 10 times higher than that in the electrolyte, it would allow the injection volume to be increased by the same factor of 10 compared to conditions in which the analyte has no velocity change. There are many ways in which the velocity can be changed, including conductivity and electric field strength, pH, and affinity with sample/electrolyte components, and these form the basis of a number of different methods for concentration. The focus on this chapter will be to highlight and discuss the various approaches to on-line concentration that have been used for carbohydrates.

3.1.1. Field-Amplified Sample Stacking

The oldest method for on-line concentration in CE is based on a difference in conductivity between the sample and the electrolyte and is called field-amplified sample stacking (FASS) (24). Due to the lower conductivity of the sample zone, a proportionally higher electric field is distributed over the sample zone than the electrolyte, thus causing ions in the sample to move quickly to the sample/electrolyte interface after which they experience a reduction in velocity (Fig. 1a). For best results, the difference in conductivity should be approximately 10, but higher differences may be more convenient and typically do not adversely impact upon the separation. Improvements in sensitivity of 10–20 are typically obtained when using FASS and nearly all separations of derivatized sugars

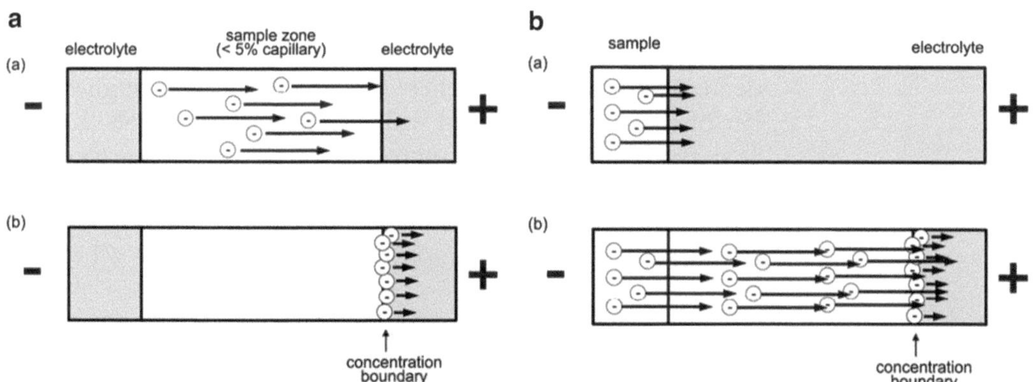

Fig. 1. **A**: Schematic representation of FASS, immediately after the voltage has been applied (**a**) and after the stacking has finished (**b**); **B**: schematic representation of FASI, immediately after the voltage has been applied (a) and after the stacking has finished (b).

are performed under FASS conditions due to the reaction mixture being diluted to stop the reaction.

1. Sample components need to be charged for FASS to be successful. If derivatizing carbohydrates, the derivatized product ideally should have a permanent negative (sulfonic acid groups such as APTS) or positive (quaternary ammonium such as p-aminophenyl ammonium chloride) charge to ensure stacking under all conditions. Weak acid and base functionalites may be suitable depending upon the conditions. If separating underivatized carbohydrates, charge can be imparted through the addition of borate to the sample and/or using a pH sufficiently high for deprotonation.
2. The sample should have a conductivity at least 10 times lower than the separation electrolyte, through either diluting the sample (with water or another low-conductivity solvent) or increasing the concentration of the separation electrolyte.
3. For optimal performance, the injection volume under stacking conditions should be restricted to 1–3% of the length to the window. This restriction arises because of differences in localized EOF between the separation electrolyte and the sample which induces pressure waves at the interface. A low-EOF environment decreases this effect and allows slightly larger volumes to be injected, up to 5% of the capillary volume, above which other factors dominate and lead to a degradation in performance.

3.1.2. Field-Amplified Sample Injection

The difference between FASS and field-amplified sample injection (FASI) is the way the analytes are injected (25, 26). FASS employs a hydrodynamic injection, while FASI uses a voltage injection, depicted schematically in Fig. 1b. The advantage of using a voltage injection is that the analytes will be injected into the capillary by a combination of both their electrophoretic mobility and the EOF (27). This means that if the physical sample volume injected into the capillary is limited to 5% of the capillary volume, which will be defined by the EOF, then there will be a significantly higher number of analyte ions injected with FASI than with FASS. Kamoda et al. used this strategy to improve the sensitivity of sugars derivatized with 3-aminobenzoic acid by 57 times (shown in Fig. 2) and applied this method to improve the detection of N-linked oligosaccharides isolated from glycoproteins separated by 2D gel electrophoresis (28).

1. FASI also requires the analytes to be charged and the same approaches can be used as for FASS discussed above.
2. FASI also suffers the same limitation of sample volume as FASS, with the difference being that in FASI the sample is physically injected into the capillary by EOF. Under a low-EOF

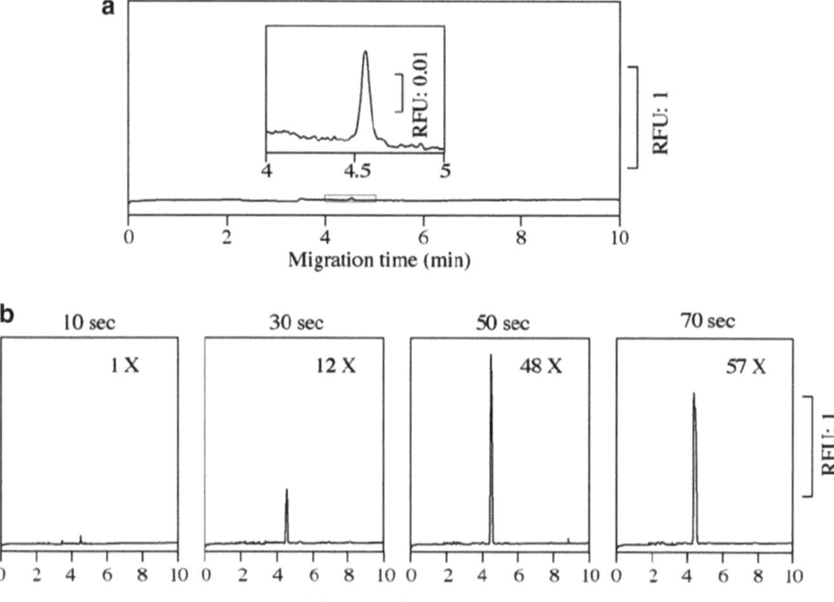

Fig. 2. Analysis of 3AA-Glc$_5$ using (**a**) FASS and (**b**) FASI. Conditions: 100 mM Tris–borate buffer, pH8.3, containing 10% PEG70000 with a DB-1 capillary, 100 μm ID, 30 cm total, (20 cm to detector), 25 kV, 25 °C, LIF excitation at 325 nm, with emission above 405 nm. (a) Pressure injection 1 psi for 10 s. (b) Voltage injection at 10 kV for 10–70 s after 1 psi, 10 s injection of water (From (28) with permission).

environment, enhancements in sensitivity >1,000 can be easily obtained.

3. FASI is a biased injection mode, meaning that analytes with a higher mobility will be injected more than analytes with a lower mobility. This can be a problem when the sample has a highly variable matrix without attempts to compensate beforehand, but with a uniform matrix, the same extent of bias is generally present with calibration standards. It does mean that there will be lowered limits of detection for ions with a higher mobility.
4. Many authors report the benefit of injecting a short hydrodynamic plug of water before the sample (29, 30) (see Note 1).

3.1.3. Large-Volume Sample Stacking

To overcome the limited injection volume of FASS, Burgi and Chien developed large-volume sample stacking (LVSS) (31) which was further refined by He and Lee without polarity switching (LVSEP) (32). This approach allows the injection of a whole capillary volume of sample, with simultaneous stacking of the analytes and removal of the sample matrix by EOF prior to separation. In LVSS, the transition from stacking/matrix removal to separation occurs by manually switching the polarity at the correct time (Fig. 3a). In LVSEP, the transition is controlled chemically through variation of the electroosmotic flow (Fig. 3b). LVSEP has recently

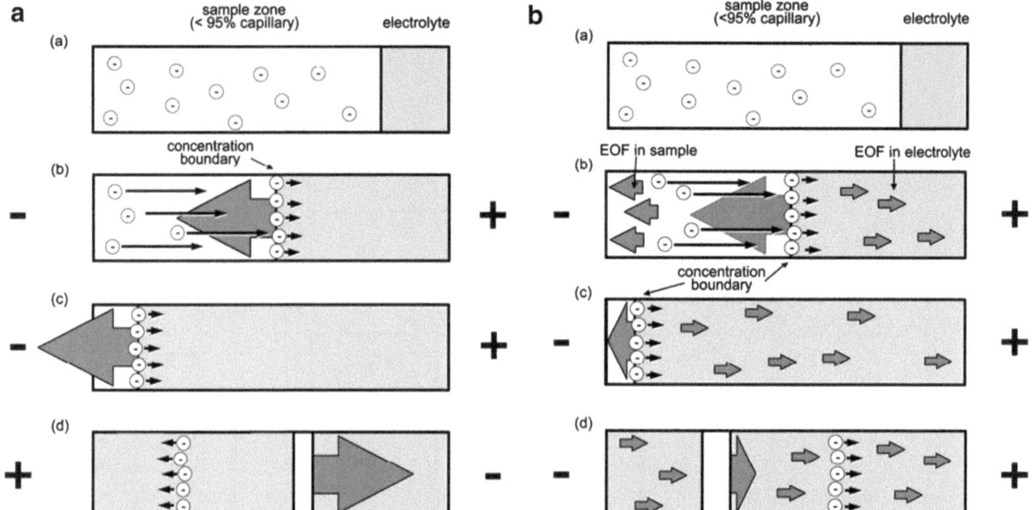

Fig. 3. **A**: LVSS with polarity switching. (a) Sample has a lower conductivity than the electrolyte and fills up to 95% of the capillary. (b) Reverse voltage is applied so that the analytes stack on the rear boundary between the sample and the electrolyte, which moves back towards the inlet due to the EOF. (c) When the optimal current is reached (80–95% of normal), the voltage is reversed. (d) The analytes separate normally in a counter EOF mode; **B**: LVSS with surfactant dissolution used to remove sample matrix. (a) Sample is prepared in a lower-conductivity solution and is injected hydrodynamically (up to 95% of the capillary volume). (b) Voltage is applied so that the analytes stack on the boundary between electrolyte and sample. This boundary moves towards the inlet as the EOF in the sample is of higher magnitude and opposite direction to that in the electrolyte. (c) As the sample matrix is removed, the speed at which the concentration boundary moves is reduced as more of the capillary has a reversed EOF. (d) When almost all of the sample matrix is removed, the concentration boundary stops and reverses its direction, with the analytes then separated as normal.

been developed to improve the separation of APTS derivatized sugars in capillaries, lowering the LOD by 500-fold to 2 pM (33). When implemented in a microchip, the same group reported enhancements up to 2,900-fold within 3 min, with the resulting separation shown in Fig. 4, but no detection limits were given (34). LVSS and LVSEP are one of the few approaches that allow the entire capillary to be filled with sample while still leaving sufficient capillary length for separation because of the matrix removal step.

1. LVSS and LVSEP require the analytes to be charged and in a low-conductivity matrix as for FASS. Again, all of the same approaches can be used.
2. The key requirement for LVSS and LVSEP is that the velocity of the analyte in the sample matrix is higher than that of the EOF which is being used to remove the sample matrix from the capillary.
3. Timing for polarity reversal in LVSS is critical. Too long a matrix removal step results in the stacked analytes exiting the capillary. Too short a removal leaves a large volume of low-conductivity solvent in the capillary which induces additional

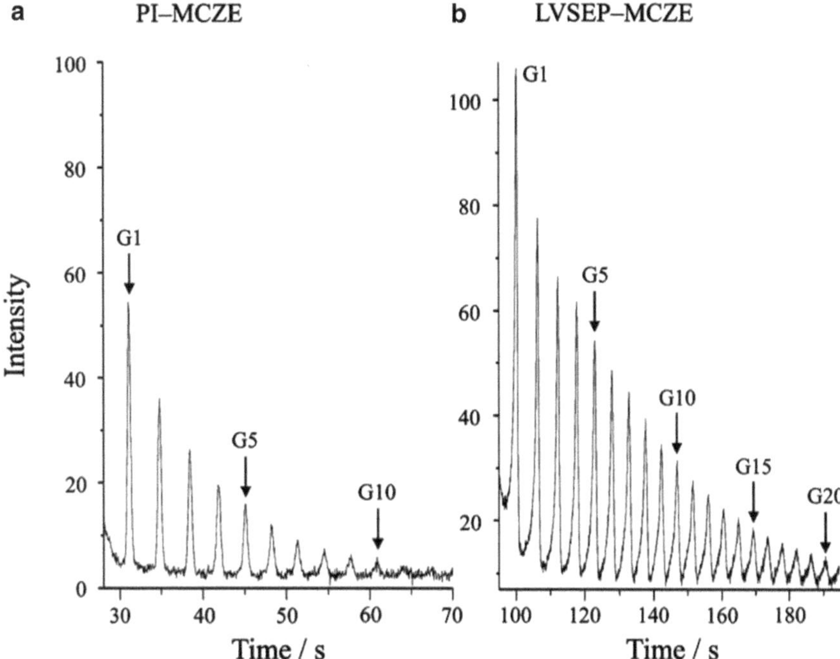

Fig. 4. Microchip electrophoresis of APTS-derivatized glucose ladder with (**a**) conventional pinched injection of 160 ppb mixture and (**b**) microchip LVSEP of 320 ppt mixture (From (34) with permission).

zone broadening. The polarity is typically switched when the current reaches 90–95% of the current when the capillary is fully filled with the separation electrolyte. Commercial CE instruments use time-defined methods steps, thus it is not possible to automatically switch the polarity in LVSS when the current reaches a predetermined value. Manual switching is therefore most frequently used (see Note 2).

4. LVSEP chemically controls the transition from stacking/matrix removal to separation. As this relies on a change in the EOF during the separation, well-controlled surface chemistry is required. It is more difficult to lose analytes out of the inlet when using LVSEP, but the repeatability is often poorer than with LVSS.

3.1.4. Isotachophoresis

Isotachophoresis (ITP) concentrates components based on differences in electric field, but unlike FASS, the electric field is created using leading and terminating electrolytes, with the leading electrolyte positioned in front of the sample in the capillary. The leading electrolyte contains a co-ion with an electrophoretic mobility higher than that of the analytes, while the terminating electrolyte has a co-ion with a lower mobility than that of the analytes. Upon application of the separation voltage, there is a high electric field over the terminator and a low electric field over the leader, which

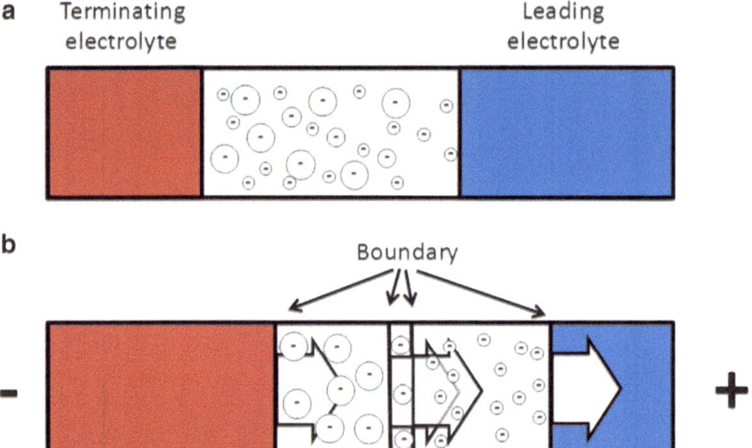

Fig. 5. Schematic of ITP concentration (**a**) immediately before application of the voltage and (**b**) once steady state has been reached. Note that the length of the ITP zones is proportional to the concentration of the ions.

causes ions of intermediate mobility to become arranged in between these two. Concurrently, there is also a concentration adjustment, with ions that exceed their steady-state concentration forming "blocks" while ions that do not exceed their steady state concentration focusing into very sharp peaks (Fig. 5). ITP is a separation technique within its own right, but for use as concentration approach, the ITP system must be transient. This is typically achieved by only having the ITP conditions maintained for a fraction of the separation, with the system transitioning to zone electrophoresis (or another suitable separation mode). Auriola et al. used transient ITP to enhance the detectability of oligosaccharides isolated from *Pseudomonas aeruginosa* with MS detection as shown in Fig. 6 (35). A 10–50-fold improvement in sensitivity was obtained, and the authors estimated that only 30 μg of cell biomass would be required for full identification of the oligosaccharides. One of the advantages of transient ITP is that it is compatible with high-conductivity samples, with high concentrations of ions being reduced to their steady-state concentration.

1. To maximize the range of analytes that can be focused by ITP, the leading ion should have as high an electrophoretic mobility as possible, and the terminator mobility should be as low as possible, and in some cases, can be supplemented with an organic solvent such as acetonitrile. More selective focusing can be achieved by reducing the mobility gap between the leading and terminating ions so that it just bridges the analytes.
2. The concentration of the leading ion defines the concentration to which the sample ions are concentrated to according to the Kohlrausch regulating function. This should be as high as

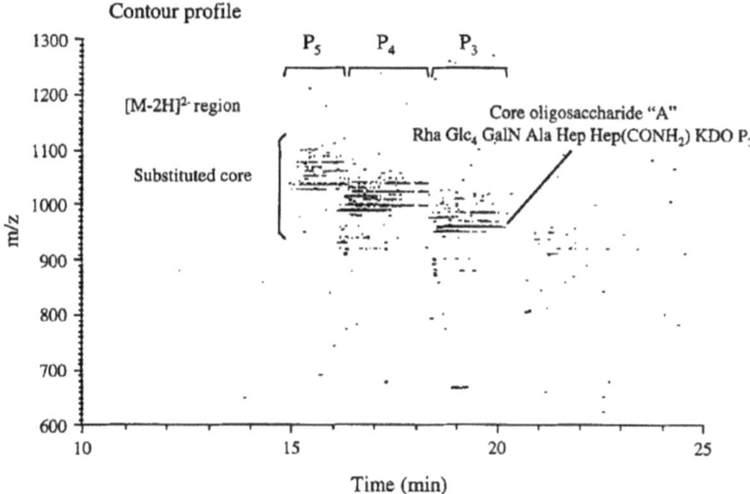

Fig. 6. Contour profile of *m/z* versus time of transient ITP with CE–MS analysis of gel-purified oligosaccharides from the direct hydrolysis of 330 mg of wet cells of Pseudomonas aeruginosa 05. Conditions: 50 mM triethylamine formate, pH 4.5; terminator: 10 mM MES, pH 4.5, −30 kV with concurrent pressure application of 75 mbar, 90 cm × 50 μm ID capillary, sheath flow of 5 μL/min of 10 mM triethylamine, pH 4.5, in 25% aqueous methanol (From (35) with permission).

possible without causing instability problems due to too high a current.

3. Ideally, it is easier to develop ITP systems with a pH between 4 and 10 to avoid the influence of hydrogen and hydroxide ions that can complicate the ITP system (see Note 3).
4. To make the ITP system transient, it is necessary to only have a small volume of leader, terminator, or both. The simplest approach is to use the separation electrolyte ions as leader (terminator) and to inject a small hydrodynamic injection of terminator after the sample (leader before the sample). The ITP system begins to dissipate when the leader (terminator) has completely migrated through the sample zone and reached the separation electrolyte (see Note 4).
5. Maximum injection volumes of ITP are typically 10–20% of the capillary length due to the need to dissipate the ITP zones and then separate the components with sufficient resolution.

3.1.5. Dynamic pH Junction

A pH difference between the sample and electrolyte can be used to concentrate analytes, as shown in Fig. 7; however, analytes must be ionizable, meaning that they are ampholytes, weak acids or weak bases (36–38). A low-pH sample/high-pH electrolyte or a high-pH sample/low-pH electrolyte can be used, with the former system used primarily for weak acids and the latter more ideal for weak bases (39). Dynamic pH junction can theoretically be applied for both native and derivatized carbohydrates but in practice has only

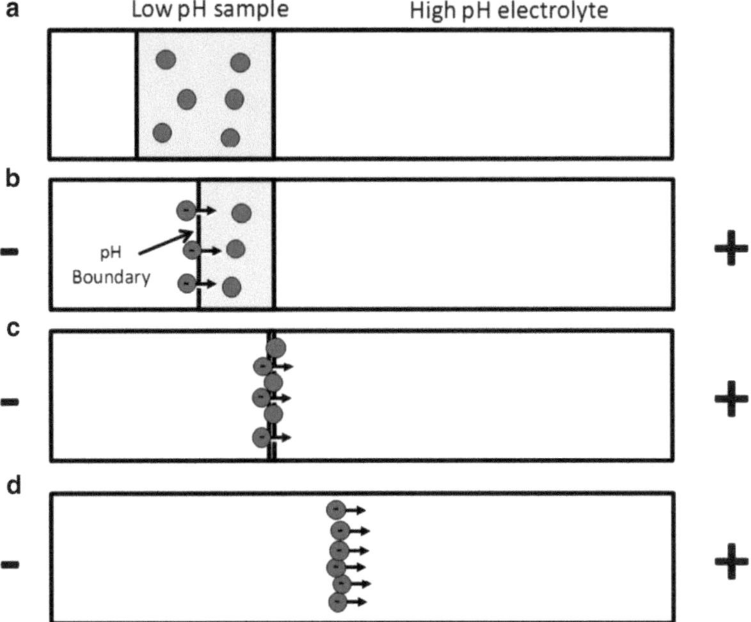

Fig. 7. Schematic of concentration using a dynamic pH junction. (**a**) Sample is injected into the capillary and analytes are neutral and have no mobility in the sample zone; (**b**) application of the voltage causes the rear pH boundary to move through the sample. When reaching analytes, these are immediately deprotonated in the high-pH electrolyte and move towards the cathode back into a low-pH environment. (**c**) Focusing of the weak acid around the pH boundary continues until the boundary has passed entirely through the sample zone; (**d**) the pH boundary dissipates and the weak acids are released to migrate by conventional zone electrophoresis.

been reported for the latter. Kazarian et al. used a low-pH sample (formic acid) and high-pH electrolyte (borate) to improve the sensitivity of carbohydrates modified with novel fluorescein reagent (40–42). This strategy was chosen because the derivatization reaction occurs at low pH and can be directly injected into the capillary, thereby avoiding the dilution step that is normally required. A 515-fold improvement in sensitivity was obtained (Fig. 8) and was later implemented in microchips. One of the advantages of stacking with a dynamic pH junction is that it is tolerant of salt and is therefore compatible with highly saline solutions.

1. The essential requirement for stacking with a dynamic pH junction is that the velocity of the analyte in the separation electrolyte must be greater than the velocity of the sample/electrolyte boundary which must be greater than the velocity in the sample (see Note 5).
2. The velocity of the sample/electrolyte boundary depends on the pH of the sample and electrolyte as well as the composition (concentration and mobility of the anions and cations). As the system is quite complex, optimization can be difficult, which

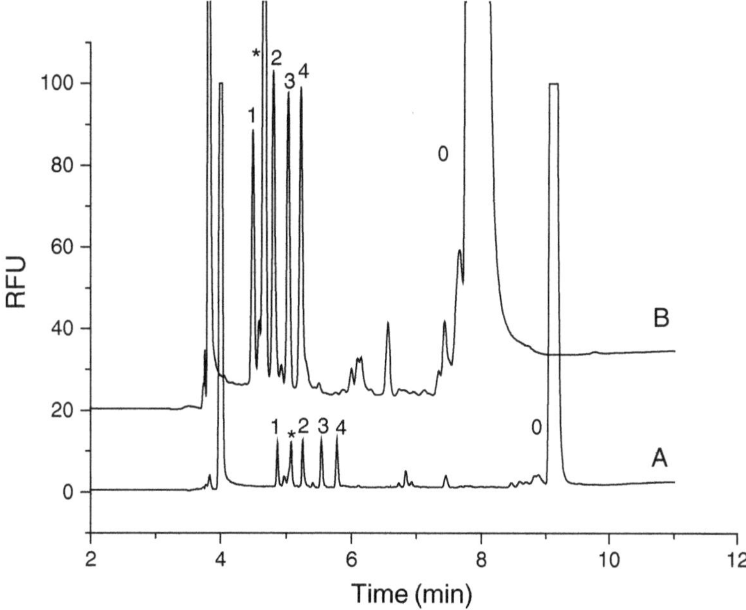

Fig. 8. Electropherograms demonstrating preconcentration via dynamic pH junction in a fused silica capillary where (A) typical CE injection of 4 s at 50 mbar, (B) dynamic pH junction at 40 s, 50 mbar. The analyte peaks were assigned as follows: 0, 0-2-(aminoethyl)fluorescein; 1, maltose; 2, glucose; 3, allose; 4, galactose; *, derivatization peak. Conditions: sample diluted 1000, fused silica capillary coated with PSS and PDADMAC where PSS is the outer most layer, voltage: +30 kV; injection at 50 mbar 4 and 40 s (Modified from (41)).

explains why there have been a limited number of systems developed (see Note 6).

3. Maximum injection volumes with dynamic pH junction are typically 10–20% of the capillary volume due to the need to also separate the concentrated analytes (see Note 7).

3.1.6. Sweeping

Sweeping is distinctly different from stacking. Sweeping relies on the analytes in the sample zone being “swept” into a sharp peak by one of the electrolyte components moving through the sample (Fig. 9). This approach was developed in the late 1990s for neutral molecules separated by micellar electrokinetic chromatography (MEKC) which, because of their lack of electrophoretic mobility, cannot be concentrated by other stacking approaches (43–46). It is most widely used with surfactants; however, any charged electrolyte component can be used for sweeping (47). For carbohydrates, Quirino and Terabe investigated the use of borate sweeping to concentrate monosaccharides (48). The extent of interaction of the analytes with the electrolyte component governs the extent to which the analyte zone is narrowed, and because of the low interaction constants, only a 10–40-fold increase in sensitivity was obtained. When combined with direct

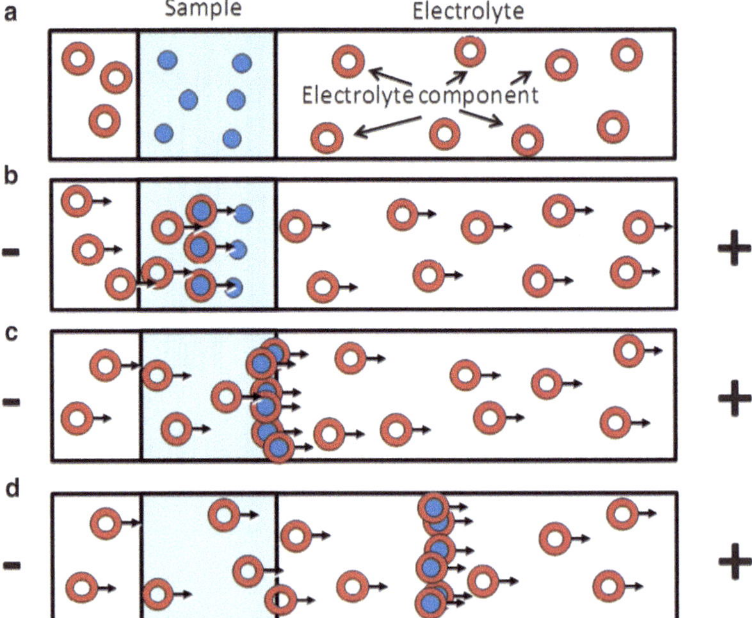

Fig. 9. Schematic of sweeping. (**a**) Sample devoid of the sweeping reagent is injected into the capillary. (**b**) Upon application of the voltage, the sweeping reagent moves through the capillary and accumulates all of the analytes from the sample zone into a sharp zone (**c**). (**d**) After the sample has been swept, the components separate by normal separation principles.

UV absorbance detection, this resulted in very high detection limits (>0.1 mM). Because sweeping is based on an entirely different concentration mechanism, it can be used in conjunction with other stacking methods to refocus analytes already concentrated through another means providing even further improvements in sensitivity (e.g., dynamic pH junction–sweeping).

1. Best results for sweeping have been observed when the sample has a conductivity similar to that of the electrolyte (see Note 8).
2. The use of other salts in the electrolyte can lead to a hybrid sweeping—isotachophoresis concentration mechanism, which is called "high-salt sample stacking" by some papers within the literature. There is no conceptual difference to this and other sweeping methods reported, although there are subtle differences in the way the sweeping reagent is stacked before it sweeps through the sample.
3. For the same reasons as discussed for other techniques, sweeping is also limited to a maximum volume of 10–20% of the capillary.

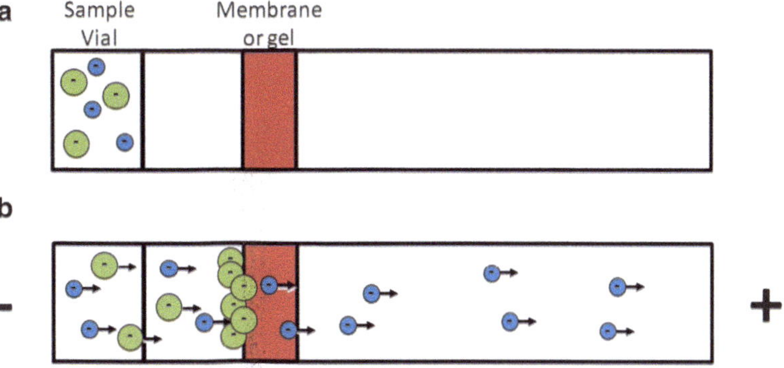

Fig. 10. Schematic of size-exclusion membrane-based concentration. (**a**) The sample contains small and large ions, and the capillary is filled with an electrolyte consisting of small ions. (**b**) Upon application of the voltage, anions move towards the cathode. Small ions freely move through the membrane while large ions cannot and are concentrated at the surface of the membrane.

3.1.7. Membrane Preconcentration

Membrane filtration has long been a method for desalting large molecules based on a size-exclusion mechanism. This idea can also be used to concentrate large compounds in electrophoresis whereby large ions are physically excluded from penetrating the capillary during injection due to a size-exclusion effect and are thus concentrated, as shown in Fig. 10. While this can be implemented in capillaries, it is conceptually easier to integrate with a subsequent separation when using a microchip than it is when using capillaries. Yamamoto et al. used a small polyacrylamide membrane to concentrate APTS-derivatized oligosaccharides (Fig. 11) and improved the sensitivity by 100,000 with a 2 min injection of sample (49). This was coupled with an electrophoretic separation in the same microchip, and a separation of 10^{-8} mol/L APTS-derivatized isomaltooligosaccharides was demonstrated.

1. Because of the planar nature of microchips, it is technically easier to integrate a size-exclusion-based concentration mechanism with a subsequent separation. It can be implemented in capillaries, but the membrane or gel must be placed at the "injection end of the capillary" and sample injected from the "detection end" to allow a subsequent separation to be performed.
2. Without expertise in polymer chemistry, it is easier to integrate a commercially available size-exclusion membrane. In capillaries, this can be done with a capillary joiner, while in microchips it can be done by placing a membrane between microchannels in two different microchip layers during bonding.

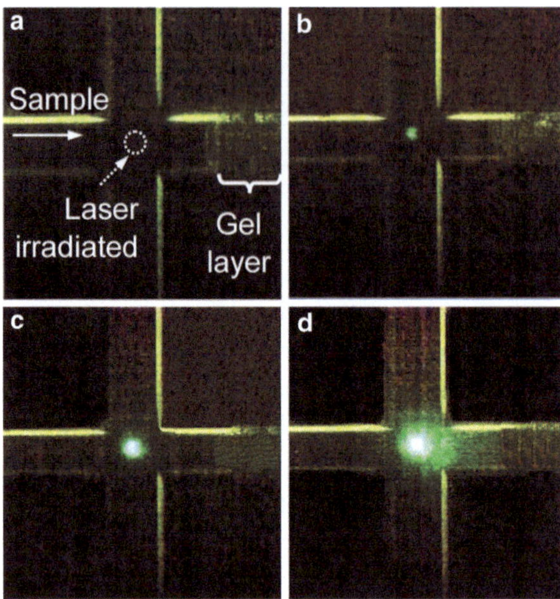

Fig. 11. Time-sequence images of APTS concentration at the channel cross in the PMMA microchannels with in situ fabricated anionic polyacrylamide preconcentrator. All channels were filled with 10 mM acetate of pH 4.2; the left channel is filled with 0.1 mM APTS and applied 100 V to the right channel reservoir. Images (**a-d**) were taken, respectively, after 85, 90, 110, and 125 s (From (49) with permission).

4. Notes

1. While there is much uncertainty within the literature about improvement in peak height, area, efficiency, and resolution, there appears to be some consensus regarding an improved repeatability of the method with a water injection most likely due to the water injection "rinsing" the capillary and electrolyte before entering the sample vial and thus reducing cross contamination of the sample.
2. As the current increases very rapidly towards the end of matrix removal, it is difficult to precisely control this transition. It is possible to improve this by setting a maximum current threshold just above the matrix removal current. This serves to reduce the voltage when the current threshold is reached, and a lower voltage reduces the speed at which stacking and matrix elimination occur. This strategy can be used to decrease the time tolerances required to accurately switch the polarity, leading to higher repeatability when fully automated.
3. It is possible to develop ITP systems at lower or higher pH, and this is usually achieved by neutralization with a counter-ionic species (e.g., to create a cationic terminator at low pH,

the hydrogen is neutralized with a counter-migrating weak acid from the leading electrolyte).

4. Optimization of the leader (terminator) plug is required with optimization of the sample injection volume to ensure that the entire sample is sufficiently stacked.
5. For a weak acid (such as an underivatized sugar), this can easily be achieved by having the pH of the sample >1 pH unit lower than the analytes' pK_a and the separation electrolyte >1 pH unit higher.
6. The most common systems have used a low-pH phosphate buffer and a high-pH borate buffer, with the borate buffer approximately 2–5 times higher concentration.
7. It is important to note that the concentration of the components and the pH of the sample and electrolyte can lead to prolonged duration of the dynamic pH junction even though injection volumes are small, thus requiring careful optimization.
8. As the sample must be devoid of the sweeping reagent, the sample cannot be simply diluted in separation electrolyte. It is usually adjusted with other electrolyte components, or even other buffered salt solutions, but it is essential that there is no sweeping reagent in the sample.

Acknowledgment

The author would like to thank the Australian Research Council for funding and provision of a QEII Fellowship (DP0984745).

References

1. Harvey DJ (2011) Derivatization of carbohydrates for analysis by chromatography; electrophoresis and mass spectrometry. J Chromatogr B 879:1196–1225
2. Kazarian AA, Hilder EF, Breadmore MC (2010) Fluorophores and chromophores for the separation of carbohydrates by CE. In: Volpi N (ed) Capillary electrophoresis of carbohydrates: from monosaccharides to complex polysaccharides. Humana Press, New York, pp 23–52
3. Evangelista RA, Liu MS, Chen FTA (1995) Characterization of 9-aminopyrene-1,4,6-trisulfonate-derivatized sugars by capillary electrophoresis with laser-induced fluorescence detection. Anal Chem 67:2239–2245
4. Lv Z, Sun Y, Wang Y, Jiang T, Yu G (2005) Ultrasensitive capillary electrophoresis of oligoguluronates with laser-induced fluorescence detection. Chromatographia 61:615–618
5. García Moreno MDV, Castro Mejías R, Natera Marín R, García Barroso C (2002) Analysis of sugar acids by capillary electrophoresis with indirect UV detection. Application to samples of must and wine. Euro Food Res Technol 215:255–259
6. Lee YH, Lin TI, Luh TY (1996) Determination of carbohydrates by high-performance capillary electrophoresis with indirect absorbance detection. J Chromatogr B 681:87–97
7. Monahan J, Gewirth AA, Nuzzo RG (2002) Indirect fluorescence detection of simple sugars

via high-pH electrophoresis in poly(dimethylsiloxane) microfluidic chips. Electrophoresis 23:2347–2354

8. Ramírez SC, Carretero AS, Blanco CC, De Castro MHB, Gutiérrez AF (2005) Indirect determination of carbohydrates in wort samples and dietetic products by capillary electrophoresis. J Sci Food Agri 85:517–521
9. Cheng X, Zhang S, Zhang H, Wang Q, He P, Fang Y (2008) Determination of carbohydrates by capillary zone electrophoresis with amperometric detection at a nano-nickel oxide modified carbon paste electrode. Food Chem 106: 830–835
10. Cao Y, Wang Y, Chen X, Ye J (2004) Study on sugar profile of rice during ageing by capillary electrophoresis with electrochemical detection. Food Chem 86:131–136
11. García CD, Henry CS (2003) Direct determination of carbohydrates, amino acids, and antibiotics by microchip electrophoresis with pulsed amperometric detection. Anal Chem 75:4778–4783
12. Huang X, Kok WT (1995) Determination of sugars by capillary electrophoresis with electrochemical detection using cuprous oxide modified electrodes. J Chromatogr A 707:335–342
13. Colón LA, Dadoo R, Zare RN (1993) Determination of carbohydrates by capillary zone electrophoresis with amperometric detection at a copper microelectrode. Anal Chem 65:476–481
14. Maxwell EJ, Ratnayake C, Jayo R, Zhong X, Chen DDY (2011) A promising capillary electrophoresis-electrospray ionization-mass spectrometry method for carbohydrate analysis. Electrophoresis 32:2161–2166
15. Grundmann M, Rothenhöfer M, Bernhardt G, Buschauer A, Matysik FM (2012) Fast counter-electroosmotic capillary electrophoresis-time-of-flight mass spectrometry of hyaluronan oligosaccharides. Anal Bioanal Chem 402(8):2617–2623
16. Campa C, Coslovi A, Flamigni A, Rossi M (2006) Overview on advances in capillary electrophoresis-mass spectrometry of carbohydrates: a tabulated review. Electrophoresis 27:2027–2050
17. Zamfir A, Seidler DG, Schönherr E, Kresse H, Peter-Katalinić J (2004) On-line sheathless capillary electrophoresis/nanoelectrospray ionization-tandem mass spectrometry for the analysis of glycosaminoglycan oligosaccharides. Electrophoresis 25:2010–2016
18. Koller A, Khandurina J, Li J, Kreps J, Schieltz D, Guttman A (2004) Analysis of high-mannose-type oligosaccharides by microliquid chromatography-mass spectrometry and capillary electrophoresis. Electrophoresis 25:2003–2009
19. Klampfl CW, Buchberger W (2001) Determination of carbohydrates by capillary electrophoresis with electrospray-mass spectrometric detection. Electrophoresis 22:2737–2742
20. Breadmore MC (2007) Recent advances in enhancing the sensitivity of electrophoresis and electrochromatography in capillaries and microchips. Electrophoresis 28:254–281
21. Breadmore MC, Dawod M, Quirino JP (2011) Recent advances in enhancing the sensitivity of electrophoresis and electrochromatography in capillaries and microchips (2008–2010). Electrophoresis 32:127–148
22. Breadmore MC, Haddad PR (2001) Approaches to enhancing the sensitivity of capillary electrophoresis methods for the determination of inorganic and small organic anions. Electrophoresis 22:2464–2489
23. Breadmore MC, Thabano JRE, Dawod M, Kazarian AA, Quirino JP, Guijt RM (2009) Recent advances in enhancing the sensitivity of electrophoresis and electrochromatography in capillaries and microchips (2006–2008). Electrophoresis 30:230–248
24. Mikkers FEP, Everaerts FM, Verheggen T (1979) High-Performance Zone Electrophoresis. J Chromatogr 169:11–20
25. Burgi DS, Chien RL (1991) Optimization in Sample Stacking for High-Performance Capillary Electrophoresis. Anal Chem 63:2042–2047
26. Burgi DS, Chien RL (1991) Application of Sample Stacking to Gravity Injection in Capillary Electrophoresis. J Microcol 3:199–202
27. Haglund H, Tiselius A (1950) Zone Electrophoresis in a Glass Powder Column—Preliminary Report. Acta Chem Scand 4:957–962
28. Kamoda S, Nakanishi Y, Kinoshita M, Ishikawa R, Kakehi K (2006) Analysis of glycoprotein-derived oligosaccharides in glycoproteins detected on two-dimensional gel by capillary electrophoresis using on-line concentration method. J Chromatogr A 1106:67–74
29. Zhang CX, Thormann W (1996) Head-column field-amplified sample stacking in binary system capillary electrophoresis: a robust approach providing over 1000-fold sensitivity enhancement. Anal Chem 68:2523–2532
30. Zhang CX, Thormann W (1998) Head-column field-amplified sample stacking in binary system capillary electrophoresis. 2. Optimization with a preinjection plug and application to micellar electrokinetic chromatography. Anal Chem 70:540–548

31. Chien RL, Burgi DS (1992) Sample Stacking of an Extremely Large Injection Volume in High-Performance Capillary Electrophoresis. Anal Chem 64:1046–1050
32. He Y, Lee HK (1999) Large-volume sample stacking in acidic buffer for analysis of small organic and inorganic anions by capillary electrophoresis. Anal Chem 71:995–1001
33. Kawai T, Watanabe M, Sueyoshi K, Kitagawa F, Otsuka K (2012) Highly sensitive oligosaccharide analysis in capillary electrophoresis using large-volume sample stacking with an electroosmotic flow pump. J Chromatogr A 1232:52–58
34. Kawai T, Sueyoshi K, Kitagawa F, Otsuka K (2010) Microchip electrophoresis of oligosaccharides using large-volume sample stacking with an electroosmotic flow pump in a single channel. Anal Chem 82:6504–6511
35. Auriola S, Thibault P, Sadovskaya I, Altman E (1998) Enhancement of sample loadings for the analysis of oligosaccharides isolated from Pseudomonas aeruginosa using transient isotachophoresis and capillary zone electrophoresis—electrospray—mass spectrometry. Electrophoresis 19:2665–2676
36. Aebersold R, Morrison HD (1990) Analysis of Dilute Peptide Samples by Capillary Zone Electrophoresis. J Chromatogr 516:79–88
37. Britz-Mckibbin P, Bebault GM, Chen DDY (2000) Velocity-difference induced focusing of nucleotides in capillary electrophoresis with a dynamic pH junction. Anal Chem 72:1729–1735
38. Britz-Mckibbin P, Chen DDY (2000) Selective focusing of catecholamines and weakly acidic compounds by capillary electrophoresis using a dynamic pH junction. Anal Chem 72: 1242–1252
39. Kazarian AA, Hilder EF, Breadmore MC (2011) Online sample pre-concentration via dynamic pH junction in capillary and microchip electrophoresis. J Sep Sci 34:2800–2821
40. Kazarian AA, Hilder EF, Breadmore MC (2008) Utilisation of pH stacking in conjunction with a highly absorbing chromophore, 5-aminofluorescein, to improve the sensitivity of capillary electrophoresis for carbohydrate analysis. J Chromatogr A 1200:84–91
41. Kazarian AA, Hilder EF, Breadmore MC (2010) Capillary electrophoretic separation of mono- and di-saccharides with dynamic pH junction and implementation in microchips. Analyst 135:1970–1978
42. Kazarian AA, Smith JA, Hilder EF, Breadmore MC, Quirino JP, Suttil J (2010) Development of a novel fluorescent tag O-2-[aminoethyl)fluorescein for the electrophoretic separation of oligosaccharides. Anal Chim Acta 662:206–213
43. Quirino JP, Kim JB, Terabe S (2002) Sweeping: concentration mechanism and applications to high-sensitivity analysis in capillary electrophoresis. J Chromatogr A 965:357–373
44. Quirino JP, Terabe S (1999) Sweeping with an enhanced electric field of neutral analyte zones in electrokinetic chromatography. J High Resol Chromatogr 22:367–372
45. Quirino JP, Terabe S (1999) Sweeping of analyte zones in electrokinetic chromatography. Anal Chem 71:1638–1644
46. Quirino JP, Terabe S, Bocek P (2000) Sweeping of neutral analytes in electrokinetic chromatography with high-salt-containing matrixes. Anal Chem 72:1934–1940
47. Aranas AT, Guidote AM Jr, Quirino JP (2009) Sweeping and new on-line sample preconcentration techniques in capillary electrophoresis. Anal Bioanal Chem 394:175–185
48. Quirino JP, Terabe S (2001) Sweeping of neutral analytes via complexation with borate in capillary zone electrophoresis. Chromatographia 53:285–289
49. Yamamoto S, Hirakawa S, Suzuki S (2008) In situ fabrication of ionic polyacrylamide-based preconcentrator on a simple poly(methyl methacrylate) microfluidic chip for capillary electrophoresis of anionic compounds. Anal Chem 80:8224–8230

Chapter 4

Determination of Monosaccharides Derivatized with 2-Aminobenzoic Acid by Capillary Electrophoresis

Mitsuru Abo, Li-ping He, Kae Sato, and Akira Okubo

Abstract

Reducing monosaccharides were derivatized with 2-aminobenzoic acid (2-AA) through reductive amination using sodium cyanoborohydride as a reductant, and the derivatives were separated by capillary zone electrophoresis with UV detection using 50 mM sodium phosphate (pH 5.5) or 150 mM sodium borate—50 mM sodium phosphate (pH 7.0) running buffer. The derivatives of monosaccharides, which are major components of various carbohydrate materials, were completely separated within 25 min.

Key words: 2-Aminobenzoic acid, Capillary zone electrophoresis, Reductive amination

1. Introduction

Determination of the monosaccharide composition of glycoconjugates is the first step in the elucidation of carbohydrate structures. Since saccharides have no characteristic UV absorption or fluorescence, they are often derivatized with UV detectable or fluorescent reagents. The reducing terminals of saccharides react with amines to produce imines, which are then transformed into secondary amines through reductive amination, using an appropriate reductant. For the analysis of monosaccharides, high separation efficiency is required because several of them are stereoisomers. Capillary electrophoresis (CE) has emerged as a highly promising technique which is capable of high-resolution separation and analyte quantification using an extremely small amount of samples.

Based on the method by Anumula (1, 2) and Bigge et al. (3), we have previously described a rapid, effective, and highly sensitive analysis method using CE (4–7). The derivatives of saccharides

Nicola Volpi and Francesca Maccari (eds.), *Capillary Electrophoresis of Biomolecules: Methods and Protocols*, Methods in Molecular Biology, vol. 984, DOI 10.1007/978-1-62703-296-4_4, © Springer Science+Business Media, LLC 2013

were separated by capillary zone electrophoresis with UV detection using 50 mM sodium phosphate buffer (pH 5.5) or 150 mM sodium borate—50 mM sodium phosphate buffer (pH 7.0) as the running electrolyte solution. In particular, saccharide derivatives with same molecular weights such as 2-AA aldohexoses (mannose and glucose) and 2-AA aldopentoses (ribose and xylose) were also well separated using the either buffers.

2. Materials

All solutions were prepared using ultrapure water (>18.2 MΩ cm at 25°C) and analytical grade reagents. Sodium cyanoborohydride ($NaBH_3CN$) was purchased from Aldrich Chemical Co. Inc. (Milwaukee, WI, USA), which is highly hygroscopic and is thus stored under dry conditions. Other reagents were purchased from Wako Pure Chemical Industries (Osaka, Japan). All carbohydrates were in the D-form.

2.1. Preparation of Running Buffer and Reagent Solution

Sodium phosphate buffer was prepared by dissolving an appropriate amount of sodium dihydrogen phosphate and disodium hydrogen phosphate in Milli-Q water, and the pH was adjusted to 5.5, 6.0, and 7.0.

Sodium borate–phosphate buffer was prepared by dissolving boric acid and sodium dihydrogen phosphate in Milli-Q water, and the pH was adjusted to 7.0 by sodium hydroxide solution. These buffers were filtered through a cellulose nitrate membrane (pore size: 0.45 μm) before use.

The reagent solution containing 0.2 M 2-AA and 1.0 M $NaBH_3CN$ was freshly prepared (see Note 1).

2.2. Measurement Conditions for Capillary Electrophoresis

Waters Quanta 4000E capillary electrophoresis system or Beckman Coulter P/ACE MDQ capillary electrophoresis system was used with a fused silica capillary of 50 μm i.d. and effective length 60 cm (total length 70 cm). The applied voltage was 20 kV and the signal was measured with on-line column detection at 214 nm. The measurement temperature was set at 25°C.

3. Methods

All procedures were carried out at room temperature except the derivatization reaction.

1. An aliquot (30 μl) of a mixture of saccharides (450 nmol each) was placed in a microtube and lyophilized.

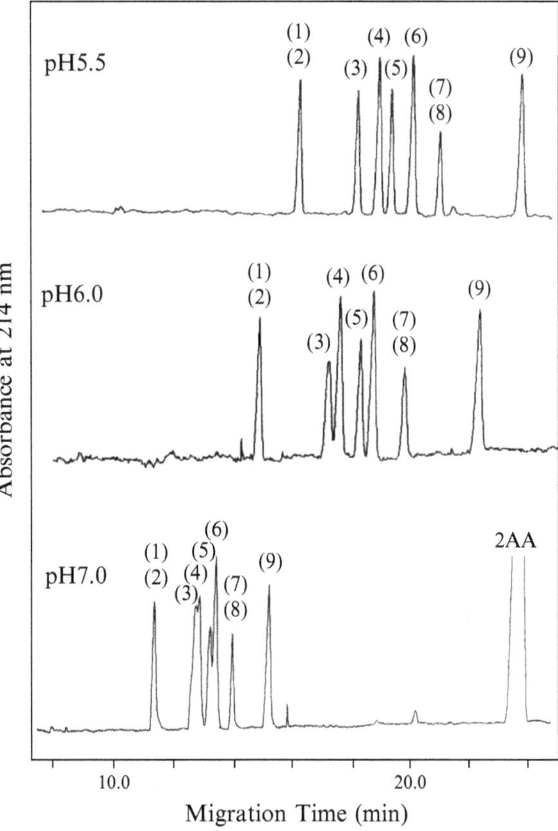

Fig. 1. Electropherograms of saccharide derivatives in 50 mM sodium phosphate buffer at pH varying from 5.5 to 7.0. Concentrations: each 60 μM. Peaks assignment: (1) 2-AA-Cellobiose; (2) 2-AA-Mal; (3) 2-AA-Man; (4) 2-AA-Glc; (5) 2-AA-Rib; (6) 2-AA-Xyl; (7) 2-AA-Threose; (8) 2-AA-Erythrose; (9) 2-AA-Glycelaldehyde.

2. 200 μl of the reagent solution (see Note 1) was added to the lyophilized sample in the microtube (see Note 2).
3. Dissolution of the saccharide(s) was facilitated by gentle vortexing and the resultant solution was then heated at 65°C for 2 h (see Note 3) with the tube capped tightly.
4. After cooling to ambient temperature, the saccharide derivatives were diluted with the running buffer to 50-fold (see Note 2) and mixed vigorously on a vortex mixer.
5. The CE measurements were performed with an operating voltage of 20 kV at 25°C and the UV detector set at 214 nm (see Note 4). Prior to each measurement, the capillary was conditioned by rinsing with 0.2 M NaOH for 2 min, water for 2 min, and the running buffer for 3 min.
6. Sodium phosphate buffer (pH 5.5) (see Note 5) or sodium borate–phosphate buffer (pH 7.0) (see Note 6) was used as the running buffer (Figs. 1 and 2).

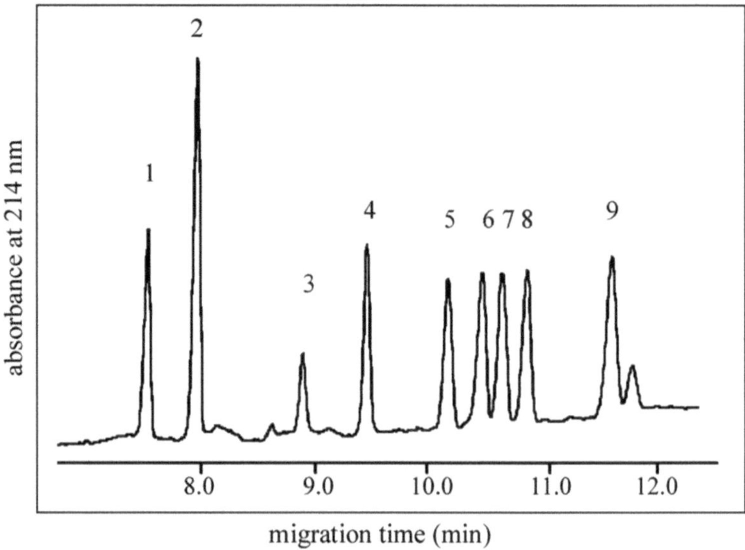

Fig. 2. Electropherogram of saccharide derivatives in 150 mM sodium borate–50 mM sodium phosphate buffer at pH 7.0. Concentrations: each 45 μM except for cellopentaose, 75 μM, and cellotetraose, 150 μM. Peaks assignment: (1) 2-AA-cellopentaose; (2) 2-AA-cellotetraose; (3) 2-AA-GlcN; (4) 2-AA-Rib; (5) 2-AA-Fuc; (6) 2-AA-Man; (7) 2-AA-Xyl; (8) 2-AA-Gal; (9) 2-AA-Glc.

7. The labeled saccharides were introduced into the capillary in hydrostatic mode (10 cm × 15 s or 100 mbar × 5 s) either directly or after dilution (see Note 2).
8. The calibration curves for all the major monosaccharides found in glycoproteins (GalN, GlcN, Fuc, Man, Gal, and Glc) were linear at the concentration ranges from 10 to 500 μM with a detection limit of 3 μM (S/N = 3).

4. Notes

1. The reagent aqueous solution containing 0.2 M 2-AA and 1.0 M $NaBH_3CN$ was freshly prepared just before derivatization. The mildness of $NaBH_3CN$, combined with its effectiveness and stability of aqueous solution, is suitable for biochemical application (8). In fact, the derivatization yields of saccharides were less than 10% when sodium borohydride ($NaBH_4$) instead of $NaBH_3CN$ was used as the reductant (data not shown).

   In addition, it has been reported that the optimum pH for iminium formation is less than 6 due to the acid catalytic reaction and the reduction of the iminium moiety by $NaBH_3CN$ was rapid at pH 6–7 (9). The p*K*a of the amino functional group of 2-AA is 4.95, and it is expected that the optimum pH of iminium formation by 2-AA is more than 4.95, avoiding the protonation of amine group. In this reaction system, the highest yields of derivatized monosaccharides, such as glucosamine (GlcN), mannose, and glucuronic acid (GlcUA), were obtained at pH 5.7 (5). The pH of the reagent solution containing 0.2 M 2-AA and 1.0 M $NaBH_3CN$ was also 5.7.

The yields of 2-AA derivatization for major monosaccharides such as Glc, Gal, Man, and Fuc were more than 90%, based on the absorbance at 214 nm derived from 2-AA.

The use of aprotic organic solvents such as dimethyl sulfoxide and tetrahydrofuran resulted in low derivatization yields (5). Water was found most suitable solvent for this reaction.

2. The reagent solution volume can be changed. When 50 μL of the reagent solution was used, 20 μL of saccharide mixture containing 1.5 mM each standards was lyophilized and then reacted with the reagent. The saccharide derivatives were diluted with the running buffer to tenfold for the CE analysis (6). When 5 μL of the reagent solution was used, the amount of saccharides should be less than 140 nmol (5).

3. For the derivatization of unstable saccharides such as uronic acid, the reaction was performed at 40°C for 16 h.

   In CE analysis, galacturonic acid and glucuronic acid migrated slower than neutral and amino sugar. The migration time of these is longer than that of unreacted 2-AA (5).

4. The detection sensitivity for 2-AA derivatives can be improved by fluorescent detection (2).

5. Below pH 5.5, good separation could not be possibly achieved because the peaks of these derivatives heavily overlapped each other (data not shown).

   The pK_1 and pK_2 for 2-AA-Glc were 2.2 and 4.8, corresponding to the dissociation of carboxyl and amino groups, respectively. Since the 2-AA derivatives of saccharides carry a negative charge at the carboxylic group in the buffer above pH 5.0, the migration order of these depends on the Stokes' radius of the derivatives. The structural consideration on the separation of 2-AA-Glc and 2-AA-Man has been described (6).

6. In a borate-containing running buffer, the separation of 2-AA derivatives was achieved by the differential degrees of complexation of saccharides with borate, which lead to the change of net charge and Stokes' radius. Since the principle of the separation using borate buffer was different from that using a phosphate buffer, this is the alternative choice to obtain a better separation.

References

1. Anumula KR (1993) Quantitative monosaccharide analysis of glycoproteins as anthranilyl derivatives by reverse phase HPLC. Glycobiology 3:511–514
2. Anumula KR (1994) Quantitative determination of monosaccharides in glycoproteins by high performance liquid chromatography with highly sensitive fluorescence detection. Anal Biochem 220:275–283
3. Bigge JC, Patel TP, Bruce JA, Goulding PN, Charles SM, Parekh RB (1995) Non-selective and efficient fluorescent labeling of glycans using 2-aminobenzamide and anthranilic acid. Anal Biochem 230:229–238
4. Sato K, Sato K, Okubo A, Yamazaki S (1997) Determination of monosaccharides derivatized with 2-Aminobenzoic Acid by Capillary Electrophoresis. Anal Biochem 251:119–121
5. Sato K, Sato K, Okubo A, Yamazaki S (1998) Optimization of derivatization with 2-aminobenzoic acid for determination of monosaccharide composition by capillary electrophoresis. Anal Biochem 262:195–197
6. He LP, Sato K, Abo M, Okubo A, Yamazaki S (2003) Separation of saccharidesderivatized with 2-aminobenzoic acid by capillary electrophoresis and their structural consideration by nuclear magnetic resonance. Anal Biochem 314:128–134
7. Chen J, He LP, Abo M, Zhang J, Sato K, Okubo A (2009) Influence of borate complexation on the electrophoretic behavior of 2-AA derivatized saccharides in capillary electrophoresis. Carbohydr Res 344:1141–1145
8. Hutchins RO (1991) Reduction of C=N to CHNH by metal hydrides. Compr Org Syn 8:36
9. Borch RF, Bernstein MD, Durst HD (1971) The cyanohydridoborate anion as a selective reducing agent. J Am Chem Soc 93:2897–2904

Chapter 5

Determination of Mono-, Di-, and Oligosaccharides by Capillary Electrophoresis with Capacitively Coupled Contactless Conductivity Detection

Claudimir Lucio do Lago, Thiago Nogueira, Lucas Blanes, and Renata Mayumi Saito

Abstract

Saccharides and chitooligosaccharides can be separated in electrophoretic conditions by raising the pH of the medium, which renders the corresponding alcoholate forms. These anionic species can be separated and detected with capacitively coupled contactless conductivity detection as negative peaks because of their low mobilities when compared to the hydroxyl mobility, which is the main co-ion in the background electrolyte. Three methods for different matrixes are presented in this chapter.

Key words: Capillary electrophoresis, Conductivity detection, Saccharides, Coffee, Chitooligosaccharides

1. Introduction

Various approaches have been created during the last two decades to allow the analysis of carbohydrates by capillary electrophoresis (CE). Two major problems are commonly associated with the analysis of these molecules by CE: (1) most of carbohydrates lack ionizable or charged functional groups to allow the electrophoretic mobilization, and (2) normally, they neither absorb nor fluoresce (1). Complexation with borate-based electrolytes, formation of metal cation complexes, and high-pH electrolytes are the main strategies used to convert carbohydrates in charged species (1). Carbohydrates analysis by CE has been performed using indirect UV or fluorescence detection, electrochemical detection, and mass spectrometry (1–5). In 2003, Carvalho et al. showed for the first

Nicola Volpi and Francesca Maccari (eds.), *Capillary Electrophoresis of Biomolecules: Methods and Protocols*, Methods in Molecular Biology, vol. 984, DOI 10.1007/978-1-62703-296-4_5, © Springer Science+Business Media, LLC 2013

time the CE analysis of carbohydrates using capacitively coupled contactless conductivity detection (C^4D) (6). Since then, several different sugars including mono-, di-, and some oligosaccharides were analyzed with CE-C^4D (7–11) and microchip-CE-C^4D (12).

C^4D was introduced in 1998 as a sensitive electronic detector for CE (13, 14), and comprehensive details have been described (15–17). Conductivity detection is known as one of the most universal detection systems for CE because practically all species present in the solution contribute to the analytical signal. The C^4D detector is composed of two radial metallic electrodes, which fit around the capillary. A high frequency is applied to one of the electrodes, and the resulting alternated current that passes through the cell to the second electrode is electronically amplified, rectified, filtered, and converted to a digital signal using an analog-to-digital converter. The detection cell is the small gap of around 1 mm between the electrodes. The C^4D detector measures differences in conductivity between migration analyte zones and the background electrolyte, which is converted into the form of a peak, providing the application of C^4D for the determination of non-absorbing species (18, 19). As a rule of thumb, if the mobility of the analyte is higher than the mobility of its co-ion in the electrolyte, the recorded peaks are positive, and if the mobility is lower, a negative peak will be recorded.

Sugars are weak acids, ionizing at strongly alkaline medium ($pH > 12$). Then at high pH, carbohydrates can be converted in anionic species called alcoholates, which result from the ionization of hydroxyl groups of the saccharides (20). The most employed electrolyte for carbohydrate analysis by CE-C^4D is sodium hydroxide (10–80 mmol/L) that allows the conversion of the carbohydrates in anionic species. The experiments are performed in reverse polarity applying (−10 to −25 kV) and adding a surfactant as cetyltrimethylammonium bromide (CTAB) to reverse the electroosmotic flow (EOF). Due to the high conductivity of the electrolyte, it is desirable to use capillaries of small diameters (50 μm or smaller) to avoid Joule's heating. Nevertheless, there is limitation for the separation potential.

Electropherograms of carbohydrates by CE-C^4D using sodium or potassium hydroxide as electrolytes are recorded as negative peaks. In this case, the main anion in the background electrolyte is hydroxyl, which has much higher mobility than the alcoholates do. Taking into account the rule above mentioned, negative peaks are expected in this case.

Carbohydrates are found in different forms and matrices. This chapter describes three different methods:

Method 1: *Determination of mono- and disaccharides in soft drinks, fruit juice, and other beverages.* Fructose, glucose, sucrose, and lactose can be determined using this method (6).

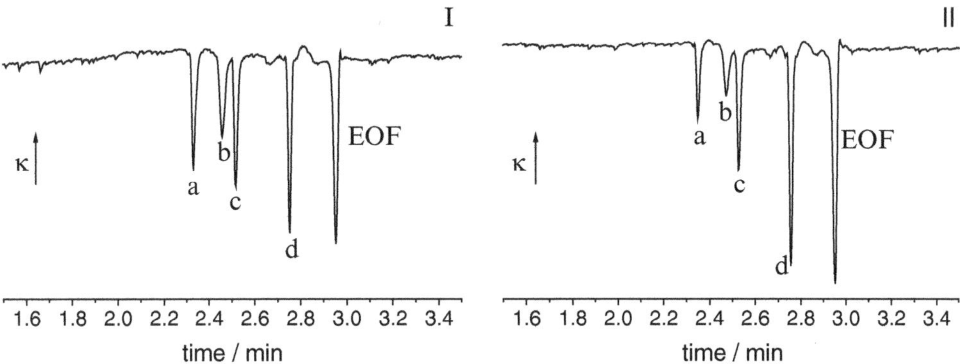

Fig. 1. Electropherograms of (I) a standard solution with 3 mmol L^{-1} of fructose (**a**), glucose (**b**), galactose (**c**), and sucrose (**d**) and (II) a cola soft drink (dilution 2:100 v/v). BGE: NaOH 10 mmol L^{-1}, Na_2HPO_4 4.5 mmol L^{-1}, and CTAB 0.2 mmol L^{-1}. A capillary of 20 μm i.d. and 44 cm long (34 cm effective length) was used.

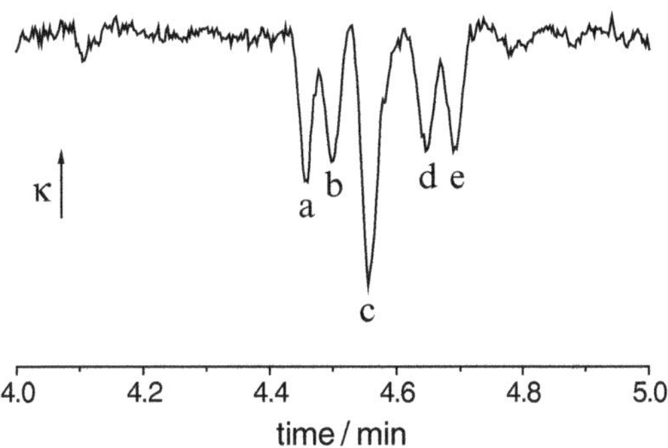

Fig. 2. Electropherogram of a standard solution with 3 mmol L^{-1} of xylose (**a**), fructose (**b**), mannose plus arabinose (**c**), glucose (**d**), and galactose (**e**). BGE: NaOH 80 mmol L^{-1}, CTAB 0.5 mmol L^{-1}, and 30% (v/v) of methanol. A capillary of 20 μm i.d. and 44 cm long (34 cm effective length) was used.

Method 2: *Determination of monosaccharides in coffee samples.* With this method, coffee adulteration can be detected through the observation of xylose and glucose in the sample (11).

Method 3: *Determination of chitooligosaccharides.* This method describes the separation of *N*-acetylglucosamine (GlcNAc), *N*′, *N*′-diacetyl-chitobiose (C2), *N*′,*N*′,*N*′-triacetyl-chitotriose (C3), tetra-*N*-acetyl-chitotetraose (C4), penta-*N*-acetylchito-pentaose (C5), and hexa-*N*-acetyl-chitohexaose (C6) (9).

Electropherograms for these three methods are shown, respectively, in Figs. 1, 2, and 3.

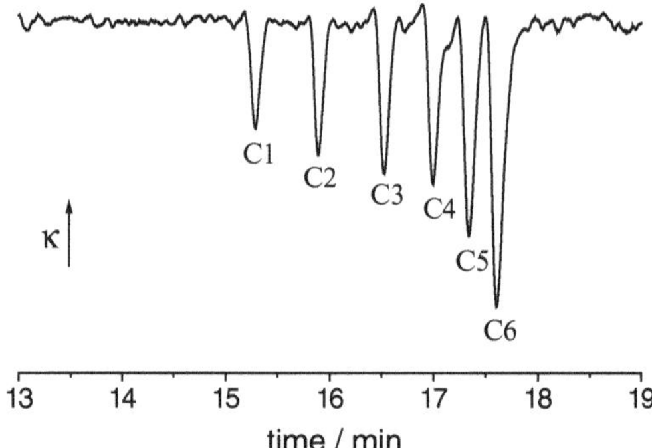

Fig. 3. Electropherogram of an aqueous solution containing 200 μmol L^{-1} of chitooligosaccharides from 1 to 6 U of *N*-acetylglucosamine, respectively, from C1 to C6. BGE: NaOH 10 mmol L^{-1}, Na_2HPO_4 4.5 mmol L^{-1}, CTAB 200 μmol L^{-1}, and 10% (v/v) acetonitrile. A capillary of 50 μm i.d. and 72.5 cm long (62.5 cm effective length) was used.

2. Materials

All the solutions should be prepared using deionized water (resistivity > 18 MΩ cm) and analytical grade reagents.

2.1. Standard Solutions Preparation

The solutions preparation is the same for all the three methods. However, the species are different for each method. The species are named below:

For method 1: Fructose, glucose, sucrose, lactose, and galactose (as internal standard)

For method 2: Arabinose, fructose, galactose, glucose, mannose, sucrose, and xylose

For method 3: *N*-acetylglucosamine (GlcNAc), *N′*, *N′*-diacetylchitobiose (C2), *N′*,*N′*,*N′*-triacetyl-chitotriose (C3), tetra-*N*-acetyl-chitotetraose (C4), penta-*N*-acetylchitopentaose (C5), hexa-*N*-acetyl-chitohexaose (C6), and methionine (as internal standard)

The solutions should be prepared as follows:

1. Prepare separately a 10 mmol L^{-1} solution dissolving the standards in deionized water.
2. Prepare at least six standard solutions containing all the analytes and the internal standard in the concentration ranging from 10 μmol L^{-1} to 100 μmol L^{-1}, diluting the respective 10 mmol L^{-1} solution in deionized water.
3. Put the standard solutions in an ultrasound bath for at least 15 min to expel dissolved CO_2.

2.2. Background Electrolyte Preparation (see Note 1)

Method 1: BGE 1—Prepare an aqueous solution of 10 mmol L^{-1} NaOH, 4.5 mmol L^{-1} Na_2HPO_4, and 0.2 mmol L^{-1} CTAB.

Method 2: BGE 2—Prepare a solution of 80 mmol L^{-1} NaOH, 0.5 mmol L^{-1} CTAB, and methanol 30% v/v.

Method 3: BGE 3—Prepare a solution of 10 mmol L^{-1} NaOH, 4.5 mmol L^{-1} Na_2HPO_4, 0.2 mmol L^{-1} CTAB, and acetonitrile 10% v/v.

2.3. Cleaning Solution

Prepare an aqueous solution of 100 mmol L^{-1} NaOH.

2.4. Capillary Column

Methods 1 and 2: Capillary 1—Fused-silica capillary of 20 μm i.d. × 375 μm o.d and 44 cm long (effective length 34 cm).

Method 3: Capillary 2—Fused-silica capillary of 50 μm i.d. × 375 μm o.d and 72.5 cm long (effective length 62.5 cm).

2.5. Conditioning of Strong Base Anion Exchanger Resin (Only for Method 2)

1. Wash *ca.* 10 g of a strong base anion exchanger resin with 100 mL of 1 mol L^{-1} NaOH solution in a glass beaker under stirring.
2. Filter the resin employing a paper filter.
3. Wash the resin with deionized water until the eluted water reaches pH near 7.

3. Methods

3.1. General Description of the Methods

Method 1: Determination of Mono- and Disaccharides in Soft Drinks, Fruit Juice, and Other Beverages

1. Pipette from 10 μL to 50 μL of the sample in a micro-tube of 1.5 mL. The ideal quantity to be pipetted will depend on the dilution factor necessary. More details are listed at Notes 8 and 9.
2. Add the 10 mmol L galactose solution (internal standard) to the sample to reach a final concentration of 3 mmol L^{-1}.
3. Insert the capillary 1 into the CE equipment.
4. Perform the conditioning of the capillary according to the proceeding described below in Subheading 3.2 using BGE 1.
5. Inject samples and the standard solutions containing all the analytes for 30 s at pressure of 1 kPa.
6. Perform the electrophoretic runs at −25 kV.
7. Execute the identification and quantitation of the analytes as described below in Subheadings 3.3 and 3.4.

Method 2: Determination of Monosaccharides in Coffee Samples

1. Prepare the sample as described in the Subheading 3.5.
2. Insert the capillary 1 into the CE equipment.
3. Perform the conditioning of the capillary according to the procedure described in Subheading 3.2 using BGE 2.
4. Inject samples and the standard solutions containing all the analytes for 30 s at pressure of 1 kPa.
5. Perform the electrophoretic runs at –25 kV.
6. Execute the identification and quantitation of the analytes as described below in Subheadings 3.3 and 3.4.

Method 3: Determination of Chitooligosaccharides

1. Add 1 μL of 10 mmol L^{-1} methionine solution (internal standard) to 10 μL of the sample.
2. Insert the capillary 2 into the CE equipment.
3. Perform the conditioning of the capillary according to the proceeding described below in Subheading 3.2 using BGE 3.
4. Inject samples and the standard solutions containing all the analytes for 30 s at pressure of 1 kPa.
5. Perform the electrophoretic runs at –15 kV.
6. Execute the identification and quantitation of the analytes as described below in Subheadings 3.3 and 3.4.

3.2. Capillary Conditioning (see Notes 2–5)

1. Insert the capillary into the CE equipment according to the recommendations of the fabricant.
2. Flush the capillary with NaOH 0.1 mol L^{-1} for 15 min.
3. Flush the capillary with water for 5 min.
4. Flush the capillary with the specific method BGE for 15 min.
5. Apply specified voltage for 10 min before the first injection (see Notes 6 and 7).

3.3. Identification of the Peaks

1. Spike the sample with one of the analytes, adding 10 μL of the respective 10 mmol L^{-1} standard solution.
2. Inject the spiked sample in the CE equipment.
3. Identify which peak had the signal increased.
4. Repeat steps 1–3 for each analyte.

3.4. Quantitation of the Analytes

1. Obtain the electropherograms for all the standard solutions.
2. Calculate the peak areas of the analyte and the internal standard in the electropherogram.
3. Plot a curve for the relative area of each species (peak area of analyte/peak area of internal standard) vs. concentration.

4. Inject the samples in the CE equipment.
5. Calculate the area of the analytes and internal standard peaks in the electropherogram of the samples.
6. Calculate the concentration of each analyte using the respective analytical curve.
7. Dilute the samples, if necessary, allowing the area of the signals be in the range of the standard solutions concentration and repeat steps from 4 to 6 (see Notes 8 and 9).

3.5. Acid Hydrolysis of Coffee Samples (see Notes 10–16)

1. Add 10 mL of 1 mol L^{-1} HCl solution to 1.00 g of dried instant coffee.
2. Heat under stirring at 90 °C in a water bath for 150 min.
3. Filter the sample using paper filter and complete to 10 mL with deionized water.
4. Add 10 g of a strong base anion exchanger previously conditioned (see Subheading 2.5).
5. Filter using paper filter.

4. Notes

1. The background electrolyte must be prepared daily.
2. To avoid the obstruction of capillary, the BGE should be filtered using a 0.45 μm membrane filter.
3. Before the conditioning of the capillary, double check if the column is not blocked, and cut both ends of the column if necessary.
4. All the solutions must be degassed to avoid the formation of bubbles inside the capillary and consequently the interruption of the current or instability in the baseline.
5. An alternative method to the ultrasound bath to expel dissolved CO_2 in the solutions is to decrease the pressure using, for example, a syringe.
6. Sometimes, even after filling the capillary with BGE and applying the running voltage, the current remains zero. In this case, perform a flush to make sure that no air bubble is inside the capillary.
7. Baseline may become unstable for high electrophoretic current (>20 μA), which requires whether to change the capillary by one of smaller bore or to decrease the concentration of the BGE. This change may be necessary because the efficiency of the heat transfer depends on the equipment.

8. Although the sample dilution of 20–100 times is often suitable, different dilution factors may be required. In this case, dilute the sample in order for the analyte peaks to fit the calibration curve.
9. If the resolution between the peaks is poor and the peak intensity is high, try a new dilution.
10. Acetonitrile removes the polyimide which covers the capillary, and then the contamination of the BGE with the polyimide can cause the variations in the migration time of the analytes. To avoid the attack of acetonitrile, the polyimide should be removed by heating both ends and cleaning the extremities with a tissue paper and methanol.
11. If the objective is the determination of carbohydrates in other matrices, we recommend starting the analysis employing the BGE 1. If two or more peaks were overlapped in the electropherograms, it will be necessary to optimize the separation. So try the following steps:
    - Increasing of the separation voltage.
    - Increasing of the concentration of the BGE.
    - Increasing of the capillary length.
    - Add an organic modifier, for example, acetonitrile, methanol, ethanol, or tetrahydrofuran.
12. When the BGE contains high amount of organic solvent (>20%, v/v), the flush should be performed using positive pressure in the injection side reservoir, instead of vacuum at the end of the column. The high volatility of the solvents may form bubbles inside the capillary when vacuum is applied.
13. The temperature inside of the equipment should be rigorously controlled since the variations can promote alteration in the migration time, baseline derivation, and peak area.
14. Always check the electrophoretic current because much information can be drawn from this parameter.
    - Current is zero suggests that the capillary contains bubble. In this case, try a flush with the BGE.
    - If the current still remains null, check if the capillary is not obstructed. In this case, try cutting both ends or change the capillary.
    - If the current is abnormally low, check if the capillary column is not partially obstructed or if the BGE was wrongly prepared.
    - If the current is abnormally high, check if the BGE was wrongly prepared.

15. The adsorption of macromolecules can affect the migration time. To minimize this trouble, perform a flush between the runs with BGE or cleaning solution followed by BGE (2 min between injections). The flush practice is important not only to minimize the effects of adsorption but to renew the solution inside the capillary.
16. The BGE in the reservoirs must be replaced frequently because the formation of products of microelectrolysis can change the pH value of BGE and cause variations in the migration time.

Acknowledgment

This work was supported by CNPq grant 471054/2008 and fellowships to TN, RMS, and CLL. LB thanks Australian Future Forensics Network (AFFIN) by financial support.

References

1. El Rassi Z (1999) Recent developments in capillary electrophoresis and capillary electrochromatography of carbohydrate species. Electrophoresis 20:3134–3144
2. Soga T, Serwe M (2000) Determination of carbohydrates in food samples by capillary electrophoresis with indirect UV detection. Food Chem 69:339–344
3. Hu Q, Zhou TS, Zhang L, Fang YZ (2001) Study of the separation and determination of monosaccharides in soluble coffee by capillary zone electrophoresis with electrochemical detection. Analyst 126:298–301
4. Campa C, Coslovi A, Flamigni A, Rossi M (2006) Overview on advances in capillary electrophoresis-mass spectrometry of carbohydrates: a tabulated review. Electrophoresis 27:2027–2050
5. Klampfl CW, Buchberger W (2001) Determination of carbohydrates by capillary electrophoresis with electrospray-mass spectrometric detection. Electrophoresis 22:2737–2742
6. Carvalho AZ, da Silva JAF, do Lago CL (2003) Determination of mono- and disaccharides by capillary electrophoresis with contactless conductivity detection. Electrophoresis 24:2138–2143
7. Richter EM, de Jesus DP, Munoz RAA, do Lago CL, Angnes L (2005) Determination of anions, cations, and sugars in coconut water by capillary electrophoresis. J Braz Chem Soc 16:1134–1139
8. Rainelli A, Hauser PC (2005) Fast electrophoresis in conventional capillaries by employing a rapid injection device and contactless conductivity detection. Anal Bioanal Chem 382: 789–794
9. Blanes L, Saito RM, Genta FA, Donega J, Terra WR, Ferreira C, do Lago CL (2008) Direct detection of underivatized chitooligosaccharides produced through chitinase action using capillary zone electrophoresis. Anal Biochem 373:99–103
10. Jaros M, Soga T, van de Goor T, Gas B (2005) Conductivity detection in capillary zone electrophoresis: inspection by PeakMaster. Electrophoresis 26:1948–1953
11. Nogueira T, do Lago CL (2009) Detection of adulterations in processed coffee with cereals and coffee husks using capillary zone electrophoresis. J Sep Sci 32:3507–3511
12. Tanyanyiwa J, Abad-Villar EM, Hauser PC (2004) Contactless conductivity detection of selected organic ions in on-chip electrophoresis. Electrophoresis 25:903–908
13. Zemann AJ, Schnell E, Volgger D, Bonn GK (1998) Contactless conductivity detection for capillary electrophoresis. Anal Chem 70: 563–567

14. da Silva JAF, do Lago CL (1998) An oscillometric detector for capillary electrophoresis. Anal Chem 70:4339–4343
15. Brito-Neto JGA, da Silva JAF, Blanes L, do Lago CL (2005) Understanding capacitively coupled contactless conductivity detection in capillary and microchip electrophoresis. Part 1. Fundam Electroanalysis 17:1198–1206
16. Brito-Neto JGA, da Silva JAF, Blanes L, do Lago CL (2005) Understanding capacitively coupled contactless conductivity detection in capillary and microchip electrophoresis. Part 2. Peak shape, stray capacitance, noise, and actual electronics. Electroanalysis 17: 1207–1214
17. Francisco KJM, do Lago CL (2009) A compact and high-resolution version of a capacitively coupled contactless conductivity detector. Electrophoresis 30:3458–3464
18. Epple R, Blanes L, Beavis A, Roux C, Doble P (2010) Analysis of amphetamine-type substances by capillary zone electrophoresis using capacitively coupled contactless conductivity detection. Electrophoresis 31:2608–2613
19. Kuban P, Hauser PC (2004) Contactless conductivity detection in capillary electrophoresis: a review. Electroanalysis 16:2009–2021
20. Rendlema JA (1973) Ionization of Carbohydrates in presence of metal-hydroxides and oxides. Adv Chem Series 117:51–69.

Chapter 6

Separation of Chitooligosaccharides in Acidic Solution by Capillary Electrophoresis

Toshiaki Hattori, Nobuhiro Anraku, and Ryo Kato

Abstract

Capillary electrophoresis (CE) of chitooligosaccharides (COS) in aqueous solution is effective for their separation from other saccharides. However, COS easily adsorb on negatively charged surfaces, such as fused silica capillaries in acidic solutions. Conventional photometric detection cannot be applied directly because saccharides do not absorb ultraviolet or visible light. Here, we describe a simple CE of COS in an acidic solution using a positively charged capillary coated with *N*-trimethoxypropyl-*N,N,N*-trimethylammonium chloride and indirect photometric detection with crystal violet nitrate background solution.

Key words: Chitooligosaccharide, Capillary electrophoresis, Positively charged coated column, *N*-trimethoxypropyl-*N,N,N*-trimethylammonium chloride, Indirect photometric detection, Crystal violet

1. Introduction

In capillary electrophoresis (CE), large-molecule polyelectrolytes such as DNA (1, 2) and synthetic polyelectrolytes (3) show almost constant electrophoretic migration because of their similar effective charges. Separation of large polyelectrolytes by CE requires a secondary process such as a gel filtration (4). The effective charges of small-molecule or small-oligomer polyelectrolytes change with the number of repeatable units in the polyelectrolyte. With a small number of units, chitooligosaccharides (COS) also have different effective charges and can be separated by simple CE. However, positively charged polymers are difficult to separate by conventional CE with fused silica capillary columns. This is because the

Nicola Volpi and Francesca Maccari (eds.), *Capillary Electrophoresis of Biomolecules: Methods and Protocols*, Methods in Molecular Biology, vol. 984, DOI 10.1007/978-1-62703-296-4_6, © Springer Science+Business Media, LLC 2013

charge on the inner wall of the capillary is negative even in acidic solution (pH >3) and positively charged species strongly adsorb to it. Consequently, it is difficult to separate these species by CE. Blanes et al. (5) developed a method for CE separation of COS in a mixed solvent of 0.01 mol/L NaOH and 10% acetonitrile–water. With this method, larger COS showed higher mobility than smaller COS. Separation of COS in the alkaline solution was successful because anionic species formed and eliminated the adsorption problem on the capillary column. However, separation of anionic COS can be affected by other saccharides. Compared to separation in an alkaline solution, CE of COS in an acidic solution requires treatment of the capillary inner wall to prevent adsorption to the column. Basic proteins can be separated using positively charged capillary columns with multiple ionic polymer layers (6). However, dynamic coatings, such as polyelectrolyte multilayer membranes, are not successful at reducing adsorption of COS. We demonstrated that a positively charged capillary covalently coated with *N*-trimethoxypropyl-*N,N,N*-trimethylammonium chloride effectively separated COS, and the electrophoretic mobility of COS was measured (7).

Saccharides, including COS, do not absorb ultraviolet or visible light. Beaudoin et al. (8) developed a CE separation method for a mixture of COS using an aminopyrene–fluorophore conjugation procedure by laser-induced fluorescence detection. The derivatization method was effective for sensitive detection of COS. However, it required a lengthy pretreatment process for each sample. Therefore, universal detectors, such as contactless conductivity (5) or indirect detection by a conventional photodetector, are effective for the CE of COS. Here, we describe a simple and rapid CE of COS using an indirect detection UV–vis detection with a suitable background solution.

2. Materials

1. COS from enzymatic hydrolysis of fully deacetylated chitosan (Katakura Chikkarin Co., Ltd. Tokyo, Japan) (see Note 1).
2. Fused silica capillary (I.D. 75 μm, O.D. 325 μm, GL Sciences Inc., Tokyo, Japan).
3. *N*-trimethoxypropyl-*N,N,N*-trimethylammonium chloride 50% methanol solution (Gelest Inc., Morrisville, PA).
4. Crystal violet (chloride salt) (Wako Pure Chemical Industries Ltd., Osaka, Japan).

5. Anion-exchange resin beads (IRA 400 J Cl, Organo Corporation, Tokyo, Japan).
6. All solutions were prepared using Milli-Q water (18 MΩ cm at 25°C, Millipore, Billerica, MA).

3. Methods

3.1. Preparation of the Positively Charged Capillary

1. Wash a fused silica capillary (80 cm long) with the following to be passed through at high pressure (100 kPa) 1 mol/L NaOH solution for 1 h, 1 mol/L HNO_3 solution for 1 h, and then pure water for 30 min.
2. Mix *N*-trimethoxypropyl-*N*,*N*,*N*-trimethylammonium chloride 50% methanol solution (0.8 g) with 1 mL of 0.1 mol/L acetate buffer solution (pH 3.5) and 5 mL of ethanol.
3. Run 3 mL of the mixed solution into the capillary (see Note 2).
4. Run 2 mL of ethanol into the capillary (see Note 2).
5. Repeat steps 3 and 4.
6. Dry the coated capillary at 100°C for 18 h.
7. Make a detection window 25 cm from one end of the capillary. The effective length of electromigration for the capillary was 55 cm (see Note 3).

3.2. Preparation of the Background Solution

1. Introduce crystal violet chloride (50 mL, 10^{-3} mol/L) at 1 mL/min to a column containing an anion-exchange resin for nitrate exchange (see Note 4).
2. Pass Milli-Q water (50 mL) through the column at the same rate.
3. Collect the crystal violet eluate (see Note 5).
4. Evaporate water from the eluate under reduced pressure.
5. Dry the crystal violet nitrate salt in a vacuum desiccator.
6. Recrystallized the dry nitrate salt in a water–ethanol solution.
7. Dry the recrystallized nitrate salt in a vacuum desiccator.
8. Dissolve crystal violet nitrate (0.03 g) in 50 mL of Milli-Q water, and adjust the pH to 3.0–4.0 using 1 mol/L HNO_3 (see Note 6).

3.3. CE Procedure

1. Set up the CE including CE apparatus with a high-voltage supply (HCZE-30P No.25, Matsusada Precision Inc., Shiga, Japan), the coated capillary column (80 cm × 50 μm ID, 55 cm effective length), a UV–vis detector (CV^4 CE absorbance detector, ISCO Inc., CA), two vials each containing a Pt electrode, a regulator and N_2 cylinder, and a computer connected

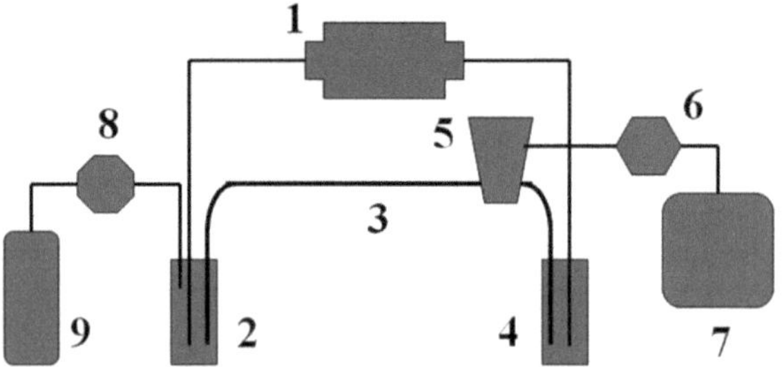

Fig. 1. CE system. 1 High-voltage power supply connected to Pt electrodes, 2 sample vial or running buffer vial, 3 positively charged coated capillary column, 4 receiving vial, 5 UV–vis detector, 6 digital multimeter, 7 computer, 8 regulator, 9 N_2 cylinder.

to a digital multimeter (PC500, Sanwa Electric Instrument Co., Ltd., Tokyo, Japan) (Fig. 1). Signals could be recorded every second on the personal computer to produce an electropherogram.

2. Immerse each electrode in a bottle of the background solution.
3. Pass the background solution through the column at high pressure (100 kPa) for 5 min.
4. Replace the background solution with a bottle of sample solution (see Note 7).
5. Pass the sample solution through the column at low pressure (4 kPa) for 3 s.
6. Replace the sample solution with a bottle of background solution.
7. Initialize detection and record the signal.
8. Apply a constant voltage (−15 kV) and monitor the visible light absorption spectrum at 540 nm.
9. Obtain electropherograms at room temperature by recording the signal every second (Fig. 2a) (see Note 8).
10. Pass the background solution through the column at high pressure (100 kPa) for 5 min after the measurements were complete (see Note 9).

4. Notes

1. COS was prepared by chitosanase from *Bacillus sp.* No. 7-M (9, 10). COS included dimers, trimers, tetramers, pentamers, hexamers, and trace amounts of monomers and higher oligomers (Katakura Chikkarin Co., Ltd. Tokyo, Japan).

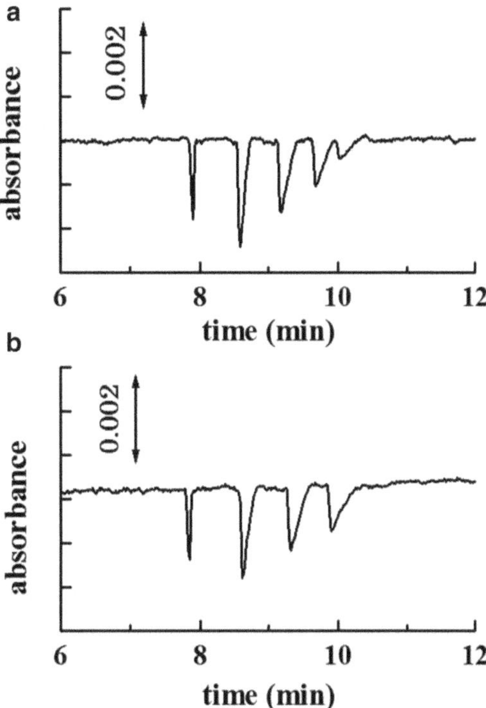

Fig. 2. Electropherograms of COS (10^{-3} eq. mol/L). (**a**) Background solution prepared from crystal violet nitrate and HNO_3 at pH 3. (**b**) Background solution prepared from crystal violet chloride and HCl at pH 3.

2. The mixed solution and ethanol were contained in a 10 mL glass syringe. After the syringe was connected to the capillary, the solution was injected into the capillary using the syringe plunger.
3. The electroosmotic flow was confirmed by electromigration of *N,N*-dimethylformamide as a neutral maker at 214 nm. The typical electroosmotic force was 6.4×10^{-4} cm^2 V^{-1} s^{-1}.
4. The column containing an anion-exchange resin for nitrate exchange was prepared as follows. Anion-exchange resin beads (IRA 400 J Cl, Organo Corporation, Japan) were soaked in 100 mL of 0.1 mol/L HNO_3 for 24 h. The soaked beads (50 mL) were packed into a 50 mL of burette (ø10mm) with glass wool at the bottom. Any air bubbles were removed. A solution of 0.1 mol/L HNO_3 (200 mL) was passed through the packed beads at 1 mL/min. Finally, water was passed through until the eluate pH was 7.
5. The absence of chloride ions after anion exchange was checked by adding a few drops of 0.1 mol/L silver nitrate to the anion-exchange solution to confirm that no white precipitate formed.

6. Chloride ion contamination must be avoided, and salt leakage from the reference electrode should be monitored using a pH meter. The amount of 1 mol/L HCl to be added was checked in a preliminary test. The background solution was checked using a pH meter.
7. COS was dissolved in 10^{-3} M of HNO_3. The peak areas were used to plot calibration curves for each oligomer (dimer, trimer, tetramer, pentamer, hexamer) (Dainichiseika Color & Chemicals, Mfg., Co. Ltd., Tokyo, Japan). To determine the amount of sample injected, a constant amount of tetraphenylphosphonium ion was added to the sample solution as an internal standard (7).
8. When using a chloride background solution, it was difficult to detect COS hexamers (Fig. 2b). A mixed background solution containing crystal violet chloride and HNO_3 was also unsuitable.
9. The capillary was washed with 10^{-3} mol/L HNO_3 before storage. However, the positively charged capillary gradually decomposed during CE runs and with storage.

References

1. Dong Q, Stellwagen E, Dagle JM, Stellwagen NC (2003) Free solution mobility of small single-stranded oligonucleotides with variable charge densities. Electrophoresis 24:3323–3329
2. Hoagland DA, Arvanitidou E, Welch C (1999) Capillary electrophoresis measurements of the free solution mobility for several model polyelectrolyte systems. Macromolecules 32:6180–6190
3. Cottet H, Gareil P (2000) From small charged molecules to oligomers: a semiempirical approach to the modeling of actual mobility in free solution. Electrophoresis 21:1493–1504
4. Altria KD (1996) Capillary electrophoresis guidebook: principles, operation, and applications. Method Mol Biol 52:9
5. Blanes L, Saito RM, Genta FA, Donegá J, Terra WR, Ferreira C, do Lago CL (2008) Direct detection of underivatized chitooligosaccharides produced through chitinase action using capillary zone electrophoresis. Anal Biochem 373:99–103
6. Katayama H, Ishihama Y, Asakawa N (1998) Stable cationic capillary coating with successive multiple ionic polymer layers for capillary electrophoresis. Anal Chem 70:5272–5277
7. Hattori T, Anraku N, Kato R (2010) Capillary electrophoresis of chitooligosaccharides in acidic solution: its simple determination method using quaternary-ammonium-modified column and indirect photometric detection with Crystal Violet. J Chromatogr B 878:477–480
8. Beaudoin M-E, Gauthier J, Boucher I, Waldron KC (2005) Capillary electrophoresis separation of a mixture of chitin and chitosan oligosaccharides derivatized using a modified fluorophore conjugation procedure. J Sep Sci 28:1390–1398
9. Izume M, Ohtakara A (1987) Preparation of d-glucosamine oligosaccharides by the enzymatic hydrolysis of chitosan. Agric Biol Chem 51:1189–1191
10. Uchida Y, Ohtakara A (1988) Chitosanase from Bacillus species. Methods Enzymol 161:501–505

Chapter 7

Capillary Electrophoresis for the Analysis of Glycosaminoglycan-Derived Disaccharides

Yuqing Chang, Bo Yang, Amanda Weyers, and Robert J. Linhardt

Abstract

Capillary electrophoresis is a common technique used for glycosaminoglycan-derived disaccharide analysis because of its high resolving power, high separation efficiency, high sensitivity, short analysis time, and straightforward operation. CE coupled to laser-induced fluorescence (LIF) detection shows an approximately 100 times higher sensitivity than traditional UV detection at 232 nm. 2-Aminoacridone (AMAC) is a widely used fluorophore for labeling unsaturated disaccharides by deductive amination, which is one of the most important method of derivatization of disaccharides for CE-LIF detection. Outlined in this chapter is a protocol of analyzing glycosaminoglycan-derived disaccharides by CE-LIF with AMAC derivatization.

Key words: Glycosaminoglycan, Disaccharide, 2-Aminoacridone, Capillary electrophoresis, Laser-induced fluorescence

1. Introduction

Heparin (HP), heparan sulfate (HS), chondroitin sulfate (CS), and dermatan sulfate (DS) are linear, highly charged polysaccharides that belong to the glycosaminoglycan (GAG) family (1), which are known to mediate in many life processes. HP is a widely used anticoagulant drug (2), and both HP and HS are implicated in cellular processes, such as regulation of enzymic catalysis, and cell–cell interaction (3, 4). CS/DS may be involved in participating and mediating cell–cell interaction and communication (5). The polysaccharide chains of HP/HS and CS/DS have closely related structures and consist of a repeating disaccharide structure, which consists of a hexosamine and a uronic acid. For HP/HS, the hexosamine could be either *N*-acetylated (GlcNAc), *N*-sulfonated

Nicola Volpi and Francesca Maccari (eds.), *Capillary Electrophoresis of Biomolecules: Methods and Protocols*, Methods in Molecular Biology, vol. 984, DOI 10.1007/978-1-62703-296-4_7, © Springer Science+Business Media, LLC 2013

(GlcNS), or unsubstituted (GlcNH), all of which can be 3- and/or 6-*O*-sulfonated; the uronic acid could be either glucuronic acid (GlcA) or iduronic acid (IdoA), both of which can be 2-*O*-sulfonated. For CS/DS, the hexosamine is *N*-acetylgalactosamine (GalNAc) and can be sulfated at 4- and/or 6-position, and the uronic acid could be sulfated at 2 position. Eight, commercially available, enzymatically prepared, HP/HS unsaturated disaccharide standards are described in Table 1, and CD/DS disaccharide standards are described in Table 2. Defining the structure of GAGs is an important factor in elucidating their structure–activity relationship. A common strategy for detailed structural analysis of GAGs involves either complete or partial depolymerization by either enzymatic or chemical means to obtain constituent disaccharides, for disaccharide analysis (6). Modern separation techniques, including high-performance liquid chromatography (HPLC) (7), gel permeation chromatography (GPC) (8, 9), polyacrylamide gel electrophoresis (PAGE) (10), and CE (11, 12), have been applied to HP/HS and CS/DS analysis to help solve many complex structures.

Capillary electrophoresis (CE) is one of the most powerful techniques for GAG analysis because of its high resolving power, high separation efficiency, high sensitivity, short analysis time,

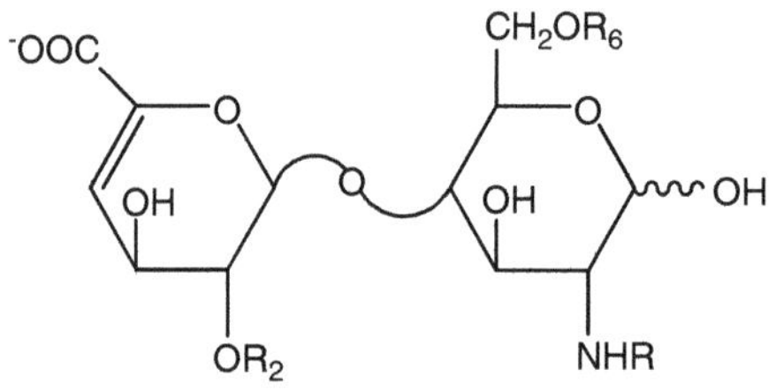

Table 1
The structures of the eight Δ-disaccharide standards from HP/HS

Reference number	Disaccharide	Formulas	R_2	R_6	R
1	TriS	ΔUA(2S)-GlcNS(6S)	SO_3^-	SO_3^-	SO_3^-
2	2S6S	ΔUA(2S)-GlcNAc(6S)	SO_3^-	SO_3^-	Ac
3	2SNS	ΔUA(2S)-GlcNS	SO_3^-	H	SO_3^-
4	NS6S	ΔUA-GlcNS(6S)	H	SO_3^-	SO_3^-
5	2S	ΔUA(2S)-GlcNAc	SO_3^-	H	Ac
6	6S	ΔUA-GlcNAc(6S)	H	SO_3^-	Ac
7	NS	ΔUA-GlcNS	H	H	SO_3^-
8	0S	ΔUA-GlcNAc	H	H	Ac

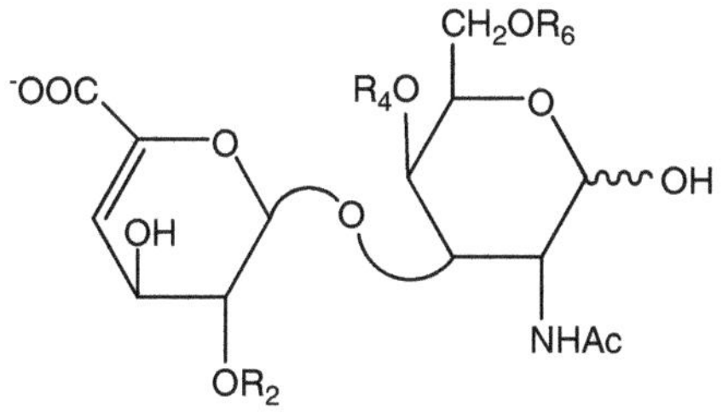

Table 2
The structures of the eight Δ-disaccharide standards from CS/DS

Reference number	Disaccharide	Formulas	R_2	R_4	R_6
9	TriS	ΔUA(2S)-GalNAc(4S)(6S)	SO_3^-	SO_3^-	SO_3^-
10	S_D	ΔUA(2S)-GalNAc(6S)	SO_3^-	H	SO_3^-
11	S_B	ΔUA(2S)-GalNAc(4S)	SO_3^-	SO_3^-	H
12	S_E	ΔUA-GalNAc(4S)(6S)	H	SO_3^-	SO_3^-
13	2S	ΔUA(2S)-GalNAc	SO_3^-	H	H
14	6S	ΔUA-GalNAc(6S)	H	H	SO_3^-
15	4S	ΔUA-GalNAc(4S)	H	SO_3^-	H
16	0S	ΔUA-GalNAc	H	H	Ac

straightforward operation (13), and compatibility with a variety of detection methods, including MS, NMR, and LIF (6). GAG-derived disaccharides can be detected by CE with UV detector at 232 nm created by the unsaturated bond in nonreducing uronic acid residue, which has an extinction coefficient of approximately 5,500 M^{-1} cm^{-1} (14). The addition of a fluorophore can greatly change the chromatographic properties of GAG-derived disaccharides and increase the sensitivity when detected by both UV and LIF detector (16). Reductive amination is one of the most frequently used derivatization method, and a number of labeling reagents have been applied to GAG analysis, such as 2-aminopyridine (2-AP), 7-aminonaphthalene-1,3-disulfonic acid (ANDS), 8-aminonaphthalene-1,3,6-trisulfonic acid (ANTS), and 2-aminoacridone (AMAC) (15). AMAC is a neutral fluorophore with $\lambda_{exc}=428$ nm and $\lambda_{em}=525$ nm, which has been previously used to HP/HS and CS/DS analysis (11, 12, 16–20). By using AMAC as labeling molecule, sensitivity for detection of GAG-derived disaccharides is greatly enhanced, and resolution is also improved (15).

2. Materials

Prepare all solutions with HPLC-grade water (purchased from Sigma-Aldrich) and analytical-grade reagents unless indicated otherwise. Prepare and store all solutions and reagents at room temperature unless indicated otherwise.

2.1. Separation of Eight HP/HS Unsaturated Disaccharide Standards and Eight CS/DS Disaccharide Standards by High-Performance Capillary Electrophoresis

2.1.1. Derivatization of Unsaturated HP/HS and CS/DS Disaccharides with AMAC

1. HP/HS and CS/DS unsaturated disaccharide standards (see Note 1).
2. AMAC (≥98%, Sigma-Aldrich). Store at −20°C.
3. Dimethyl sulfoxide (DMSO).
4. Labeling solution. 0.1 M AMAC in glacial acetic acid–DMSO (3:17, v/v). Store at 4°C.
5. Reducing agent. 1 M $NaBH_3CN$:82.84 mg sodium cyanoborohydride ($NaBH_3CN$) in 1 mL water (see Note 2).
6. Reconstitution solution. DMSO–water (1:1, v/v). Store at 4°C.
7. Lyophilizer.
8. Centrifuge.
9. Water bath set to 45°C.

2.1.2. Separation and Calibration Curve of Eight HP/HS Disaccharide Standards and Eight CS/DS Disaccharide Standards

1. HPCE system (Agilent Technologies).
2. ZetaLif (Picometrics, France) detector (λ_{exc} = 488 nm).
3. Uncoated fused-silica capillary (50 μm i.d., 85 cm total length, 70 cm effective length).
4. Preconditioning solution. 1 M sodium hydroxide (NaOH) (see Note 3).
5. 0.22 μm Steritop filters Millipore.
6. Running buffer. 50 mM phosphate buffer, pH 3.5 (see Note 4). Dissolve 0.6000 g monosodium phosphate (NaH_2PO_4) in about 90 mL of HPLC-grade water and titrate with 1 M HCl to pH 3.5. Make up volume to 100 mL.

2.2. Analysis of GAG-Derived Disaccharides in Biological Sample

2.2.1. Extract and Enzymic Degradation of GAGs

1. PBS buffer: Dissolve 8 g sodium chloride (NaCl), 0.20 g potassium chloride, 1.44 g disodium phosphate (Na_2HPO_4), and 0.24 g monopotassium phosphate (KH_2PO_4) in 800 mL of water. pH should be adjusted to 7.4 with dilute HCl. Make up volume with water to 1 L.
2. Defatting Solutions
   (a) 2:1 chloroform–methanol: Mix two parts chloroform to one part methanol (v/v)

(b) 1:1 chloroform–methanol: Mix equal parts of chloroform and methanol

(c) 1:2 chloroform–methanol: Mix one part chloroform with two parts methanol

3. Actinase E solution: 2 mg/mL solution of actinase E (Kaken Pharmaceutical, Japan). Add water to dry enzyme powder and let it dissolve slowly. This should be prepared immediately before use.
4. Water bath set to 55°C.
5. Tissue homogenizer: LabGEN 7 (Cole-Parmer, Illinois).
6. Strong anion-exchange column: Vivapure Q IEX H ion exchange columns (Sartorius Stedim, France)
7. Urea/chaps solution: Dissolve 480.48 g urea and 20 g chaps (3-[(3-cholamidopropyl)dimethylammonio]-1-propanesulfonate) in 1 L of water.
8. 200 mM NaCl solution: 11.69 g NaCl in 1 L water.
9. 16% NaCl solution: 160 g NaCl in 1 L water.
10. Dialysis membrane (3, 10, 30 kDa MWCO Spectra/Por Dialysis Membrane, Spectrum Labs, California)
11. Centrifugal filters (3 and 3.5 kDa MWCO Amicon Ultra Centrifugal Filter Units, Millipore, Massachusetts)
12. Heparin lyase I (EC 4.2.2.7), heparin lyase II (no EC assigned), and heparin lyase III (EC 4.2.2.8). Cloning, *E. coli* expression, and purification of the recombinant heparin lyase I, II, and III from F. heparinum were performed in our laboratory as described (21–23). These enzymes are commercially available (Sigma Chemical, Missouri).

3. Methods

Carry out all procedures at room temperature unless otherwise specified.

3.1. Separation of Eight HP/HS Unsaturated Disaccharide Standards and Eight CS/DS Unsaturated Disaccharide Standards

3.1.1. Derivatization of Unsaturated HP/HS and CS/DS with AMAC

1. Lyophilize unsaturated disaccharide standards.
2. Add 5 μL (per 10 nmol of disaccharides) of 0.1 M AMAC in glacial acetic acid–DMSO (3:17, v/v) solution to the lyophilized disaccharides (see Note 5).
3. Add 5 μL (per 10 nmol of disaccharides) of reducing agent 1 M $NaBH_3CN$ to the reaction mixture (see Note 6).
4. Incubate the reaction mixture in water bath at 45°C for 4 h.
5. Make mixtures of HP/HS and CS/DS unsaturated disaccharide standards and make up the reaction mixture with re-constitution solution to desired volume (see Note 7).

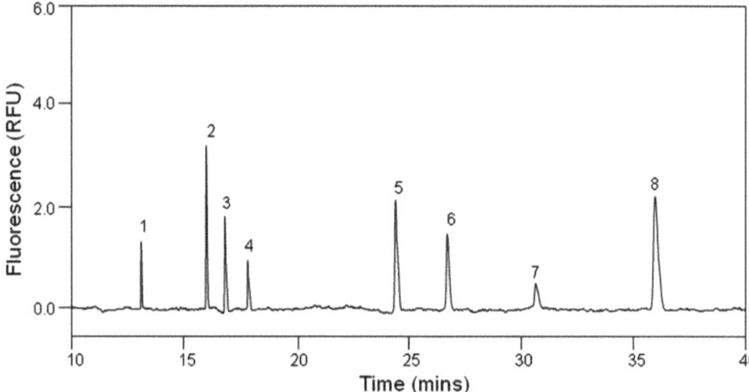

Fig. 1. Electrophoregram of eight HP/HS Δ-disaccharides. Analysis was performed at 25°C, pressure injection of 50 mbar × 10 s, using 50 mM phosphate buffer, pH 3.5, under 30 kV with reversed polarity.

3.1.2. Separation and Calibration Curve of Eight HP/HS Unsaturated Disaccharide Standards and Eight CS/DS Unsaturated Disaccharide Standards

1. Precondition the capillary before each run (see Note 8).
   (a) Flush with 1 M NaOH for 2 min
   (b) Flush with water for 2 min
   (c) Flush with running buffer for 3 min
2. Inject the sample by pressure mode for 50 mbar × 10 s at reversed polarity.
3. Separation is taken under 30 kV. The separation profiles of eight HP/HS unsaturated disaccharides and eight CS/DS unsaturated disaccharides are shown in Figs. 1 and 2.
4. Run a series of disaccharide standard solutions of different concentrations. Draw calibration curve for HP/HS and CS/DS (see Table 3).

3.2. Analysis of GAG-Derivatized Disaccharides in Biological Sample

3.2.1. Extract and Enzymic Degradation of GAGs

1. Cut tissues into 2–3 cm square pieces. Wash tissues in PBS buffer to remove excess blood. Freeze-dry to remove excess water.
2. Defat tissues using defatting solutions A–C. Immerse tissue successively in each solution, starting with A, for 12–24 h. Pour off solution (see Note 9). Allow tissue to fully dry (to remove all organic solvent) before continuing.
3. Submerse dry, defatted tissue in actinase E solution and incubate at 55°C for at least 12 h. Use 5–10 mL solution for every 1 g of tissue. After the first 12 h, homogenize tissue with tissue homogenizer. If undigested tissue remains, additional enzyme may be added and incubate for additional time as needed.
4. Filter digested material through 0.22 μm filter to remove unsolubilized material. Add urea and chaps to make the resulting filtered solution 8 M urea and 2 wt.% chaps (0.48 g urea and 0.02 g chaps for every mL of solution).

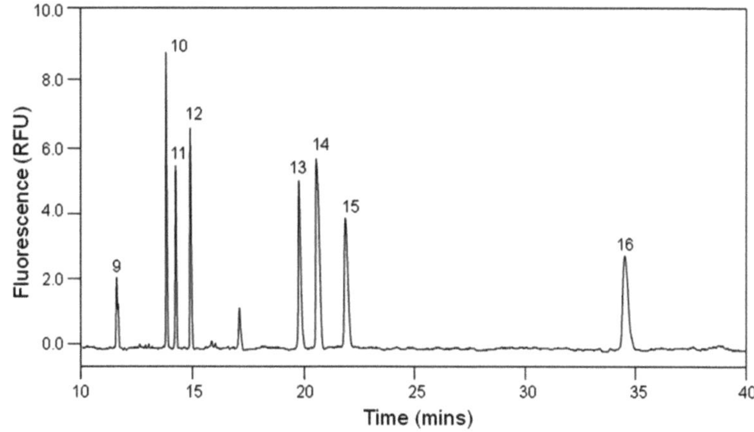

Fig. 2. Electrophoregram of eight CS/DS Δ-disaccharides. Analysis was performed at 25°C, pressure injection of 50 mbar × 10 s, using 50 mM phosphate buffer, pH 3.5, under 30 kV with reversed polarity.

Table 3
Linearity equations for AMAC-derivatized HP/HS and CS/D Δ-disaccharides

Disaccharide type	Reference number	Disaccharide	Linearity equations
HP/HS[a]	1	TriS	$Y=0.72718\ X+0.06179$, $R^2=0.98859$
	2	2S6S	$Y=2.34914\ X+0.28360$, $R^2=0.99523$
	3	NS2S	$Y=1.25743\ X+0.30320$, $R^2=0.98814$
	4	NS6S	$Y=0.83603\ X+0.17811$, $R^2=0.98841$
	5	2S	$Y=6.05419\ X+0.55524$, $R^2=0.99884$
	6	6S	$Y=4.33948\ X+0.34730$, $R^2=0.99313$
	7	NS	$Y=5.02764\ X+0.75179$, $R^2=0.98820$
	8	0S	$Y=7.66238\ X+1.03039$, $R^2=0.99517$
CS/DS[b]	9	TriS	$Y=1.86003\ X+1.87668$, $R^2=0.99517$
	10	S_D	$Y=1.68848\ X+8.25444$, $R^2=0.97362$
	11	S_B	$Y=1.01521\ X+3.28328$, $R^2=0.97650$
	12	S_E	$Y=1.63450\ X+1.91876$, $R^2=0.99964$
	13	2S	$Y=3.03737\ X+7.05318$, $R^2=0.99149$
	14	6S	$Y=3.14119\ X+6.78632$, $R^2=0.99826$
	15	4S	$Y=2.69588\ X+6.61650$, $R^2=0.99273$
	16	0S	$Y=1.55614\ X+8.42892$, $R^2=0.98708$

[a]Tested range is from 0.2 to 5 ng/μL
[b]Tested range is from 0.3 to 25 ng/μL

5. Isolate GAGs from tissue using strong anion-exchange (SAX) column. GAGs were isolated following a modified version of the manufacturer's protocol. The following should be briefly done: Wash column 2–3 times with 1 column volume (c.v.) of water. Wash column with 1 c.v. of urea/chaps solution. Load filtered, digested sample (see Note 10). Wash sample 3 times

with 1 c.v. of 200 mM NaCl. Elute sample with 2 washes of 0.5 c.v. 16% NaCl.

6. For every 1 mL of purified GAGs isolated above, add 4 mL methanol (methanol to make 80% of total solution). Mix well, place in 4°C fridge overnight. Centrifuge at 5,000 × *g* for 30 min and pour off supernatant to isolate precipitated GAG.
7. Isolated GAG is then desalted by dissolving in a minimal amount of water and dialyzed (against 1 L water for every 10 mg of GAG) or loaded onto a spin column (3 kDa MWCO) and washed with 5 column volumes of water.
8. GAG samples (5 μg) were incubated with the chondroitinase ABC (5 mU) and chondroitinase ACII (2 mU) at 37°C for 10 h. The enzymatic products were recovered by centrifugal filtration (30 kDa MWCO). CS/DS disaccharides, passed through the filter, were freeze-dried and ready for CE-LIF analysis.
9. The heparinase I, II, and III (5 mU each) were added into the remainder and incubated at 37°C for 10 h. The products were again recovered by centrifugal filtration (10 kDa MWCO), and the HP/HS disaccharides were similarly collected and freeze-dried and ready for CE-LIF analysis.

3.2.2. Derivatization of GAG-Derived Disaccharides with AMAC

1. Add 5 μL (per estimated 10 nmol of disaccharides) of 0.1 M AMAC in glacial acetic acid–DMSO (3:17, v/v) solution to the lyophilized disaccharides.
2. Add 5 μL (per estimated 10 nmol of disaccharides) of reducing agent 1 M $NaBH_3CN$ to the reaction mixture.
3. Incubate the reaction mixture at 45°C for 4 h.
4. Make up the reaction mixture with reconstitution solution to desired volume.

3.2.3. Determination of GAG-Derived Disaccharides

1. Precondition the capillary and inject the sample with the same procedure as suggested in Subheading 3.1.2. Electrophoregrams are shown in Figs. 3 and 4; peaks were identified by either coinjection or comparing with standard unsaturated disaccharide profiles. Presence of GAG-derivatized disaccharides was also proven by mass spectrometry.

4. Notes

1. Unsaturated disaccharide standards of CS/DS (ΔDi-0S: ΔUA-GalNAc, ΔDi-4S: ΔUA-GalNAc4S, ΔDi-6S: ΔUAGalNAc6S, ΔDi-2S: ΔUA2S-GalNAc, ΔDi-diS_B: ΔUA2S-GalNAc4S, ΔDi-diS_D: ΔUA2S-GalNAc6S, ΔDi-diS_E: ΔUA-GalNAc4S6S,

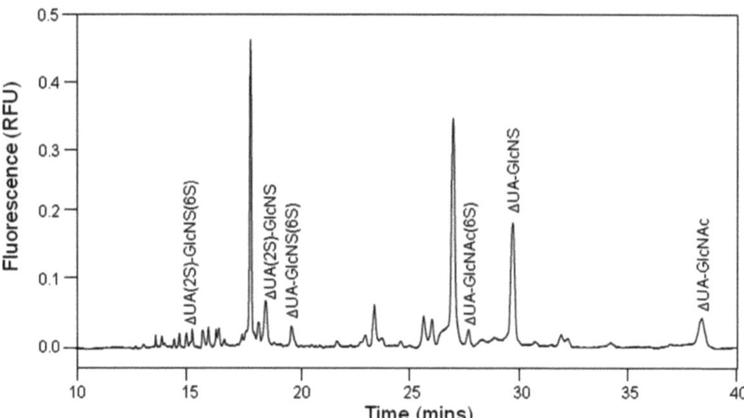

Fig. 3. Electrophoregram of HP/HS-derivatized disaccharides from camel liver. Analysis was performed at 25°C, pressure injection of 50 mbar × 10 s, using 50 mM phosphate buffer, pH 3.5, under 30 kV with reversed polarity. Unlabeled small and large peaks are fluorescent compounds not corresponding to the eight HP/HS-derivatized disaccharide standards.

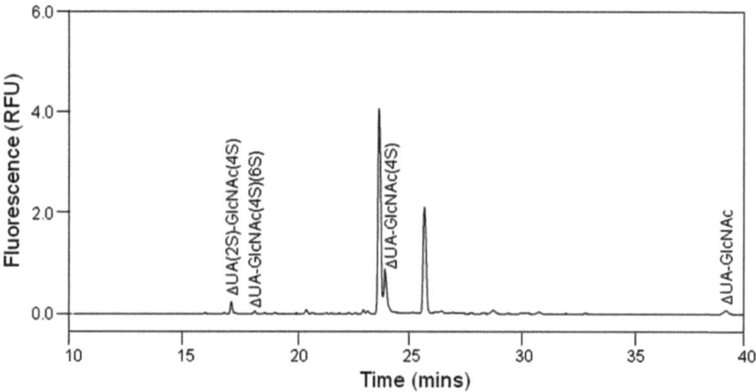

Fig. 4. Electrophoregram of CS/DS-derivatized disaccharides from camel liver. Analysis was performed at 25°C, pressure injection of 50 mbar × 10 s, using 50 mM phosphate buffer, pH 3.5, under 30 kV with reversed polarity. Unlabeled small and large peaks are fluorescent compounds not corresponding to the eight CS/DS-derivatized disaccharide standards.

ΔDi-triS: ΔUA2SGalNAc4S6S). Unsaturated disaccharide standards of heparin/HS (ΔDi-0S: ΔUA-GlcNAc, ΔDi-NS: ΔUA-GlcNS, ΔDi-6S: ΔUA-GlcNAc6S, ΔDi-2S: ΔUA2S-GlcNAc, ΔDi-2SNS: ΔUA2S-GlcNS, ΔDi-NS6S: ΔUA-GlcNS6S, ΔDi-2S6S: ΔUA2S-GlcNAc6S, ΔDi-triS: ΔUA2S-GlcNS6S) were obtained from SEIKAGAKU CORPORATION (Japan). Make 1 μg/μL standard solution with HPLC-grade water and store at −20°C.

2. The reducing agent solution must be made fresh prior to use.

3. All solutions used for capillary electrophoresis must be filtered through a 0.22 μm membrane filter and degassed before use.
4. Be aware that pH will change with temperature and buffer additives. This separation buffer is suggested by Militsopoulou et al. (11). It is worth noting that below pH 4.0, the ionization of silanols is low and the electroosmotic flow mobility is insignificant. So when conventional electrode polarity is reversed, negatively charged disaccharides are drawn to the anode only under the influence of electrophoresis.
5. The coupling reaction between unsaturated disaccharides and the fluorophore proceeds through reductive amination, which involves the reducing end of disaccharide and amine group of AMAC. 100–500 times of excess free AMAC tag is used to produce high derivatization yield.
6. The purpose to add $NaBH_3CN$ to reaction mixture is to stabilize the conjugate formed from reductive amination between unsaturated disaccharides and AMAC.
7. Under the suggested separation conditions, excess AMAC receives positive charge, which means that AMAC will not enter the capillary at reversed polarity mode. This is proven by injecting AMAC solution only into the capillary and applying the same separation condition. No peak is found in the electrophoregram.
8. Condition of capillary is critical to migration time of analytes and peak shape. Generally, new capillary will be washed with MeOH, 1 M HCl, 1 M NaOH, and operating buffer, with water flushing at each interval, until the baseline is good enough for analysis. After each day, the capillary should be washed with 1 M NaOH and water for 5 min and dried with air.
9. If necessary, defatting solutions may need to be filtered as they are removed to retain small tissue particles. As a general rule of thumb, use about 10–50 mL of each defatting solution for every 1 g of dry tissue weight; more can be used for fattier tissues.
10. Each column will have a max binding capacity, which will be the absolute amount of GAG the column can isolate; overloading the column will result in sample loss. The amount of each digested, filtered tissue to load onto each column should be based on the anticipated GAG content of the tissue which can be found in literature (typically on the order of 10–50 mg/g dry tissue).

References

1. Volpi N, Maccari F, Linhardt RJ (2008) Capillary electrophoresis of complex natural polysaccharides. Electrophoresis 29:3095–3106
2. Linhardt RJ (1991) Heparin: an important drug enters its seventh decade. Chem Ind 2:45–50
3. Grag HG, Linhardt RJ, Hales CA (2005) Chemistry and Biology of Heparin and Heparan Sulfate. Elsevier B. V, New York, NY
4. Linhardt RJ, Turnbull JE, Wang HM, Loganathan D, Gallagher JT (1990) Examination of the substrate specificity of heparin and heparan sulfate lyases. Biochemistry 29:2611–2617
5. Alicia M, Hitchcock AM, Yates KE, Shortkroff S, Costello CE, Zaia J (2007) Optimized extraction of glycosaminoglycans from normal and osteoarthritic cartilage for glycomics profiling. Glycobiology 17:25–35
6. Yang B, Solakyildirim K, Chang Y, Linhardt RJ (2011) Hyphenated techniques for the analysis of heparin and heparan sulfate. Anal Bioanal Chem 399:541–557
7. Rice KG, Kim YS, Grant AC, Merchant ZM, Linhardt RJ (1985) High-performance liquid chromatographic separation of heparin derived oligosaccharides. Anal Biochem 150: 325–331
8. Hileman RE, Smith AE, Toida T, Linhardt RJ (1997) Preparation and structure of heparin lyase-derived heparan sulfate oligosaccharides. Glycobiology 7:231–239
9. Chuang WL, McAllister H, Rabenstein L (2001) Chromatographic methods for product-profile analysis and isolation of oligosaccharides produced by heparinase-catalyzed depolymerization of heparin. J Chromatogr A 932:65–74
10. Rice KG, Rottink MK, Linhardt RJ (1987) Fractionation of heparin-derived oligosaccharides by gradient polyacrylamide-gel electrophoresis. Biochem J 244:515–522
11. Militsopoulou M, Lamari FN, Hjerpe A, Karamanos NK (2002) Determination of twelve heparin- and heparan sulfate-derived disaccharides as 2-aminoacridone derivatives by capillary zone electrophoresis using ultraviolet and laser-induced fluorescence detection. Electrophoresis 23:1104–1109
12. Mitropoulou TN, Lamari F, Syrokou A, Hjerpe A, Karamanos NK (2001) Identification of oligomeric domains within dermatan sulfate chains using differential enzymic treatments, derivatization with 2-aminoacridone and capillary electrophoresis. Electrophoresis 22: 2458–2463
13. Mao W, Thanawiroon C, Linhardt RJ (2002) Capillary electrophoresis for the analysis of glycosaminoglycans and glycosaminoglycan-derived oligosaccharides. Biomed Chromatogr 16:77–94
14. Skidmore MA, Guimond SE, Dumax-Vorzet AF, Atrih A, Yates EA, Turnbull JE (2006) High sensitivity separation and detection of heparan sulfate disaccharides. J Chromatogr A 1135:52–56
15. Lamari FN, Kuhn R, Karamanos NK (2003) Derivatization of carbohydrates for chromatographic, electrophoretic and mass spectrometric structure analysis. J Chromatogr B 793:15–36
16. Militsopoulou M, Lecomte C, Bayle C, Couderc F, Karamanos NK (2003) Laser-induced fluorescence as a powerful detection tool for capillary electrophoretic analysis of heparin/heparan sulfate disaccharides. Biomed Chromatogr 17:39–41
17. Mastrogianni O, Lamari F, Syrokou A, Militsopoulou M, Hjerpe A, Karamanos NK (2001) Microemulsion electrokinetic capillary chromatography of sulfated disaccharides derived from glycosaminoglycans. Electrophoresis 22: 2743–2745
18. Hitchcock AM, Bowman MJ, Staples GO, Zaia J (2008) Improved workup for glycosaminoglycan disaccharide analysis using CE with LIF detection. Electrophoresis 29:4538–4548
19. Zinellu A, Pisanu S, Zinellu E, Lepedda AJ, Cherchi GM, Sotgia S, Carru C, Deiana L, Formato M (2007) A novel LIF-CE method for the separation of hyaluronan- and chondroitin sulfate-derived disaccharides: application to structural and quantitative analyses of human plasma low- and high-charged chondroitin sulfate isomers. Electrophoresis 28:2439–2447
20. Viola M, Vigetti D, Karousou E, Bartolini B, Genasetti A, Rizzi M, Clerici M, Pallotti F, Luca GD, Passi A (2008) New electrophoretic and chromatographic techniques for analysis of heparin and heparan sulfate. Electrophoresis 29:3168–3174
21. Yoshida E, Arakawa S, Matsunaga T, Toriumi S, Tokuyama S, Morikawa K, Tahara Y (2002) Cloning, sequencing, and expression of the gene from bacillus circulans that codes for a heparinase that degrades both heparin and heparan sulfate. Biosci Biotechnol Biochem 66:1873–1879
22. Shaya D, Tocilj A, Li Y, Myette J, Venkataraman G, Sasisekharan R, Cygler M (2006) Crystal structure of heparinase II from Pedobacter heparinus and its complex with a disaccharide product. J Biol Chem 281:15525–15535
23. Godavarti R, Davis M, Venkataraman G, Cooney C, Langer R, Sasisekharan R (1996) Heparinase III from Flavobacterium heparinum: cloning and recombinant expression in Escherichia coli. Biochem Biophys Res Commun 225:751–758

Chapter 8

High-Throughput Capillary Electrophoresis–Mass Spectrometry: From Analysis of Amino Acids to Analysis of Protein Complexes

Mehdi Moini

Abstract

Recent advances in capillary electrophoresis–mass spectrometry (CE-MS) interfacing using porous tip is leading to commercialization of CE-MS with a sheathless interface for the first time. The new sheathless interface in conjunction with CE capillary coatings using self-coating background electrolytes (BGE) has significantly simplified CE-MS analysis of complex mixtures. CE-MS, with its high separation efficiency, compound identification capability, and ability to rapidly separate compounds with a wide range of mass and charge while consuming only nanoliters of samples, has become a valuable analytical technique for the analysis of complex biological mixtures. These advances have allowed a single capillary to analyze a range of compounds including amino acids, their D/L enantiomers, protein digests, intact proteins, and protein complexes. With these capabilities, CE-MS is poised to become the multipurpose tool of separation scientists. More recently, an eight-capillary CE in conjunction with an 8-inlet mass spectrometry has allowed 8 CE-MS analyses to be performed concurrently, significantly increasing throughput.

Key words: Capillary electrophoresis, Electrospray ionization, Mass spectrometry, Amino acids, Enantiomers, Biological clocks, Dating, Self-coating background electrolyte, Peptides

1. Introduction

A quarter century after its first introduction, capillary electrophoresis–mass spectrometry (CE-MS), with its high separation efficiency, compound identification capability, and ability to rapidly separate compounds with a wide range of mass and charge while consuming only nanoliters of samples, has become a valuable analytical technique for the analysis of complex biological mixtures. Capillary electrophoresis–mass spectrometry is the only analytical

Nicola Volpi and Francesca Maccari (eds.), *Capillary Electrophoresis of Biomolecules: Methods and Protocols*, Methods in Molecular Biology, vol. 984, DOI 10.1007/978-1-62703-296-4_8, © Springer Science+Business Media, LLC 2013

technique with which one can analyze amino acids, their racemic mixtures, peptides, proteins, intact protein complexes, and intact cells simply by changing the composition of the background electrolyte (BGE). This allows a single capillary to analyze a range of omics, such as metabolomics, proteomics (both top-down and bottom-up), functional proteomics, and cellomics. In this regard, CE-MS is considered the multipurpose tool of the separation scientists. CE-MS is especially useful for the analysis of charged compounds which are not easily retainable under reverse-phase high-performance liquid chromatography (HPLC) and in this regard is considered a complementary technique to HPLC. Comparison between CE-MS and HPLC-MS in proteomics area has been discussed (1, 2). In addition to its multipurpose characteristics, CE-MS also offers many other important advantages compared to HPLC-MS such as the following:

- High separation efficiency because of its flat flow profile
- High-speed separation allowing sample to be analyzed in minutes
- Low-pressure operation allowing injection of nL or less of the samples or injection of the intact cells
- Ease of changing BGE (1–2 min)
- Higher sensitivity under ESI due to lower flow rates
- Ease of operation (no fittings, no leaks, etc.)

While these advantages were well known to the practitioners in the field, the widespread use of CE-MS and its commercialization remained elusive due to the difficulties of interfacing CE to MS. This changed in 2007 when a simple, reproducible CE to MS interface design using a porous tip was introduced. The new design provided a path for the commercialization of sheathless interface, which in turn will provide a turn-key sheathless CE-MS instrument for wide usage within the scientific community and later in clinical laboratories.

1.1. CE-MS Interface Design

An important feature of any CE/MS interface is the method by which electrical current is provided to the CE outlet/ESI electrode. Over the past 25 years, a variety of CE/ESI-MS interfaces have been introduced, various reviews of which have been published and will not be discussed here (3–10). These interfaces are divided into three general categories: sheath-flow, sheathless, and split-flow interfaces. Sheathless interfaces are especially popular due to their high sensitivity of detection (resulting from the absence of sheath liquid to dilute the CE effluent), and most interfaces introduced in the past 25 years deal with developing this type of interfaces; however, novel sheath-flow designs that minimize the dilution of the analyte by sheath liquid are still being pursued and published (11, 12). In split-flow interfaces the electrical connection to the CE capillary outlet is achieved by diverting part of the CE background electrolyte (BGE) out of the capillary through an opening near the capillary outlet. The CE buffer exiting the opening contacts a sheath metal tube which acts as the CE outlet/ESI shared electrode. In cases in which the ESI source uses a metal needle, the voltage contact to the CE buffer is achieved by simply inserting the outlet of the CE capillary, which contains an opening, into the existing ESI needle. As a result of the concentration-sensitive nature of ESI, splitting a small percentage of the CE flow has minimal effect on the sensitivity of detection. In addition, because the liquid is flowing through the opening and out of the capillary, there is no dead volume associated with this interface. Moreover, bubble formation due to redox reactions of water at the electrode does not affect CE/ESI-MS performance, because the actual metal/liquid contact occurs outside of the CE capillary. Despite these advantages, split-flow interface is a multistep interface design since it requires a separate step for sharpening the capillary outlet outer diameter (o.d.) tip for efficient ESI. In addition, reaching a consistent split ratio is currently difficult to achieve (13).

1.1.1. Sheathless CE-MS Interface Using a Porous Tip

To address the deficiencies mentioned above, in 2007 we introduced porous tip interface as a robust and reproducible, single-step sheathless design. In this design, the electrical connection to the capillary outlet is simply achieved by etching ~4 cm of the capillary outlet using a solution of 49% HF until it becomes porous. The etching process also simultaneously reduces the outlet tip for efficient ionization, eliminating the separate tip-sharpening process of the older designs. Small ion transport through the porous section of the capillary in contact with the conductive solution in the ESI needle provides voltage to the solution inside the capillary for ESI. In CE, in addition to providing voltage for ESI, ion transport through the porous tip closes the CE electrical circuit (14). Figure 1 shows the rate of HF etching for 150-μm-o.d., 20-μm-i.d. fused-silica capillary, with its polyimide coating (~10 μm) removed prior to etching. The procedure for HF etching is provided in Appendix 1. As shown, to etch porous a nominal 150-μm-o.d., 20-μm-i.d. capillary requires about 30 min. Once etched, the capillary is inserted inside an existing ESI needle filled with a conductive solution (0.1–1 M formic or acetic acid). Figure 2 shows the overall schematic of the porous tip interface and exemplifies the uniform spray that is formed using porous tip design. A sharp tip, good electrical connections, and minimal bubble formation are the main reasons for the uniformity of the electrospray plume under porous tip design. An additional advantage of the porous tip is that the inlet and outlet inner diameters are the same, which reduces the chance of the tip getting clogged since any particulate that gets into the porous tip capillary will exit from the other end of the capillary. This is in contrast to the use of a nanospray interface using pulled

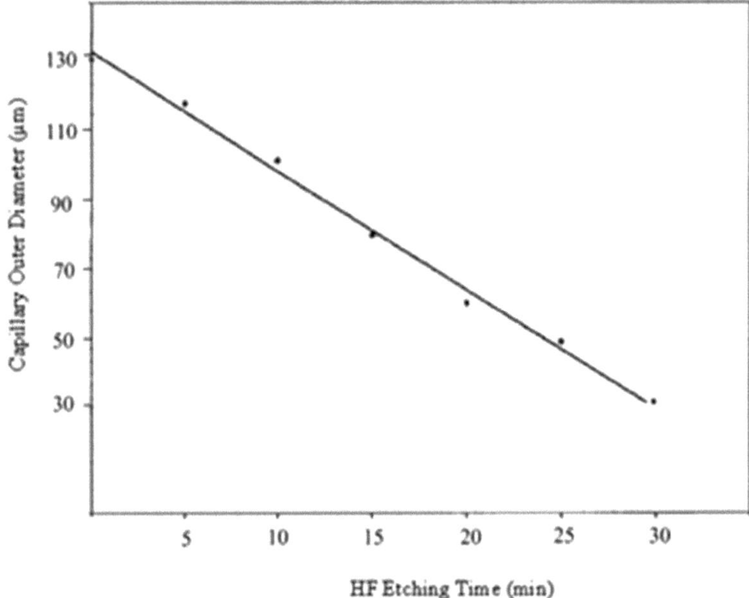

Fig. 1. Etching rate for a nominal 150-μm-o.d. fused-silica capillary in 49% HF.

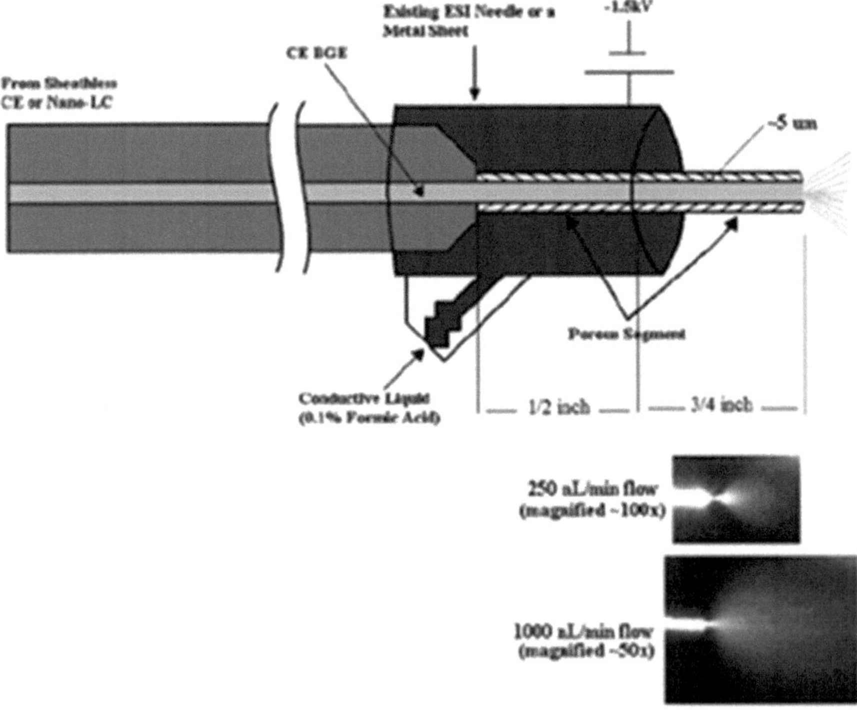

Fig. 2. Schematics of the porous tip. Insets, ESI plume from a nLC operating at 250 nL/min (magnified by ~100×) and 1 μL/min (magnified by ~50×) (reprinted with permission from (Moini, M. (2007) Simplifying CE-MS Operation. 2. Interfacing Low-Flow Separation Techniques to Mass Spectrometry Using a Porous Tip, *Anal. Chem.* 79, 4241–4246). Copyright (2007) American Chemical Society).

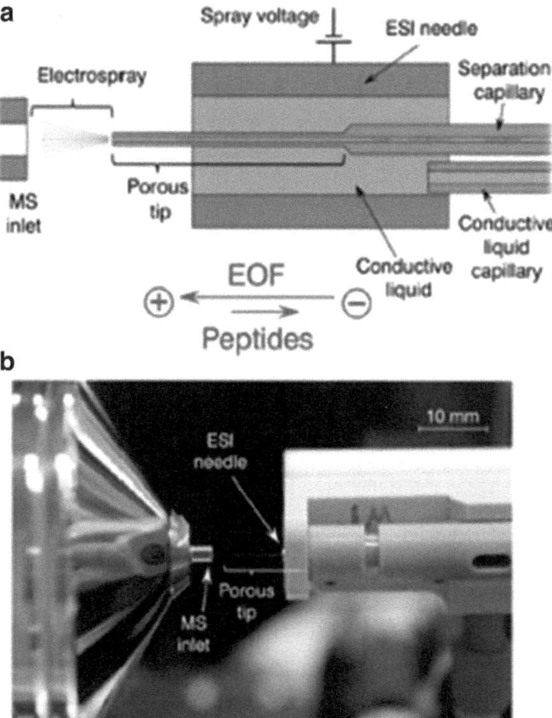

Fig. 3. The high-sensitivity porous sprayer interface (**a**) schematic and (**b**) photograph of the prototype interface ref 2 (reprinted with permission from (Faserl, K., Sarg, B., Kremser, L., Lindner, H. (2011) Optimization and Evaluation of a Sheathless Capillary Electrophoresis–Electrospray Ionization Mass Spectrometry Platform for Peptide Analysis: Comparison to Liquid Chromatography–Electrospray Ionization Mass Spectrometry, *Anal. Chem.* 83, 7297–7305). Copyright (2011) American Chemical Society).

tips, as its tip is usually drawn to less than 10 μm i.d., which can trap particulates with diameters larger than the capillary opening at the inlet. Since in the pull-tip design, only the tip is sharp, tip clogging will result in the loss of the spray tip and the analysis. Damage to a porous tip, however, can be easily repaired by etching away 1–2 mm of the tip off using the HF solution. Because the porous tip has about the same o.d. and i.d. for the ~4-cm porous tip, removing 2 mm of the tip will not affect spray performance. Figure 3 shows the prototype of a commercial version of this interface design (2).

Another advantage of the porous tip design is that it can be applied to a variety of capillary o.d.s and i.d.s, as demonstrated in Figs. 4 and 5, where an enolase digest has been analyzed using capillaries with the same inner diameters but different outer diameters (Fig. 4) and with the same capillary o.d. but different i.d.s and lengths (Fig. 5).

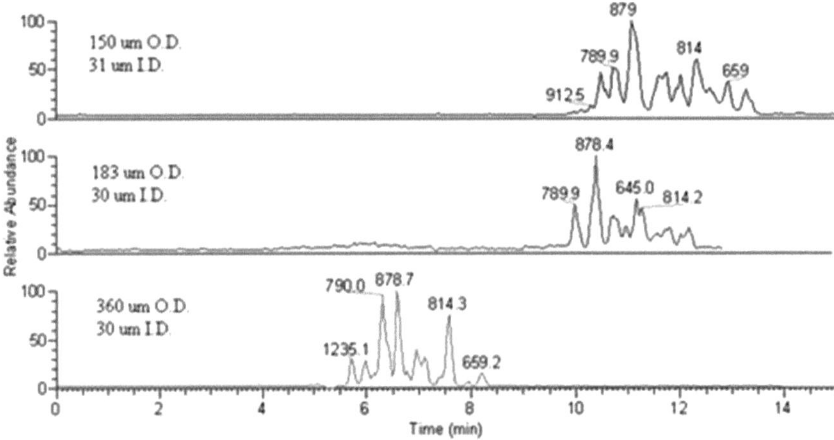

Fig. 4. CE-MS of enolase digest utilizing a porous tip on different capillary o.d.s.

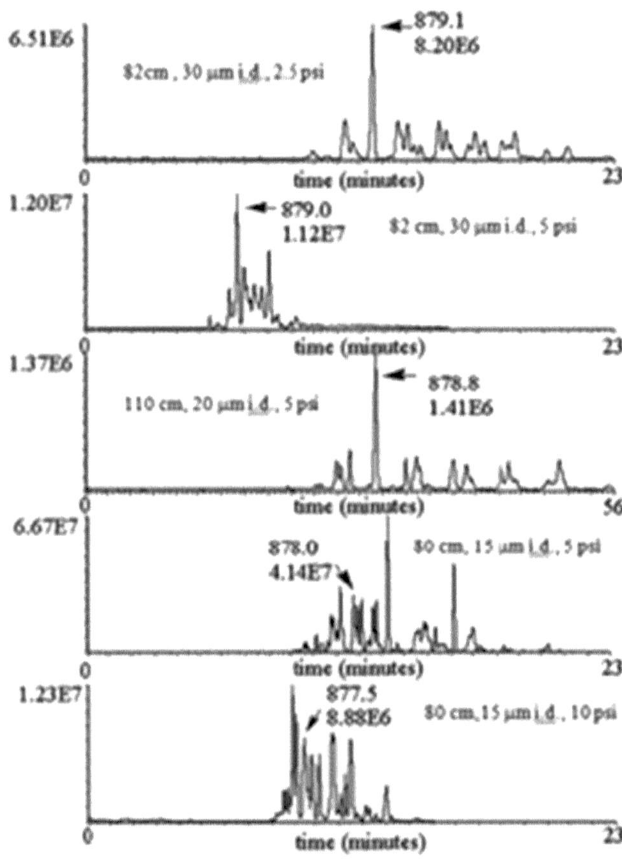

Fig. 5. CE-MS of enolase digest utilizing a porous tip on capillaries with the same o.d.s. (150 µm) but different i.d.s.

2. Applications of CE-MS

In the past decade we have applied CE-MS to the analysis of a wide range of biological mixtures from amino acid mixtures to the analysis of the chemical contents of a single cell. The idea is to develop sensitive, fast, and simple CE-MS methods that do not require analyte or capillary derivatization, which are both time-consuming and expensive. To achieve our goal we mainly rely on the addition of additives to the CE BGE that, for example, minimize analyte–wall interaction or enhance sensitivity and separation.

2.1. Analysis of Underivatized Amino Acid

CE-MS is an ideal analytical technique for the analysis of underivatized amino acid analysis. This is because gas chromatography–mass spectrometry is unable to analyze underivatized amino acids, which are thermally labile. Moreover, except for larger amino acids with hydrophobic side chain, reverse-phase HPLC-MS with conventional ion-pairing reagents (0.1% TFA and 0.1% formic acid) cannot analyze most amino acids because under these conditions amino acids are positively charged and they are co-eluted with void volume. However, CE-MS can separate and detect underivatized amino acids with ease. Figure 6 shows the CE/ESI-MS electropherogram of the separation of the 20 standard protein amino acids using a 130-cm-long, underivatized CE capillary in conjunction with 1 M formic acid as the BGE. For this experiment, ~0.4 nL of the 20 amino acid standard solution (~400 fmol of each amino acid) was injected (15). The use of an underivatized capillary is a big advantage since capillary derivatization techniques are usually long, labor-intensive, expensive, and short-lived. While under 1 M formic acid as the BGE, all amino acids were separated and detected, however, their relative intensities varied almost 3 orders of magnitude, even though an equimolar solution of amino acids had been used. The variation in relative intensities of amino acids is due to variation in ionization efficiencies of amino acids under ESI (16, 17).

2.1.1. Enhanced Sensitivity of Amino Acids by Adding 18-Crown-6 to the BGE During Amino Acid Analysis

To enhance the detection sensitivity of the amino acids that showed low sensitivity with 1 M formic acid as the BGE (Fig. 6) as well as to remedy the disparity between detection limits of amino acids, we have used 18-crown-6 (18-C-6) with a concentration between 20 and 30 mM as an additive and complexation reagent to the BGE. Once added to the BGE, 18-C-6 formed stable complexes with amino acids, which exhibited high ionization efficiencies and low (attomole levels) detection limit for most amino acids (Fig. 7).

2.1.2. Separation of Amino Acid Enantiomers Using (+) 18-Crown-6-Tetracarboxylic Acid (18-C-6-TCA)

While the use of 18-C-6 enhanced the sensitivity of detection of most amino acids, it was unable to separate amino acid enantiomers because it lacks a chiral center. To achieve high sensitivity of detection for amino acid and to achieve amino acid enantiomer separation, enantiomeric pure (+) 18-C-6-TCA instead of 18-C-6

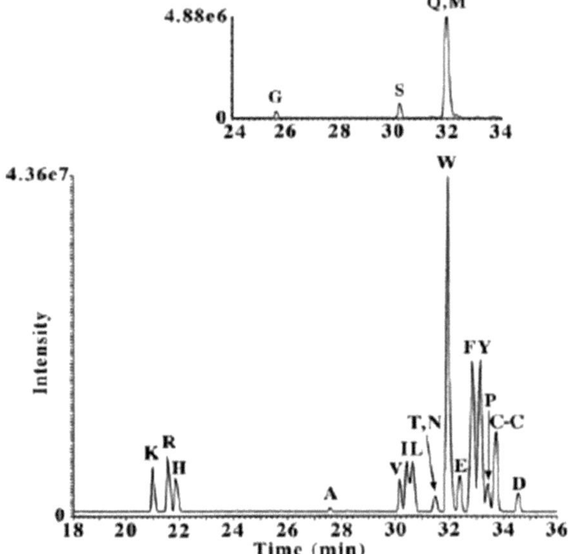

Fig. 6. CE-MS analysis of underivatized amino acids using 1 M formic acid as the BGE (reprinted with permission from (Schultz, C. L., Moini, M. (2003) The analysis of underivatized amino acids and their D/L enantiomers using sheathless CE-MS, *Anal. Chem.* 75, 1508–1513). Copyright (2003) American Chemical Society).

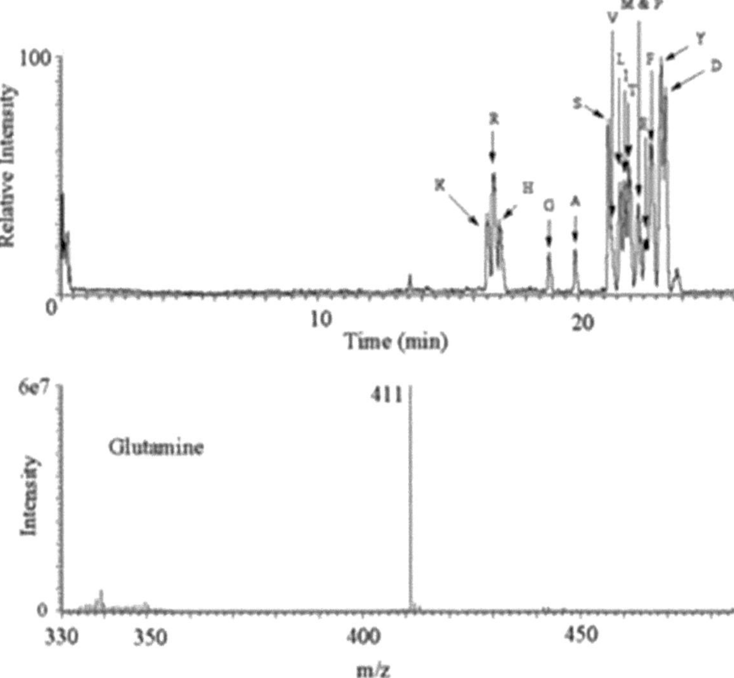

Fig. 7. CE-MS of amino acids using 18-crown-6 as the BGE additive (*top panel*). Mass spectrum of the protonated glutamine (MW 146)/18-crown-6 (MW 264) complex (*bottom panel*).

was utilized as the background electrolyte and used in conjunction with an underivatized, 130-cm-long, 20-μm-i.d., 150-μm-o.d. fused-silica capillary (18). During the enantiomer separation, while interaction between the protonated amine and the oxygens of the ethylene bridge of 18-C-6-TCA is responsible for complex formation, it is apparently the carboxyl group substituents of 18-C-6-TCA that allow for enantiorecognition (18). Both D and L enantiomers form stable complexes with 18-C-6-TCA; however, in most cases, the D enantiomers separate from the L enantiomers because D enantiomers spend more time with 18-C-6-TCA (except for serine and threonine, in which the L enantiomers migrated last). By monitoring the m/z range of the amino acid/18-C-6-TCA complexes (m/z 515–700), most of the standard amino acids and many of their enantiomers were separated and detected with high separation efficiency and high sensitivity (nanomolar concentration detection limits) in one run (Fig. 8). The solutions of 18-C-6-TCA also worked well as the CE/ESI-MS BGE for low-level detection of several neurotransmitters and some of their D/L enantiomers as well as for the analysis of amino acids at endogenous levels in lysed red blood cells.

2.1.3. Amino Acid Racemization (AAR) as a Biological Clock

AAR is a posttranslational modification (i.e., modification of the protein structure after its initial synthesis), which stems from the intrinsic instability of certain AA residues that leads to racemization. On earth, proteins are synthesized from L-amino acids. However, after incorporation into proteins, bound L-AAs start to racemize to their D-form under a reversible first-order kinetic reaction until equilibrium is reached (19). Each amino acid has a different intrinsic AAR rate, which is a function of the physical, chemical, and biological (protein sequence and its secondary, tertiary, and quaternary structures) states of the specimen, and acts as an independent biological clock. To measure the rate of conversion of L- to D-amino acids in proteins, the proteinaceous specimens are digested by hydrochloric acid to free amino acids, D- and L-amino acids are separated by CE-MS, and the ratio of the D/L is measured to estimate the age of the specimen (20). Among AAR rates, the aspartic acid racemization rate is fast enough to be used for more recent specimens (<2,500 years); however, the AAR rates of other amino acids are used to date older specimens (21). The CE-MS technique discussed in Subheading 2.1.2 can separate and detect amino acids and their D/L enantiomers in one run, in ~20 min, and using only ~1 nL of sample. Unlike the analysis of amino acids by gas chromatography or HPLC with UV detectors, CE-MS uses underivatized amino acids and requires no sample preparation, which significantly reduces analysis time.

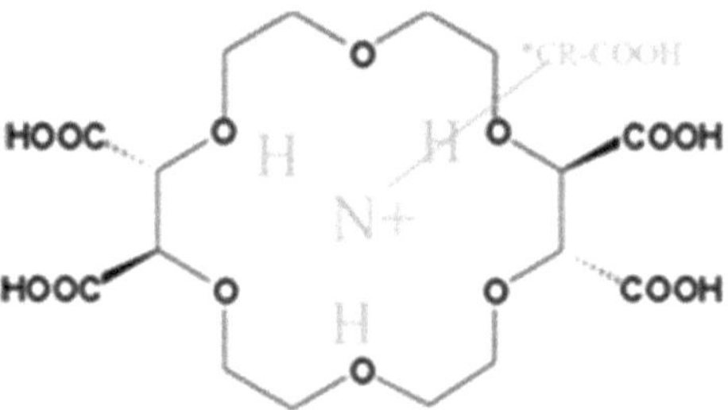

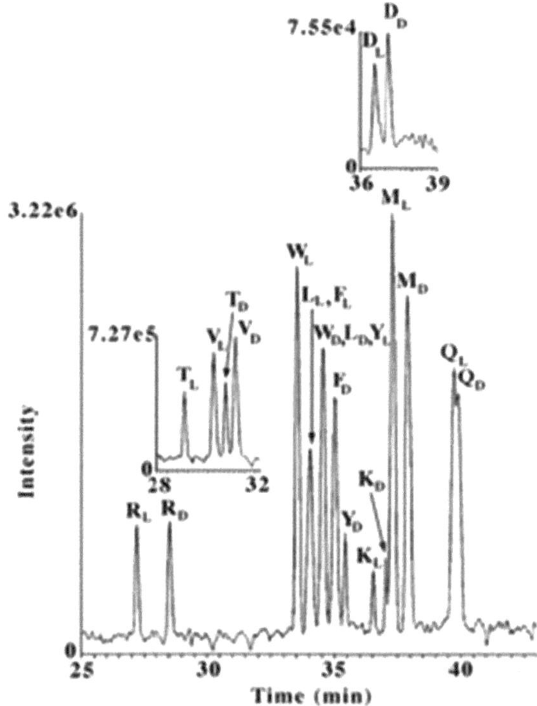

Fig. 8. CE-MS of 11 amino acid enantiomers in one run using a 30-mM aqueous solution of 18-C-6-TCA as the CE BGE (reprinted with permission from (Schultz, C. L., Moini, M. (2003) The analysis of underivatized amino acids and their D/L enantiomers using sheathless CE-MS, *Anal. Chem.* 75, 1508–1513). Copyright (2003) American Chemical Society).

2.1.4. Application of Amino Acid Racemization in Dating Museums' Proteinaceous Specimens

Because of the existence of a large number of precious proteinaceous specimens such as silk and wool textiles, leather and animal gut objects, bone and tissues, ink, paper, paint, coatings, binders (and associated adhesives), and paleo-organic matter in museum collections, the identification of the nature of the degradative state of such historic objects is often critical to their preservation. To preserve these objects it is desirable to understand the fundamental factors that cause their degradation, to identify the deterioration markers that determine their degradation stage and their age, and to study the environmental factors that affect their deterioration. In most cases, however, there is insufficient uniform material available from museums' samples for accurate ^{14}C dating since this technique usually requires a few milligrams of specimen. For precious objects, even a few milligrams are a relatively large amount of material, and in most cases it is almost impossible to obtain (22). In this area our goal was to develop a high-throughput CE-MS technique for dating proteinaceous specimens while consuming minimal amount of samples. As a proof of concept, we applied this technique for dating silk textiles fabricated from the cocoons of the silkworm Bombyx mori by measuring the D/L ratio of the aspartic acid for several well-dated textiles ranging in age from the present to ~2,500 years ago (20). The silk samples included fresh silk (2010); untreated silk from the Oka Studio, Freer–Sackler Museum, Smithsonian Institution (SI), Washington, DC (~1990); Sheridan flag, National Museum of American History, SI (1883–1888); Mexican War flag, National Museum of American History, SI (1845–1846); a man's suit coat, Museum of the City of New York (1740); a silk textile from Istanbul, Textile Museum, Washington, DC (1551–1599); a silk tapestry from the Fontainebleau Series, Kunsthistorisches Museum, Vienna, Austria (1540s); tiraz silk yarns from Egypt, Textile Museum, Washington, DC (993); and silks from the Warring States Period, China, brown (top) and red (bottom), Metropolitan Museum of Art, New York City (475–221 b.c.). The results are shown in Fig. 9, with an insert showing the relative increase of the D vs. L enantiomer of aspartic acid/18-C-6-TCA complex (m/z 574). From Fig. 9, it is clear that the data points fit very well with a theoretical curve obtained from reversible first-order chemical kinetics (solid curve).

2.1.5. Effect of Environmental Factors on Racemization

The effect of UV radiation on modern degummed silk, which was UV-radiated from 0 to 1,920 h, was also investigated (Fig. 10). It was observed that the D/L ratio increased from ~1.5 to ~3.5%. The results indicate that after ~2,000 h of radiation, D/L ratio change is equivalent to natural aging of ~100 years. As shown, UV radiation has a direct effect on silk aging.

3. Application of CE-MS to the Analysis of Peptides and Proteins

The most important factor for successful CE-MS analysis of peptides and proteins is prevention of peptides and protein adsorption on fused-silica inner wall (23). Fused silica is the most commonly used capillary material in CE due to its good thermal conductivity, ultraviolet transparency, and uniform inner diameter (24). However, the use of bare fused-silica capillary in CE for peptide and protein analysis is problematic because of the chemical composition of the

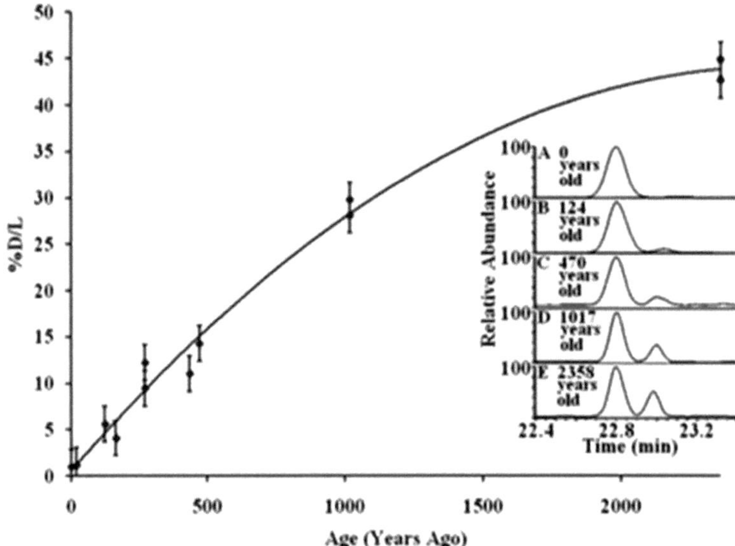

Fig. 9. Ratios of D- vs. l-aspartic acid for several well-dated museums' specimens. *X*-axis on the inset is relative intensity (reprinted with permission from (Moini, M., Klauenberg, K., Ballard, M. (2011) Dating Silk By Capillary Electrophoresis Mass Spectrometry, *Anal Chem.* 83, 7577–7581). Copyright (2011) American Chemical Society).

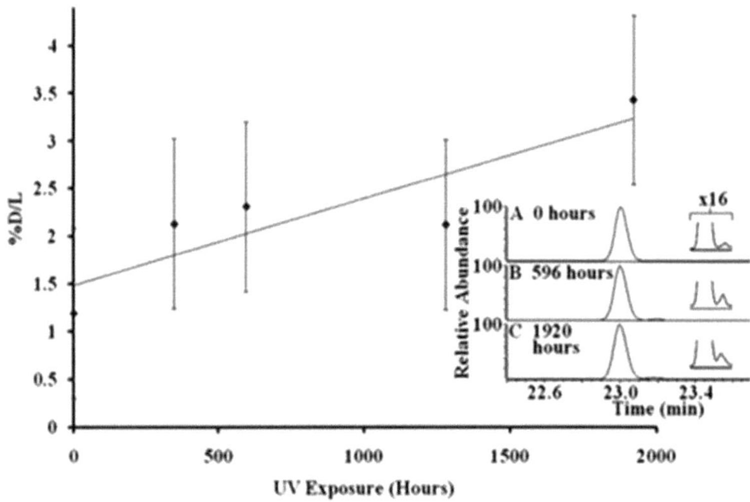

Fig. 10. Effect of UV radiation on the ratio of D- vs. l-aspartic acid of a modern silk sample (reprinted with permission from (Moini, M., Klauenberg, K., Ballard, M. (2011) Dating Silk By Capillary Electrophoresis Mass Spectrometry, *Anal Chem.* 83, 7577–7581). Copyright (2011) American Chemical Society).

inner wall of the capillary. It exposes a hydrophobic, negatively charged, and pH-dependent surface that can interact with the positively charged analyte (25). The analyte–wall interactions lead to irreproducibility, impaired efficiencies, and peak broadening. Because of the heterogeneity of silanol groups on the inner walls of the fused-silica capillary, there exists a spectrum of micro p*K*a values within the capillary centered on 6.3 (Fig. 11) (26). Since the

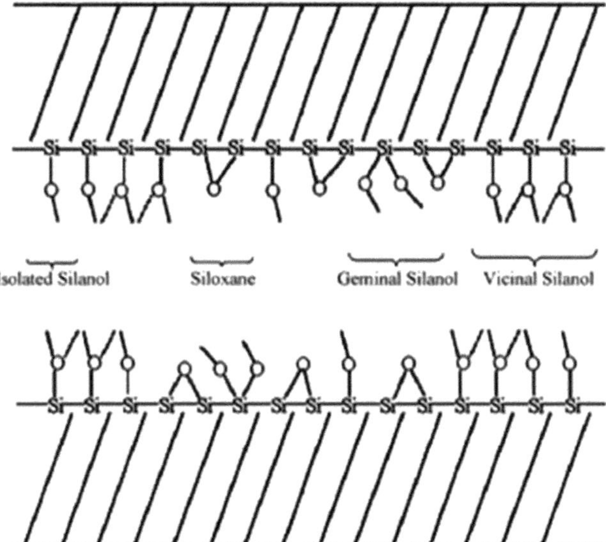

Fig. 11. Schematic diagram of the inner wall of an underivatized fused-silica capillary with silanol groups exposed (reprinted with permission from (Garza, S., Chang, S., Moini, M. (2007) Simplifying capillary electrophoresis–mass spectrometry operation: Eliminating capillary derivatization by using self-coating background electrolytes, *J. of Chromatogr. A* 1159, 14–21). Copyright (2007) Elsevier).

acid–base ionization constant (p*K*a) of silanol groups is 2–9 (based on change of electroosmotic flow (EOF) versus pH experiment), an aqueous BGE with a pH > p*K*a will impose a net negative charge on the silanol groups. Acidic BGEs with pH < 2 have been used to protonate these silanol groups to prevent analyte–wall interaction (4). However, highly acidic BGEs have disadvantages such as producing a high CE current which can cause electrical discharge through the capillary wall leading to capillary breakage, especially for thin-wall capillaries (e.g., 150 μm o.d.). In addition, highly acidic conditions are not suitable for analyzing protein complexes and other higher-order structures. To minimize analyte–wall interaction without the use of highly acidic BGE, a series of derivatization techniques have been developed to positively charge the capillary inner wall under mildly acidic BGE conditions suitable for ESI. Two general derivatization techniques are used: covalent and dynamic coating of the fused-silica capillary inner wall. Covalent coatings involve chemical reactions between the silanol groups on the capillary wall and the coating material (27). An example of a covalent coating is the trimethoxyaminopropylsilane (APS) treatment of the capillary. In the APS derivatization technique, the methoxy groups of APS and the –OH silanol groups attach to the inner wall and react together to leave positively charged amino groups inside the capillary wall under acidic conditions, which in turn reverses the EOF. Although covalent derivatization provides optimum separation for proteins and peptides under mildly acidic

BGE, the derivatization process is time-consuming, labor-intensive, and short-lived (a few runs). Moreover, the derivatization mostly fails to give complete coverage of the silica inner surface and is usually unstable at high pH. Furthermore, covalent coatings typically require in situ reaction steps that are hard to control and affect the homogeneity and reproducibility of the coated capillaries. Another limitation is that in situ derivatization can plug the capillary by producing a very viscous polymer solution that sometimes cannot be completely washed out. Alternatively, dynamic coatings require the temporary physical adsorption of the polymer to the capillary wall. Dynamic coatings consist of treating the capillary wall with a solution containing the coating agent, performing a few separations, removing the coating using a solution of NaOH, rinsing with water, and recoating the capillary again once the wall coating degrades. This type of coating can be divided into two general categories: adsorption of the polymer coating to the capillary wall prior to the CE analysis (precoating) or using the polymer reagent as an additive in the BGE (self-coating). The main disadvantage of precoating techniques is their gradual degradation. While multilayer dynamic coatings have enhanced the coating stability (28), the main advantage of the self-coating BGEs is their ease of use. The use of polymers as dynamic coatings for CE-UV analysis has recently been reviewed (29).

3.1. Simplifying CE-MS Operation by Using Self-Coating Background Electrolytes

Polybrene (PB) and polyE 323 (PE) are cationic polymers that are able to non-covalently attach to the negatively charged wall of fused-silica capillary. The excess positive charges on the newly created surface generate a stable anodic EOF that is independent of pH within the pH range of 4–8. The molecular structures of both PB and PE are shown in Fig. 12. The structure of PE is purposely constructed to contain mixed bonding characteristics. For example, the nitrogen atoms in the backbone of PE are separated by three-atom length, whereas nitrogen atoms in PB are separated by six carbon atoms. The length of the spacer arm between the nitrogen atoms in the backbone can affect the polymer's flexibility and hydrophobicity. Also, hydroxyl groups in PE increase polymer immobilization on the capillary wall by hydrogen bonding (30). PB and PE are very effective in reversing the charge on the capillary wall, which prevents peptide and protein attachment to the capillary wall, thereby significantly improving separation efficiency. Figure 13 shows the comparison of CE-MS analyses of peptide standard using an uncoated capillary (top panel), PB and PE precoated capillary, and self-coated capillary using 33–660-nM PB and PE (from top to bottom). Figures 14 and 15 show the separation efficiency for more complex mixtures including a mixture of 6-protein digest (Fig. 14) and for a tryptic digest of 55 ribosomal proteins. As shown both PB and PE are

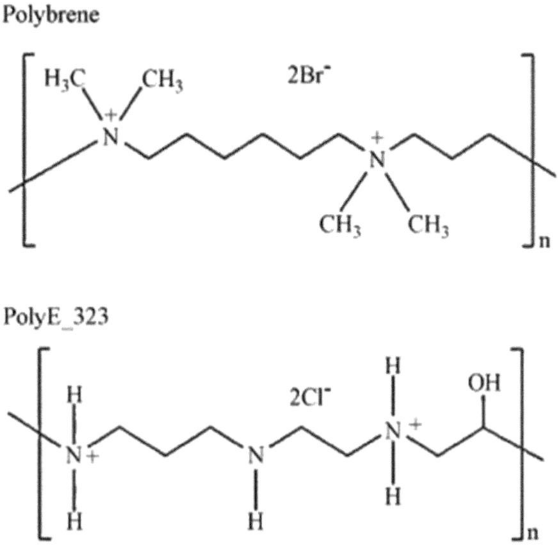

Polybrene

PolyE_323

Fig. 12. Molecular structures of polybrene (PB) and polyE 323 (PE) (reprinted with permission from (Garza, S., Chang, S., Moini, M. (2007) Simplifying capillary electrophoresis–mass spectrometry operation: Eliminating capillary derivatization by using self-coating background electrolytes, *J. of Chromatogr. A* 1159, 14–21). Copyright (2007) Elsevier).

excellent additives as a self-coating BGE for the CE/ESI-MS analysis of peptides and proteins, and they can significantly simplify CE-MS operations (23, 31, 32).

3.2. High Sequence Coverage Proteomics Using $(CE\text{-}MS/MS)^n$

Capillary electrophoresis coupled with tandem mass spectrometry (CE-MS/MS) has been used to successfully identify proteins from a variety of organisms and can provide high sequence coverage of a protein digest, often approaching total coverage for a single protein digest (33–38). However, application of CE-MS/MS to the analysis of a digest of a complex protein mixture has been limited. This is because even under CE's high separation efficiency, each CE peak still contains several peptides. Since the number of peptides analyzed by MS under full scan and MS/MS modes is limited by the mass spectrometer's rate of switching between these two modes, sharp electrophoretic peaks containing several peptides result in undersampling; that is, only a fraction of the peptides are analyzed under MS/MS mode. In addition, due to the electrospray ionization mechanism, when manipulating experimental conditions, such as pH and composition of the CE BGE, only a specific number of peptides will have ion intensities detectable by the mass spectrometer. For example, when a protein digest of a complex protein mixture of *E. coli* ribosomal proteins was analyzed by CE-MS, a single electrophoretic peak of 6 s wide contained more than 20 peptides with a wide range of intensities (Fig. 15). This usually results in low sequence coverage since many

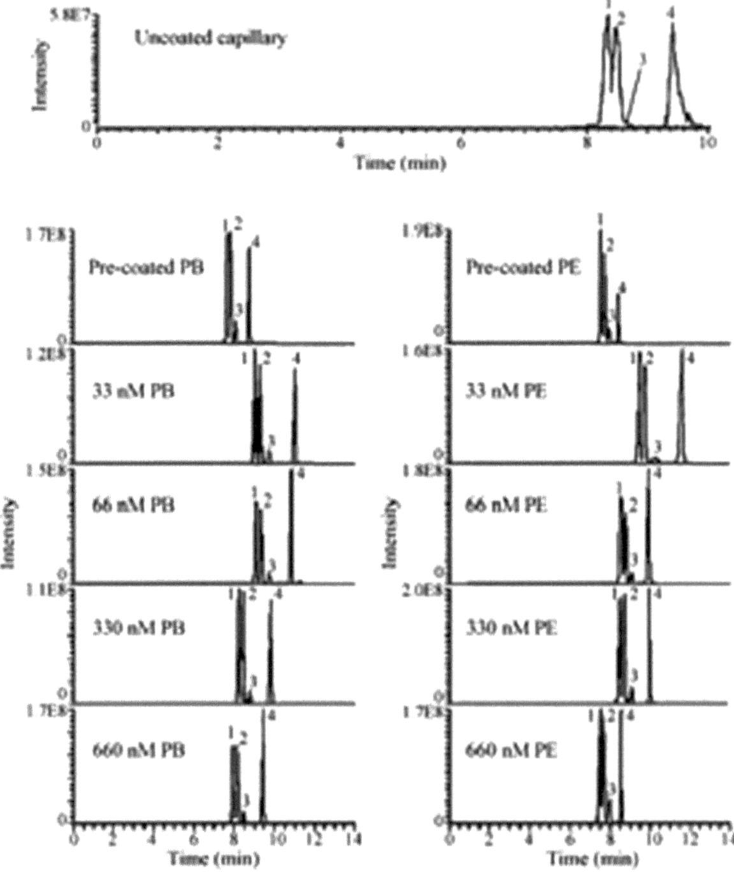

Fig. 13. Comparison of CE-MS analyses of peptide standard using an uncoated capillary (*top panel*), PB and PE precoated capillary, and self-coated capillary using 33–660-nM PB and PE (from *top* to *bottom*) (reprinted with permission from (Garza, S., Chang, S., Moini, M. (2007) Simplifying capillary electrophoresis–mass spectrometry operation: Eliminating capillary derivatization by using self-coating background electrolytes, *J. of Chromatogr. A* 1159, 14–21). Copyright (2007) Elsevier).

peptides are not sequenced by MS/MS. There are several reasons for the low coverage including the following: (1) undersampling (the existence of a large number of peptides that must be analyzed by MS/MS and the limited rate at which the mass spectrometer is capable of switching between MS and MS/MS modes); (2) wide concentration dynamic range of the proteins in a complex protein mixture, in which low-level peptides may go undetected by MS/MS; and (3) wide range of electrospray ionization efficiency of peptides under each mobile-phase composition, resulting in a wide variation in ion intensities, even for peptides with equimolar concentration. Under this condition, the mass spectrometer usually does not have sufficient time to analyze all the peptides by MS/MS, which results in undersampling and low protein coverage.

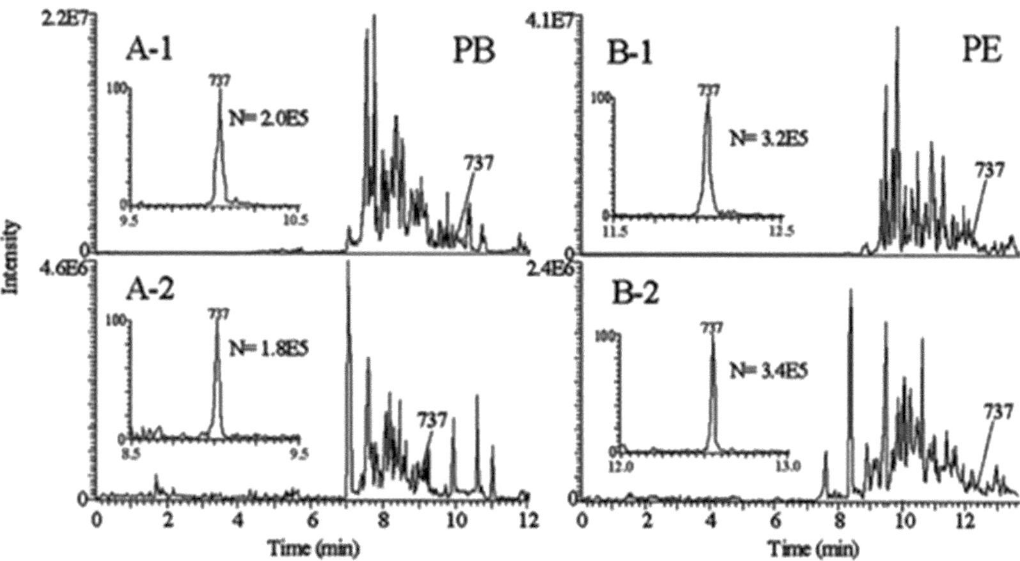

Fig. 14. Separation efficiency of CE-MS of 6-protein mix digest (reprinted with permission from (Garza, S., Chang, S., Moini, M. (2007) Simplifying capillary electrophoresis–mass spectrometry operation: Eliminating capillary derivatization by using self-coating background electrolytes, *J. of Chromatogr. A* 1159, 14–21). Copyright (2007) Elsevier).

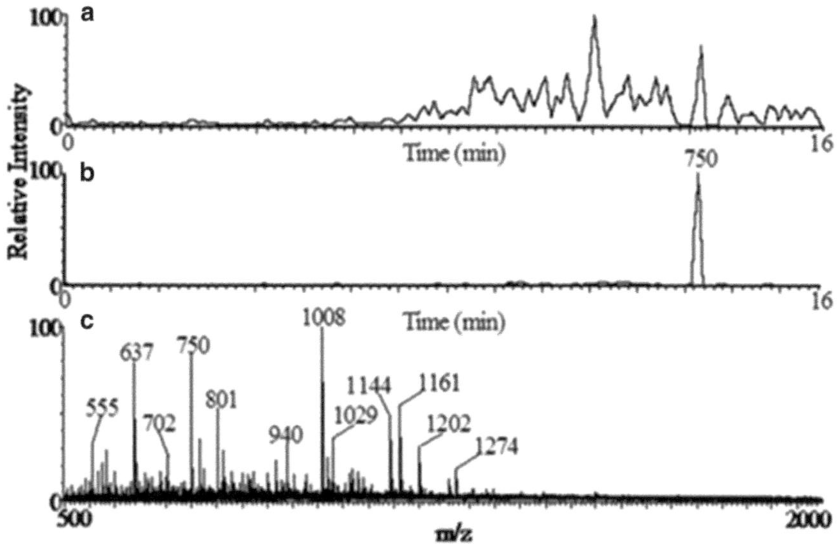

Fig. 15. Panel **a**: CE-MS/MS analysis of the tryptic digest of 55 ribosomal proteins. Panel **b**: separation efficiency for a typical CE peak. Panel **c**: mass spectrum of the peak from panel **b** (reprinted with permission from (Garza, S., Moini, M. (2006) Analysis of Complex Protein Mixtures with Improved Sequence Coverage Using (CE–MS/MS)^n, *Anal. Chem.* 78, 7309–7316). Copyright (2006) American Chemical Society).

Protein fractionation prior to MS analysis is yet another technique that can be used to solve the undersampling issue. In this technique, a complex protein mixture is first fractionated into several simpler mixtures using solubility or isoelectric point properties of the proteins. If the protein fractions are simple enough, they can be dissociated directly in the gas phase via a top-down technique such as electron-capture dissociation (39) or electron-transfer dissociation (40). More complex protein fractions are usually further separated; for example, by using SDS gels, individual bands are chemically or enzymatically digested and then analyzed by nLC-MS/MS (41, 42). Because these bands usually contain ~1 pmol or less of sample, to achieve a sensitive analysis, it is often necessary to inject the entire sample followed by analysis utilizing nano-LC-MS or nLC-MS/MS (1, 41, 42). Under this condition, the multiple-injection technique under the same nLC-MS/MS is impractical. In addition, since each LC separation takes ~1 h for complex protein mixtures, analysis would be very time-consuming. Multiple sample injections followed by MS/MS analysis, however, are an ideal match for CE-MS/MS analysis. This is because each CE-MS/MS analysis consumes only nanoliters of sample (~1/100 of a nano-HPLC run), and each analysis takes ~10 min (<1/6 of the nLC-MS/MS) (1). Because the sample consumption associated with CE is insignificant and analysis time is short, an experiment can be designed in which all CE and MS parameters are considered variables that can be manipulated after each injection followed by CE-MS/MS analysis to provide maximum sequence coverage (1). To address this low sequence coverage, we introduce a novel technique, $(\text{CE-MS/MS})^n$, which utilizes the most significant advantages of CE-MS/MS, including economy of sample size, fast analysis time, and high separation efficiency, to increase the sequence coverage of a complex protein mixture. Based on these characteristics, $(\text{CE-MS/MS})^n$ can be performed in which multiple CE-MS/MS subanalyses (injections followed by analyses) are analyzed and experimental variables are manipulated during each CE-MS/MS subanalysis in order to maximize sequence coverage. $(\text{CE-MS/MS})^n$ is a practical technique since each CE-MS/MS subanalysis consumes <10 nL, and each CE-MS/MS subanalysis takes ~10 min; therefore, several subanalyses can be performed in ~1 h consuming only nanoliters of the sample. Two techniques have been introduced to address the undersampling: (1) $(\text{CE-MS/MS})^n$ using dynamic exclusion (in this technique, several CE-MS/MS analyses (injection followed by separation) were performed in one run using the dynamic exclusion capability of the mass spectrometer until all peptide peaks were analyzed by MS/MS) and (2) gas-phase fractionation (in this technique, $(\text{CE-MS/MS})^n$ is performed by scanning a narrow mass range (every ~100 m/z) during each CE-MS/MS subanalysis without using dynamic exclusion). Under this condition, in each subanalysis, the number of peptides available for MS/MS analysis is significantly reduced, and peptides with the same nominal masses are analyzed, thereby increasing sequence coverage. Additionally, to address the lack of detection of low-level peptides in a mixture containing a wide concentration dynamic range, the

Table 1
Sequence coverage achieved by nano-LC vs. (CE-MS/MS)^6 for 6-protein digest

Protein	Nano-LC-MS coverage (%)	(CE-MS/MS)^6 coverage (%)
Carbonic anhydrase I (human)	31	81
Bovine serum albumin (bovine)	25	59
Phosphorylase B (rat)	23	54
Cathepsin C (bovine)	41	83
Carbonic anhydrase II (bovine)	12	59
Trypsinogen (bovine)	43	79

concentration of the sample was systematically increased in each subanalysis (while utilizing dynamic exclusion) so that low-intensity peptides would rise above the mass spectrometer threshold and, consequently, undergo MS/MS analysis. Moreover, to alter the ionization efficiency of peptides with low electrospray ionization efficiency and to change the migration behavior of comigrating peptides under a specific liquid composition, the CE background electrolyte was modified in several subanalyses to further improve sequence coverage. The (CE-MS/MS)^6 was applied to the analysis of the tryptic digests of the six-protein mixture (Table 1), and the results were compared with the analysis of the same mixture using a nano-HPLC-MS (Fig. 16). The sequence coverage achieved with the nano-HPLC and the (CE-MS/MS)^6 is summarized in Table 1. As shown in Table 1, (CE-MS/MS)^6 produced much better sequence coverage using only a few nanoliters of the protein digest, while the HPLC analysis used 1 μL of the same solution. The results clearly show the advantage of the (CE-MS/MS)^n for high sequence coverage proteomics while consuming a minimal amount of the samples.

3.3. CE-MS Analysis of Intact Proteins of Whole-Cell Lysate and Subcellular Proteins

Perhaps one of the most important and underutilized advantages of the CE-MS is its capability to analyze intact proteins and protein complexes. While most hyphenated chromatography–mass spectrometry techniques have a hard time analyzing complex protein mixtures, the dynamically coated capillaries discussed above have no problem analyzing proteins and protein complexes. Separation and MS analysis of intact proteins are especially useful for top-down proteomics (43), in which intact proteins in complex mixtures are

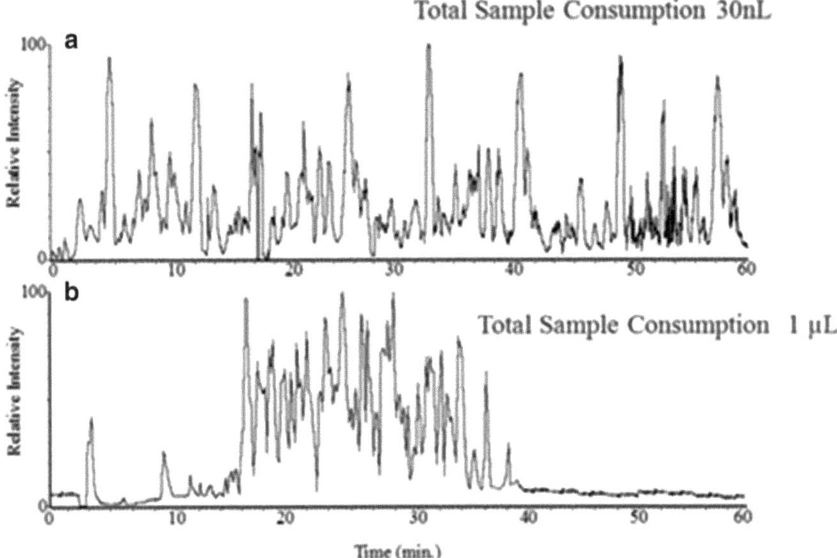

Fig. 16. Comparison between (CE-MS/MS)[6] and nano-HPLC-MS analyses of the tryptic digests of a six-protein mixture (reprinted with permission from (Garza, S., Moini, M. (2006) Analysis of Complex Protein Mixtures with Improved Sequence Coverage Using (CE–MS/MS)^n, *Anal. Chem.* 78, 7309–7316). Copyright (2006) American Chemical Society).

analyzed by MS and then undergo collision-induced dissociation (CID), electron-capture dissociation (ECD), or electron-transfer dissociation (ETD) where protein identification can be achieved without enzymatic digestion (44). While perhaps the ultimate goal for CE-MS analysis of complex protein mixtures is to analyze whole-cell lysates, currently CE-MS analysis of protein mixtures is limited to analyzing subcellular protein contents such as ribosomal proteins. Limitation for the analysis of whole-cell lysate is due to (1) limited peak capacity of CE-MS which is currently ~100, (2) insolubility of a large number of whole-cell lysate proteins in the MS-friendly background electrolyte, (3) a wide concentration range of proteins in the cell, and (4) limited volume of the sample that can be injected into CE capillary. For example, when the whole-cell lysate of *E. coli* was injected into a capillary containing 0.1% acetic acid and 25–50% organic solvents, protein precipitation resulted in capillary blockage (45). While the use of a basic BGE (20 mM ammonium bicarbonate (NH_4HCO_3) in water/methanol (75/25 v/v) with pH 8 (using NH_4OH)) reduced precipitation, because of low sensitivity of detection of proteins under this BGE, only highly abundant proteins were detected. To increase the number of proteins detected, the whole-cell lysate was fractioned into acidic and basic fractions by dissolving the whole-cell lysate in the acidic BGE. The supernatant was then analyzed using the acidic BGE and the APS-derivatized capillary. The precipitate of the acidic cell lysate was then dissolved in the 20 mM ammonium bicarbonate and analyzed by CE-MS, using the basic BGE and the HPMC-coated capillary. For each analysis, 1 nL of the acidic or basic crude lysate solution was injected. Figure 1d displays an example of the mass spectrum observed for each peak of the electropherogram. As shown, because of the large concentration dynamic range of proteins in *E. coli* cells and the limited injection volume of the CE, only 58 proteins were detected in these experiments. At the cell lysate concentration used, many low-level proteins either were below the CE-MS detection limit or were masked by comigrating proteins of larger quantity. Moreover, some proteins were not detected due to their poor solubility. The results clearly show that 1-D CE is not capable of completely analyzing such a complex protein solution with a large concentration dynamic range. Therefore, the utility of CE-MS in analyzing a less-complicated protein mixture was investigated using *E. coli* ribosomal proteins. Figure 17 shows the CE-MS electropherograms for the analysis of ribosomal proteins using 0.1% acetic acid in water/acetonitrile (50/50 v/v) (A) and 0.1% acetic acid (B) as the BGE, utilizing a 15-mm-i.d., 120-cm-long capillary. In each experiment, 1 nL of the protein solution (1.7 ng of total ribosomal protein) was injected. As shown in Fig. 17, respectively, 44 and 39 different ribosomal proteins were detected in each run. Some of the proteins not detected using the BGE containing 0.1% acetic acid in water/acetonitrile (50/50 v/v) were detected using 0.1% acetic acid. Overall 55 out of the 56 (45, 46) proteins were detected in two runs, consuming only 2 nL of the sample (3.4 ng of ribosomal protein). The results clearly show that CE-MS is a viable technique for top-down analysis in subcellular proteomics.

3.3.1. Application of CE-MS to the Analysis of the Major Proteins of Red Blood Cells (RBCs) Under Denaturing Condition

In addition to subcellular proteomics mentioned above, CE-MS is especially useful for the analysis of the major proteins of mammalian cells. Most mammalian cells are specialized cells and are usually packed with a few major proteins that are involved with the specific functions of that cell. One such cell is human RBC which is specialized in transport of oxygen and carbon dioxide. RBC contains three major proteins: hemoglobin (Hb), carbonic anhydrase I (CAI), and CAII. In its endogenous state, hemoglobin (Hb) exists as a tetramer (~450 amol/cell), which consists of two dimers, each composed of an α-chain (average MW 15 126) and a β-chain (average MW 15 865). Each chain contains a heme group (47). Carbonic anhydrase I (CAI, average MW 28 780) and carbonic anhydrase II (CAII, average MW 29 156) are the next most abundant proteins (48, 49), with the respective quantities of ~7 and ~0.8 amol in each adult RBC (50, 51). In addition to the common reasons for monitoring hemoglobin (e.g., for identification of variants), there is also a need for the identification of carbonic anhydrase isoforms. For example, a significant decrease in CAI levels could suggest hemolytic anemia or hyperthyroid Graves' disease (52, 53), CAII

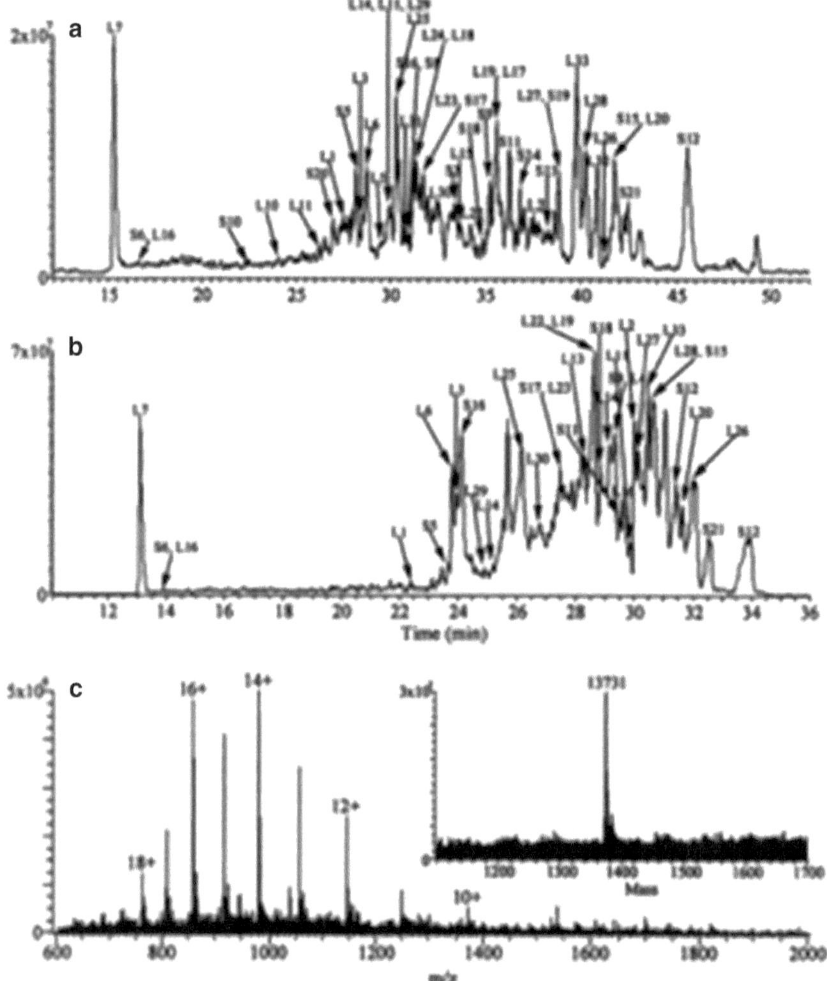

Fig. 17. (**a**) CE/ESI-MS base peak electropherogram of the *E. coli* ribosomal proteins using 0.1% acetic acid in 50% acetonitrile as the BGE. (**b**) Same as (**a**) except that 0.1% acetic acid in water was used as the BGE. Approximately 1.7 ng of total ribosomal proteins was injected in each analysis. (**c**) Mass spectrum of the S11 peak in (**a**). The inset shows the deconvoluted mass spectrum (13 731 Da) of the peak of (**c**) (reprinted with permission from (Moini, M., Huang, H. (2004) Application of capillary electrophoresis/electrospray ionization-mass spectrometry to subcellular proteomics of *Escherichia coli* ribosomal proteins, *Electrophoresis* 25, 1981–1987). Copyright (2004) John Wiley & Sons, Ltd).

deficiency results in renal tubular acidosis (54), while osteoporosis, symmetrical cerebral calcification, and mental retardation have been noted in individuals with very low levels of CAII in their RBCs (55). Moreover, a decrease in the ratio of CAI/CAII is observed in individuals deficient in glucose-6-phosphate dehydrogenase. Therefore, it is important to develop sensitive analytical techniques for the detection and identification of hemoglobin and carbonic anhydrase isoforms, preferably at single-cell levels. There are two ways to detect the chemical contents of RBCs: (1) under denaturing conditions, where protein–protein and protein–metal complexes are dissociated, and (2) under native

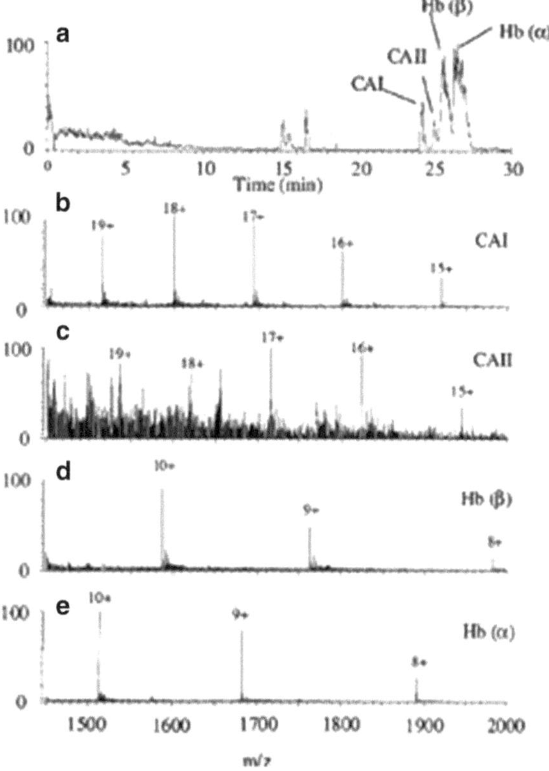

Fig. 18. (**a**) Base peak electropherogram of the lysed blood at a CAII concentration of ~88 amol/nL using the long column. Approximately 0.5 nL of lysed blood was injected. (**b**) Mass spectrum of the peak marked CAI of panel **a**. (**c**) Mass spectrum of the peak marked CAII of panel **a**. (**d**) Mass spectrum of the peak labeled Hb (β) of panel **a**. (**e**) Mass spectrum of the peak labeled Hb (α) of panel **a**. The 0.1% acetic acid solution containing 50% acetonitrile was used for this experiment. Unmarked peaks on panel **a** were not identified (reprinted with permission from (Moini, M., Demars, S.M., Huang, H. (2002) Analysis of Carbonic Anhydrase in Human Red Blood Cells Using Capillary Electrophoresis/Electrospray Ionization-Mass Spectrometry, *Anal. Chem.* 74, 3772–3776). Copyright (2002) American Chemical Society).

conditions in which protein–protein and protein–metal complexes remain intact. In this section CE-MS analysis of RBCs under denaturing conditions will be discussed.

Under denaturing conditions, all four major proteins of RBCs are detectable by CE-MS (Fig. 18); however, since CAI (~7 amol/cell) and CAII (~0.8 amol/cell) are, respectively, ~100× and ~1,000× less than α-chain and β-chain of hemoglobin (~900 amol/cell), they must be well separated from hemoglobin before they can be detected (56). Otherwise, the background chemical noise associated with the ~1,000× molar excess of hemoglobin masks the CA signals. Because of this large molar excess ratio of hemoglobin, successful detection of CAI and CAII requires complete separation of these two proteins from hemoglobin. This was achieved by using a long (125-cm) capillary and operating under reverse polarity mode (−30 kV at the inlet), which allowed CAI and CAII to migrate before hemoglobin chains and be detected. Under this condition, the detection limit of CAII in lysed blood was ~44 amol. On the other hand, the limit of detection for CAI, which was completely separated from Hb, was ~20 amol, which was limited by the LCQ MS sensitivity. For example, when a single intact RBC was analyzed with this capillary, CAI (~7 amol/RBC) was not detected. This narrow dynamic range is a limiting factor in CE/ESI-MS analysis of samples containing components in a wide range of concentrations. At the single-cell level, a more sensitive mass spectrometer is required to detect components present in low quantities (<20 amol). At the multi-cell level, or when lysed cells are used, a long capillary is needed to separate the components of the mixture. However, it is important to note that this level of sensitivity was achieved by using an older generation of mass spectrometer. Modern mass spectrometers will probably be able to detect CAI at single-cell level. A drawback of using a long capillary is, however, longer analysis time. Off-line separation of the major constituents of samples prior to CE/ESI-MS is another alternative for the detection of minor components of cells. However, this technique is time-consuming and may result in loss of sample due to additional sample handling. The results demonstrate that CE/ESI-MS is a powerful tool for the analysis of the chemical contents of single cells at low attomole levels.

3.4. Analysis of Protein Complexes by CE-MS

Analysis of the protein contents of cells under native conditions when proteins maintain their non-covalent association with other proteins, molecules, and metals is an important task. This is because the genome itself does not explain many cellular functions and processes. Proteins have specialized functions in the cellular environment and may non-covalently aggregate with DNA, RNA, cofactors, ligands, and other proteins to produce protein complexes. Protein complexes found inside and outside of biological cells take part in many biochemical pathways and perform many different functions. The effects of drugs and other ligands on a particular protein complex can also be investigated, which can become a powerful tool for the biotechnology industry. Mass spectrometry analysis of protein complexes in cells under native conditions is essential for their identification since the formation of the complex is largely affected by pH and other environmental factors. To date, MS analysis of a mixture of protein complexes is mostly achieved by off-line separation and purification of protein complexes followed by MS analysis using infusion electrospray ionization (ESI). By utilizing this technique, several types of protein complexes have been analyzed including the following: protein–DNA, protein–ligand, and protein–protein complexes (57–66). The main disadvantage of infusion ESI-MS for the analysis of protein complexes is that unbound analytes and salts present in the solution can associate with the protein complexes during the ESI process, resulting in degradation of peak resolution when the resolving power of the instrument is not high enough to separate the individual components as well as shift to higher average MW for the protein complexes as a result of these nonspecific adducts. Moreover, separation and sample cleanup is labor-intensive and time-consuming.

Capillary electrophoresis is an ideal separation technique for the analysis of protein complexes. Two important advantages of CE that are essential for the analysis of protein complexes are the following: (1) compatibility with a variety of background electrolytes at a wide pH range, which allows separation of intact protein complexes under native condition (for example, by switching from 0.1% acetic acid solution (pH ~3.5) to ammonium acetate with pH ~7 as the background electrolyte, protein complexes were measured using the same capillary that is normally used for the analysis of intact proteins under denaturing conditions), and (2) separation of endogenous protein complexes from analytes not associated with the complex that could otherwise be attached to the complex during electrospray ionization process. By separating these nonspecific analytes, CE-MS provides a more accurate representation of the constituents of the complex without the use of hydrophobic media that can disintegrate protein complexes. In recent years, we have applied this technique to the analysis of the protein complexes of RBCs directly from cell lysates and for the measurement of the protein–metal stoichiometry.

3.4.1. Application of CE-MS to the Analysis of Major Protein–Protein and Protein–Metal Complexes of Erythrocytes Directly from Cell Lysate

In this section we discuss CE-MS analysis of the RBC under native conditions, in which protein–protein and protein–metal complexes remain intact (67). Under native conditions, hemoglobin (average MW 64446) is a protein complex in RBCs that exists as a tetramer (Hb-tetramer) consisting of four non-covalently bonded protein subunits: two α-chains and two β-chains with each chain attached to a heme group (MW 616). CAI and CAII are two other proteins in RBCs that are each complexed with one zinc cofactor. This Zn cofactor is very tightly bounded to both CAI (CAI-Zn, MW 28 845) and CAII (CAII-Zn, MW 29 221) at pH ~7. Figure 1 shows the electropherogram (panel a) of a lysed blood kept in ammonium acetate for 4 h before analysis and the corresponding mass spectra (panels b–d) of the CAII-Zn, CAI-Zn, and Hb-tetramer protein complexes of panel a, respectively. The order of migration, as shown in Fig. 1, is CAI-Zn, CAII-Zn, and Hb-tetramer. This order of migration is different from our previous CE-MS analysis of lysed RBCs (56) because of the major experimental differences between CE-MS analyses of RBC under denature condition compared with analysis under native conditions. To analyze intact protein complexes, we used ammonium acetate at pH~7 (the BGE

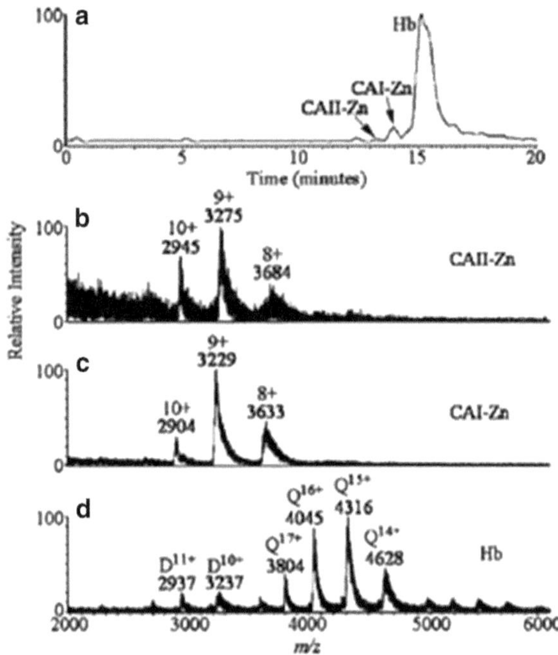

Fig. 19. (**a**) Base peak electropherogram of blood lysed and diluted 100× in 1.7 mM ammonium acetate (pH 7.4) and kept in the solution for 4 h before CE/ESI-MS analysis with 5 nL of sample injected. (**b**–**d**) Corresponding mass spectra of the peaks of panel a marked with CAII-Zn, CAI-Zn, and Hb. The Hb-dimer present in (**d**) is found to have 2 hemes attached (MW 32 322) (reprinted with permission from (Nguyen, A., Moini, M. (2008) Analysis of Major Protein–Protein and Protein–Metal Complexes of Erythrocytes Directly from Cell Lysate Utilizing Capillary Electrophoresis Mass Spectrometry, *Anal. Chem.* 80, 7169–7173). Copyright (2008) American Chemical Society).

also contained polybrene as a self-coating reagent, to prevent analyte–wall interaction.) In addition, the experiments were performed under forward polarity mode so the CE outlet electrode acted as the cathode with respect to the CE inlet electrode and acted as an anode with respect to ESI. By acting as both anode and cathode, the outlet electrode maintains a neutral pH at the end of the capillary, thereby preventing dissociation of the tetramer. In addition, under forward polarity mode, CAII-Zn and CAI-Zn migrated ahead of Hb-tetramer, avoiding being masked by, respectively, 562× and 64× molar excess of Hb-tetramer. Moreover, to achieve high separation efficiency, injection volume was limited to ~5 nL (of the 100× diluted lysed blood) resulting in high separation efficiency. Under these conditions, CAII-Zn, CAI-Zn, and Hb were almost baseline separated. Figure 19 shows the base peak electropherogram of blood lysed and diluted 100× in 1.7 mM ammonium acetate (pH 7.4) and kept in the solution for 4 h before CE/ESI-MS analysis (panel a). Panels b–d show the corresponding mass spectra of the peaks of panel a marked with CAII-Zn, CAI-Zn, and Hb. The Hb-dimer present in (D) is found to have 2 hemes attached (MW 32 322). The mass spectrum of panel d also contained a small amount of (Hb-dimer + 2(heme)) as indicated by D^{10+} and D^{11+}, as well as monomers. However, the extracted ion electropherograms of the monomers and the dimer were almost identical to that of the electrophoretic peak profile of the Hb-tetramer, indicating that they were most likely from the dissociation of Hb-tetramer at the end of the CE capillary and during the desolvation/ionization process. No separate electrophoretic peak for Hb-dimer or Hb-monomer was observed, even when injecting 4× more samples (20 nL rather than 5 nL) and using a blood sample left in ammonium acetate for 6 h (Fig. 2). These results implied that, at physiological pH, the amount of Hb-dimer in lysed blood was too low to be detected by CE/ESI-MS. This is consistent with the binding constant (K) of the Hb-dimer ↔ Hb-tetramer, which is reported to be $\sim 4 \times 10^5$ M^{-1} for fully oxygenated blood (K increases as less oxygen binds) at pH 7.4. In summary, the use of polybrene as a self-coating reagent in conjunction with ammonium acetate at pH ~7.4, narrow capillary for high separation efficiency, and forward polarity CE to avoid acid production at the tip of the capillary were overriding experimental factors for successful analysis of protein complexes. Diluting the lysed blood sample in ammonium acetate for a minimum of 6 h before injecting the sample into the CE was essential for obtaining the mass accuracy consistent with their theoretical average molecular weights. A separate electrophoretic peak for the Hb-dimer was only detected when the pH of the BGE was lowered from 7.4 to ~6.6. The tetramer was completely dissociated to dimer when the BGE pH was dropped to 5. At pH 5, the mass spectrum of the Hb-tetramer was similar to CE-MS analysis under reverse polarity mode, indicating that under reverse polarity mode the pH at the ESI tip had decreased from pH 7.4 to ~5, the pH at which Hb-tetramer completely dissociates to dimer (60).

3.4.2. Application of CE-MS to Metal Displacement and Stoichiometry of Protein–Metal Complexes

Another example of the application for CE-MS analysis to protein complexes is the study of metal displacement and stoichiometry of protein–metal complexes under native conditions. Currently, most MS analyses of protein complexes use direct-infusion ESI-MS. While this technique offers much higher sensitivity of detection compared to nuclear magnetic resonance (NMR) and X-ray diffraction techniques, since protein–metal complexes under native conditions usually are dissolved in salt solutions, their direct ESI-MS analysis requires off-line sample cleanup prior to MS analysis to avoid sample suppression during ESI. Moreover, direct infusion of the salty solution promotes nonspecific salt adduct formation by the protein–metal complexes under ESI-MS, which complicates the identification and stoichiometry measurements of the protein–metal complexes. Because of the high mass of protein–metal complexes and lack of sufficient resolution by most mass spectrometers to separate nonspecific from specific metal–protein

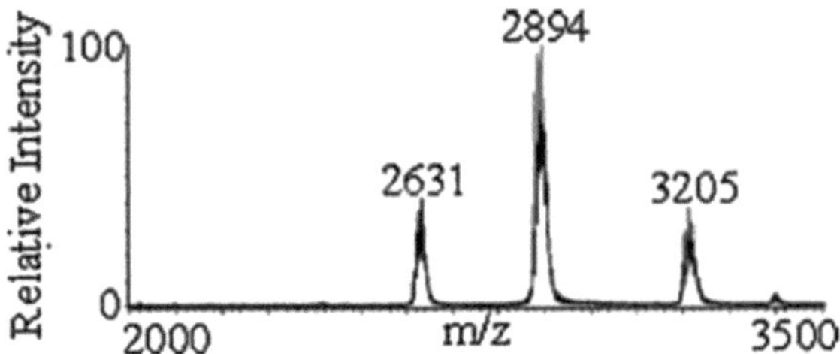

Fig. 20. Mass spectrum of the AiiA-2Co^{2+} complex at pH 6.9 obtained using infusion ESI-MS (reprinted with permission from (Garza, S., Thomas, P.W., Fast, W., Moini, M. (2010) Metal displacement and stoichiometry of protein–metal complexes under native conditions using capillary electrophoresis/mass spectrometry, *Rapid Commun. Mass Spectrom.* 24, 2730–2734). Copyright (2010) John Wiley & Sons, Ltd).

complexes, accurate protein–metal stoichiometry measurements require some form of sample cleanup prior to ESI-MS analysis. CE-MS in conjunction with a medium-resolution (10,000) mass spectrometer is an efficient, sensitive, and fast method for the measurement of the stoichiometry of the protein–metal complexes under physiological conditions (pH ~7) without the complications of the infusion ESI-MS. To demonstrate this, recently we compared the metal displacement of Co^{2+} to Cd^{2+} using both ESI-MS and CE-ESI-MS using the procedure explained above for the analysis on intact protein–protein and protein–metal complexes of RBC directly from cell lysates (68). Co^{2+} and Cd^{2+} are two metal ions necessary for activation in the monomeric AHL lactonase produced by B. thuringiensis. The AiiA from B. thuringiensis (average MW 28633) is known to have quorum-quenching activity in the presence of two bound metal ions. The AiiA–metal complex is most stable at a pH ~7 in the presence of buffer salts; however, the presence of this buffer during the ESI process suppresses the ionization of the complex. CE/ESI-MS of the AiiA-2Co^{2+} complex was performed using 10 mM ammonium acetate at pH values of 4.8 and 6.9; however, at pH of 4.8, the protein–metal complex dissociated and only the protein was detected. In contrast, at pH 6.9 the protein–metal complex remained intact. Similar to our RBC study, to obtain an accurate molecular weight free from the nonspecific salt adduct, it was necessary to dilute the sample in a solution of 10 mM ammonium acetate (pH 6.9) before performing CE/MS analysis. Figure 20 shows ESI-MS of AiiA-Co^{2+} using infusion ESI. As shown, even though the complex was diluted in ammonium acetate solution before infusion ESI-MS, the mass spectrum of the protein–metal complex still contains a number of unresolved peaks caused by multiple nonspecific metal (mostly sodium) adductions. Since the resolution of a typical TOF mass analyzer is usually not high enough to differentiate the isotope clusters of protein–metal complexes from proteins bound to

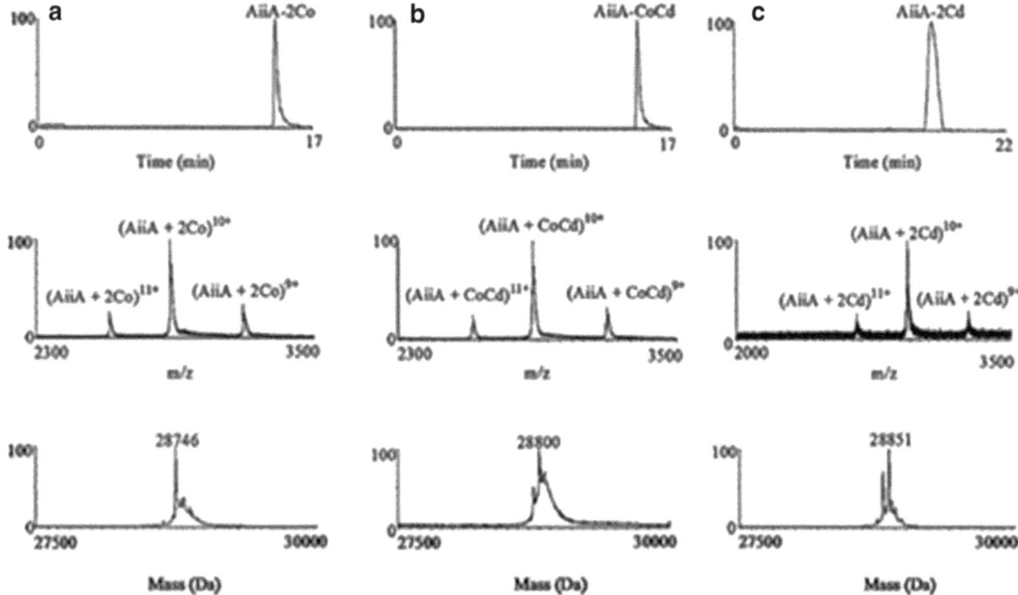

Fig. 21. The electropherograms (*top panels*), mass spectra (*middle panels*), and deconvoluted mass spectra (*bottom panels*) of AiiA–metal complexes when a 1:1 ratio (**a**), a 10:1 ratio (**b**), and a 100:1 ratio (**c**) of Cd^{2+}:Co^{2+} solutions were added to the protein solution (reprinted with permission from (Garza, S., Thomas, P.W., Fast, W., Moini, M. (2010) Metal displacement and stoichiometry of protein–metal complexes under native conditions using capillary electrophoresis/mass spectrometry, *Rapid Commun. Mass Spectrom.* 24, 2730–2734). Copyright (2010) John Wiley & Sons, Ltd).

nonspecific and specific metal ions, it becomes difficult to accurately identify the composition and stoichiometry of the protein–metal complexes. Therefore, to utilize direct-infusion MS, it is necessary to significantly, but not completely, reduce the salt content by using off-line salt-removal techniques. In contrast, under CE/MS conditions, the mass spectrum of the electrophoretic peak of the same sample contained sharp peaks (Fig. 21, panel a), corresponding to the *m/z* value of the protein–metal complex within the experimental error of ~0.02%.

The AiiA protein was originally grown in a medium containing $CoCl_2$. The mass spectrum of the protein complex shows that the AiiA is binuclear, with two cobalt atoms bound to the protein. To determine the relative binding stoichiometry of the Cd^{2+}, different ratios of $CdCl_2$ (1:1, 10:1, and 100:1) were added to the AiiA-Co^{2+} complex in order to substitute the two Co^{2+} ions with two Cd^{2+} ions. In Fig. 21, panels a–c show the CE/MS analyses of these three ratios, respectively. The top panels illustrate the electropherogram of the protein–metal complexes, the middle panels show the corresponding mass spectra, and the bottom panels show the deconvoluted mass spectra. From Fig. 21, the relative binding stoichiometry of each metal can be determined. The results indicate that, at a 1:1 ratio of Co^{2+} to Cd^{2+}, both Co^{2+} ions are tightly bound to the AiiA protein. However, as more Cd^{2+} is added (10:1 ratio), one Co^{2+} ion is completely replaced by one Cd^{2+} ion. At a 100:1 ratio of Cd^{2+} to Co^{2+}, both Co^{2+} ions are replaced by two Cd^{2+} ions. Because the results were obtained by CE/MS, all free metal ions were separated from the protein–metal complex and did not participate in the complex formation during ESI, allowing a more accurate determination of the ratios of the metal concentration needed for metal displacement.

As shown in this section, the analysis of intact protein–protein and protein–metal complexes was accomplished using CE/ESI-MS. Intact protein complexes were analyzed with no prior sample preparation. Accurate MWs (~0.01% error) were obtained for all major protein–protein and protein–metal complexes. It was observed that more accurate MW was obtained when protein complexes were diluted in ammonium acetate buffer and left in that solution for a few hours. Ion exchange between ammonium ions in the solvent and nonspecific sodium adducts of the protein–protein and protein–metal complexes slightly unraveled the complexes. CE/ESI-MS further enhanced this charge exchange and separated the nonspecific salt adducts from protein complexes, resulting in formation of higher charge-state protein complexes free from nonspecific salt adducts. Deconvolution of these higher charge-state ions provided accurate (~0.01% error) average MW of intact protein–protein and protein–metal complexes for better identification of the constituents of the complexes. Operation of the CE under forward polarity mode was essential for detection of intact protein–protein and protein–metal complexes. This was because of the electrochemical nature of CE and ESI processes. These results demonstrate that CE/ESI-MS is an effective tool for the analysis of the mixture of protein complexes and can accurately provide average MW for unambiguous identification of small (<100,000 Da) protein complexes.

4. High-Throughput CE-MS Analysis Using a CE with Eight Capillaries in Conjunction with a Mass Spectrometer with Eight Inlets

High-throughput analysis of complex mixtures has been a primary need for industries such as pharmaceutical, biotechnology, and combinatorial chemistry, where a large number of reaction by-products must be analyzed. In recent years, the effort to increase throughput using separation in conjunction with mass spectrometry has been focused on reducing the separation time. In CE-MS, shorter analysis time is also achieved by using ultrafast CE-MS (UFCE-MS) by reducing capillary inner diameter (i.d.) (<10 μm) and capillary length (<30 cm) while increasing separation voltage (>1,000 V/cm) (69). To increase throughput beyond this technique, we recently introduce multi-CE capillaries in conjunction with gated multi-inlet mass spectrometry (70). In this technique the MS atmospheric pressure sampling inlet (nozzle, bore, sampling orifice, or heated capillary) is modified by replacing its single inlet with multiple atmospheric pressure inlets. This allowed for the parallel introduction of multiple liquid streams into the MS using one electrosprayer for each inlet so that the chemical contents of all liquid streams are analyzed concurrently using a single MS. In our original multi-sprayer, multi-inlet design (71, 72), all inlets were open to the atmosphere at all times and the ESI sprayers either sprayed continuously or were turned on one at a time using a high-voltage switch. Since all of the inlets were open to the atmosphere at all times, the pressure in the first stage of the vacuum system was too high for the original pumping system to evacuate when multiple inlets were used. To maintain operating pressure within the MS, the inlet i.d. was reduced and multiple pumps were used. In order to maintain the original inlet i.d. and pumping system while still maintaining the original sensitivity (per inlet), a gated multi-sprayer, multi-inlet design was developed. As a proof of concept we used a CE with 8 parallel CE capillaries in conjunction with a mass spectrometer equipped with 8 sprayers and 8 inlets for high-throughput and concurrent analysis of 8 complex mixtures.

For multi-CE, multi-inlet TOF-MS analysis, an in-house fabricated CE instrument capable of handling eight samples and eight buffer vials was used (Fig. 22). The CE instrument was capable of pressure programming during a voltage separation. Pressure and voltage were supplied by a commercial CE instrument (Beckman Coulters, Fullerton, CA, USA). Eight 50-μm-i.d., 150-μm-o.d. CE capillaries, ranging between 50 and 80 cm in length, were interfaced to the multi-inlet MS using a sheathless CE/ESI-MS interface. Approximately 5 nL of the peptide standard was injected into each capillary using the pressure injection mode. The CE was operated in reverse polarity mode (−15 kV applied at the CE inlet) and a 0.1% acetic acid solution was used as the BGE. Figure 23 shows the schematic of the gated eight-sprayer, eight-inlet TOF-MS. For proof of concept a peptide mixture containing 5 peptides was injected into all eight-CE capillaries. Figure 24 (top panel) shows the total ion electropherogram of the analysis of the peptide standard using eight-CE capillaries. Figure 24 (bottom panel) shows the reconstructed total ion electropherogram for one of the capillaries. The low resolution achieved in this electropherogram was because of the wider i.d. (50 μm) capillary in conjunction with low separation voltage (15 kV) used in this experiment. The use of a wider capillary and lower separation voltage was necessary to obtain peak widths that were compatible with the acquisition time (4.8-s turnaround time) of our current multi-inlet instrument. This preliminary result demonstrates that gated multi-inlet mass

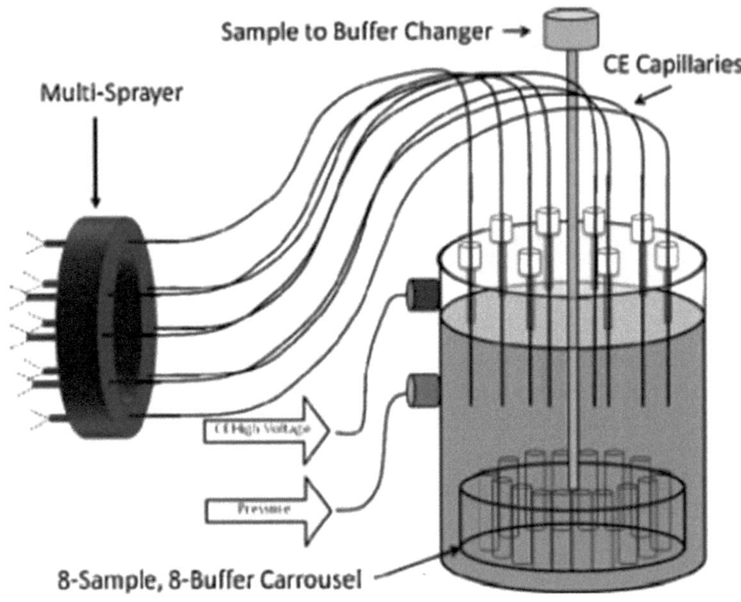

Fig. 22. Schematic of the eight-capillary CE instrument capable of providing pressure programming under voltage separation (reprinted with permission from (Moini, M., Jiang, L., Bootwala, S. (2011) High-throughput analysis using gated multi-inlet mass spectrometry, *Rapid Commun. Mass Spectrom.* 25, 789–794). Copyright (2011) John Wiley & Sons, Ltd).

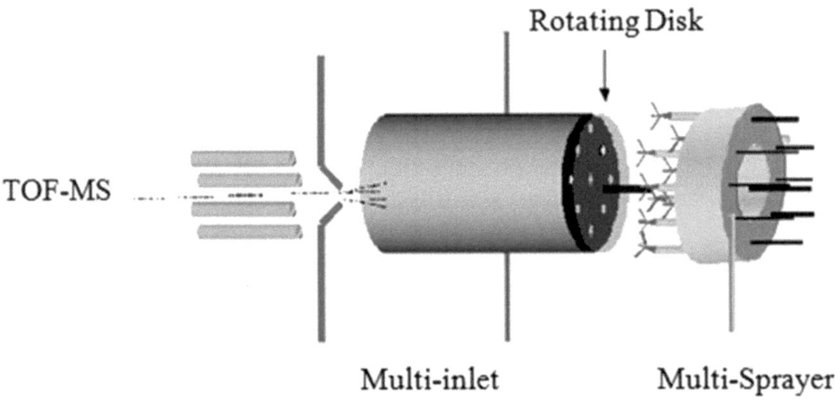

Fig. 23. Schematic of the gated eight-sprayer, eight-inlet TOF-MS (reprinted with permission from (Moini, M., Jiang, L., Bootwala, S. (2011) High-throughput analysis using gated multi-inlet mass spectrometry, *Rapid Commun. Mass Spectrom.* 25, 789–794). Copyright (2011) John Wiley & Sons, Ltd).

spectrometry is a viable technique for high-throughput analysis, where multiple streams of liquids from several CE capillaries can be analyzed using a single mass spectrometer. Up to eight peptide mixtures from up to eight CE capillaries were analyzed in about 30 min.

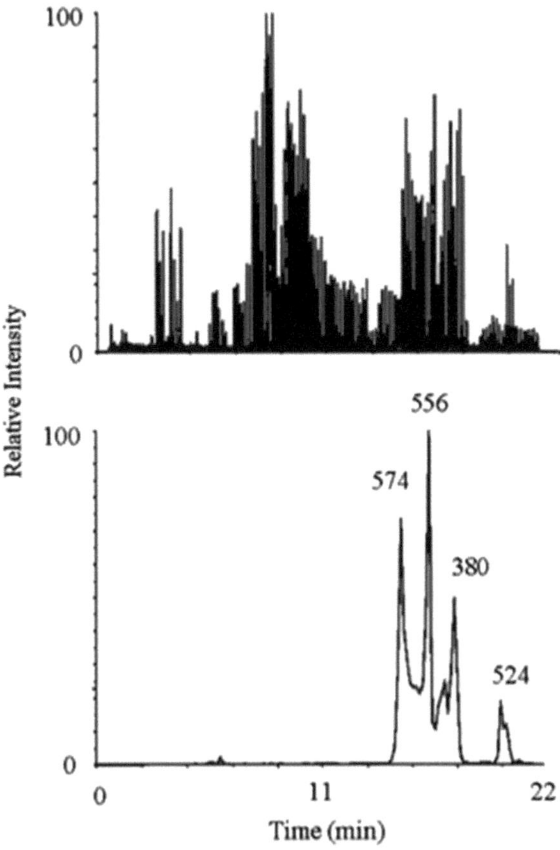

Fig. 24. Combined ion electropherogram (CIE) of the eight-CE capillary system (*top*). The same peptide standard was injected in each capillary. Extracted total ion electropherogram (TIE) of one capillary/inlet is shown at the *bottom panel*. The *m/z* 574 (1+), 556 (1+), 380 (1+), and 524 (2+) are, respectively, Met-enkephalin, Leu-enkephalin, Val-Tyr-Val, and angiotensin II (reprinted with permission from (Moini, M., Jiang, L., Bootwala, S. (2011) High-throughput analysis using gated multi-inlet mass spectrometry, *Rapid Commun. Mass Spectrom.* 25, 789–794). Copyright (2011) John Wiley & Sons, Ltd).

5. Appendix 1. Standard Operating Procedure for Making Porous Capillary Using HF

Substance: hydrogen fluoride, HF

Identification numbers: CAS number 7664-39-3

Main risks and target organs: Hydrogen fluoride is highly corrosive to all tissues. Skin: burns, necrosis; underlying bone may be decalcified. Eyes: burns

The course of the skin burns from hydrogen fluoride depends on concentration; more dilute solutions cause potentially serious but delayed (24 h) systemic effects. Absorption of the fluoride ion is a significant hazard mainly due to hypocalcaemia. Hydrogen fluoride burns to the eye are corrosive and require immediate flushing and ophthalmologic consultation.

5.1. Storage

Hydrogen fluoride should be stored in a cool, dry, well-ventilated area in tightly sealed containers. Containers of hydrogen fluoride should be protected from physical damage and should be stored separately from metals, concrete, glass, strong bases, sodium hydroxide, potassium hydroxide, and ceramics. Glass containers should not be used for storage as hydrogen fluoride will dissolve glass. Explicit warnings in appropriate languages and graphics should be maintained on all containers and work areas.

5.2. Personal Hygiene Procedures

If hydrogen fluoride contacts the skin, workers should flush the affected areas immediately with plenty of water for 15 min, followed by washing with soap and water.

Clothing contaminated with hydrogen fluoride should be removed immediately, and provisions should be made for the safe removal of the chemical from the clothing. Persons laundering the clothes should be informed of the hazardous properties of hydrogen fluoride, particularly its potential for causing irritation.

A worker who handles hydrogen fluoride should thoroughly wash hands, forearms, and face with soap and water before eating, using tobacco products, using toilet facilities, applying cosmetics, or taking medication.

Workers should not eat, drink, use tobacco products, apply cosmetics, or take medication in areas where hydrogen fluoride or a solution containing hydrogen fluoride is handled, processed, or stored.

5.2.1. Initial Medical Treatment

Thoroughly irrigate with water. Massage calcium gluconate gel into the burned skin for a minimum of 30 min and for as long as the pain persists; at times pain persists up to 4 h. If no gel is available, use a calcium (20 g Ca^{2+} in about 2 L of water) or magnesium solution for soaking.

5.2.2. Specific Preventive Measures

Appropriate latex or rubber gloves should be used when working with hydrogen fluoride. Frequent checks of glove integrity or changes of the gloves should be performed. Any suspected exposure should be immediately decontaminated. Eye goggles, aprons, sleeves, and other protective devices should be used in accord with the conditions of use. Concentrated and anhydrous hydrogen fluoride should only be used in the hood.

Use nitrile gloves, apron, face mask, and rubber gloves when working with HF solution and nitrile gloves and safety glasses when dealing with etched capillaries. Work under the hood when dealing with HF. Replace gloves each time your hands need to be outside the hood. Change gloves without touching skin.

5.2.3. Procedure for Making the Porous Tip Using the Teflon Container

1. Prepare a 500-ml saturated solution of calcium carbonate in a beaker and 500 mL of DI water in another beaker. These solutions will be used for neutralization of HF-contaminated glassware and fused-silica capillaries.

2. Fill Teflon container with fresh 48% HF, which is in a plastic container under a well-ventilated hood. To do this you may have to discard the old HF in the container by removing ~100 μL of HF solution from the HF container and slowly pouring it into the carbonate solution. Wash the container with DI water by filling the container with water and discard the water using disposable pipettes. Pour methanol into the empty Teflon container and discard it into the carbonate container.
3. Wait a few minutes for the leftover methanol in the HF container to evaporate.
4. Gradually (300–400 μL at a time) fill the dry Teflon container with fresh HF using a plastic disposable pipette (or a glass pipette if no plastic pipette is available). Fill the HF container to the top.
5. Replace gloves and etch a capillary by first burning about 4–5 cm of the capillary using a lighter. Second, remove the burnt debris at the capillary tip by using Chemwipes and methanol.
6. Attach the inlet of the capillary to the air hose in the hood using an Upchurch union.
7. Insert the bare fused-silica end of the capillary (the outlet) into the HF container all the way without touching the bottom of the container, and adjust it to prevent it from touching the wall of the HF container.
8. Wait appropriate amount of time for the outlet tip to become porous.
9. Using fresh gloves, remove the capillary outlet from the HF container and neutralize it by immersing it into the carbonate solution for a few seconds and then in water for a few seconds. Disconnect the fused-silica capillary from the air pressure and wash it under the faucet.
10. Replace the HF container top.
11. Decontaminate the area using the carbonated solution.

6. Appendix 2. Methods and Techniques for Analyzing Red Blood Cells

The chemical contents of RBCs were analyzed using three different sampling methods:

(1) A solution of lysed RBCs was injected into the capillary using the pressure injection mode. The solution was prepared by diluting 5 μL of fresh blood from a healthy adult male to 50 μL using saline solution, followed by centrifugation and removal

of the supernatant, leaving only intact cells without the plasma. After adding 50 μL of water to lyse the cells, the lysed cells were then diluted either 5× or 200× with water to final dilutions of 50× and 2,000×, respectively.

(2) Intact RBCs (one or three cells) were drawn into the capillary inlet using suction and a 200× microscope for observation.

(3) Intact RBCs suspended in a saline solution were injected into the CE capillary using the pressure injection mode of the CE instrument. The solution of suspended intact RBCs was prepared in a manner almost identical to method 1, except that saline solution was used instead of water to maintain intact cells. To homogenize the solution, the sample vial was hand-shaken immediately before sampling. It should be noted that it was necessary to use fresh blood to separate the Hb chains; when a 1-week-old blood sample was analyzed using the 0.1% acetic acid solution, only one peak with an average molecular weight of ~65,000 was observed. For intact cell analysis, best overall performance (separation and sensitivity) was achieved using the 0.1% acetic acid solution without any organic additives. However, when lysed blood was analyzed, the best overall performance was achieved using a 0.1% acetic acid solution containing 50% acetonitrile. Under these conditions, the presence of acetonitrile in the CE BGE reduced the EOF, and consequently, the four major proteins of RBCs were all separated (Fig. 18). Moreover, addition of acetonitrile also reduced the background chemical noise and adduct formation due to the incomplete desolvation of liquid droplets and, therefore, improved the S/N of CAI (Fig. 18, panel b). The improvement of S/N due to the latter BGE was further confirmed by analyzing a standard mixture of CAI and CAII (0.1 mg/mL) using a 0.1% acetic acid solution and a 0.1% acetic acid solution containing 50% acetonitrile. It was observed that while the mass spectra of these compounds had similar intensity on the absolute scale, both noise level and adduct formation were significantly lower for the BGE containing 50% acetonitrile. Other possible factors for the higher sensitivity of these proteins include better solubility and less adsorption to the capillary wall when the acetonitrile-containing BGE was used.

7. Appendix 3. Successful CE-MS analysis of protein complexes depends on three major factors

1. The use of ammonium acetate at pH 7.4. Ammonium acetate at pH ~7 was ideal for the analysis of protein complexes that we have studied. This is because it keeps the complexes intact and it is a mass-friendly background electrolyte. It was made by titrating a 0.01% acetic acid solution with a 0.01% ammonium

hydroxide solution until desired pHs were reached. At pH of 7.4 the concentration of the ammonium acetate (NH4OAc) buffer was ~1.7 mM. Polybrene (PB) was added to these buffers as the self-coating reagent 2 with a final concentration of PB at ~0.1%.

2. Soaking the protein complex in ammonium acetate for a few hours before CE-MS analysis. More accurate MWs (~0.01% error) were obtained for all major protein–protein and protein–metal complexes when they were diluted in ammonium acetate buffer and left in that solution for a few hours. The longer the protein complexes stayed in ammonium acetate solution before CE-MS analysis, the more intense and narrow the higher charge states of the protein complexes became. This was because longer times improved the ion exchange between the nonspecific, nonvolatile salts (mostly sodium) that were associated with the interior of the protein complexes under native conditions and the volatile ammonium ions of the BGE. Ion exchange between ammonium ions in the solvent and nonspecific sodium adducts of the protein–protein and protein–metal complexes slightly unraveled the complexes. CE/ESI-MS further enhanced this charge exchange and separated the nonspecific salt adducts from protein complexes, resulting in formation of higher charge-state protein complexes free from nonspecific salt adducts. Deconvolution of these higher charge-state ions provided more accurate (~0.01% error) average MW of intact protein–protein and protein–metal complexes for better identification of the constituents of the complexes.

3. Operation of the CE under forward polarity. Operation of the CE under forward polarity mode was essential for detection of intact protein–protein and protein–metal complexes. Intact protein complexes with minimal fragmentation can only be observed in forward polarity. This was because of the electrochemical nature of CE and ESI processes. As we have explained before, CE and ESI-MS represent two electrical circuits, each with two sets of electrodes, CE inlet and outlet electrodes, ESI emitter, and MS inlet electrodes. The CE/ESI-MS overlays these two separate circuits so that the CE outlet electrode and the ES emitter electrode are shared between the two circuits (CE outlet/ESI shared electrode). Therefore, under CE/ESI-MS, at the shared electrode two electrochemical reactions occur simultaneously. Depending on the polarity and magnitude of the voltage of the shared electrode compared with that of the CE inlet and MS inlet electrodes, electrochemical reactions at the shared electrode can be both reductive, both oxidative, or one reductive and the other oxidative. For example, under positive electrospray ionization mode, where +1.5 kV is applied to the shared electrode and 0–100 V is applied to the MS inlet electrode, two possibilities exist: (1) Under reverse

polarity CE, where −30 kV is applied to the CE inlet electrode, the shared electrode is anodic with respect to both CE inlet electrodes and the MS inlet electrode. (2) Under forward polarity CE, where +30 kV is applied to the CE electrode, the shared electrode is cathodic with respect to CE inlet electrode and anodic with respect to MS inlet electrode. At the anode, the major oxidation reaction for aqueous buffer solutions is typically the electrochemical oxidation of water ($2H_2O \leftrightarrow 4\ H^+ + O_2 + 4e^-$). The extent of the electrochemical reactions and therefore the pH change depends on the magnitude of the current that flows through each circuit. Under experimental conditions used here, CE current was ~1 μA, while ESI current was in the nanoampere range. Under reverse polarity mode in which the electrochemical reactions at the shared electrode are anodic for both circuits, the pH of the solution decreases enough to dissociate all the tetramer. The pH at the ESI tip under reverse polarity is estimated to decrease from pH 7.4 to 5, the pH at which some protein complexes dissociate to dimer. However, under forward polarity mode, the shared electrode is only anodic in the ESI circuit, which has a lower current, and its effect is mostly canceled by the cathodic reaction at this electrode in the CE circuit causing minimal dissociation of the tetramer. Since this dissociation (in both cases) is happening close to the outlet of the CE capillary, all dissociation products comigrate and show up as a single electrophoretic peak.

References

1. Garza S, Moini M (2006) Analysis of Complex Protein Mixtures with Improved Sequence Coverage Using (CE-MS/MS)^n. Anal Chem 78:7309–7316
2. Faserl K, Sarg B, Kremser L, Lindner H (2011) Optimization and Evaluation of a Sheathless Capillary Electrophoresis–Electrospray Ionization Mass Spectrometry Platform for Peptide Analysis: Comparison to Liquid Chromatography–Electrospray Ionization Mass Spectrometry. Anal Chem 83:7297–7305
3. Moini M (2004) Capillary Electrophoresis–Electrospray Ionization Mass Spectrometry of Amino Acids, Peptides, and Proteins. In: Strege MA, Lagu AL (eds) *Capillary Electrophoresis of Proteins and Peptides*. Humana Press, Totowa, NJ, pp 253–290
4. Moini M (2002) Capillary electrophoresis mass spectrometry and its application to the analysis of biological mixtures. Anal and Bioanal Chem 373:466–480
5. Geiger M, Hogerton AL, Bowser MT (2012) Capillary Electrophoresis. Anal Chem 84:577–596
6. Simpson DC, Smith RD (2005) Combining capillary electrophoresis with mass spectrometry for applications in proteomics. Electrophoresis 26:1291–1305
7. Busnel J-M, Schoenmaker B, Ramautar R, Carrasco-Pancorbo A, Ratnayake C, Feitelson JS, Chapman JD, Deelder AM, Mayboroda OA (2010) High Capacity Capillary Electrophoresis-Electrospray Ionization Mass Spectrometry: Coupling a Porous Sheathless Interface with Transient-Isotachophoresis. Anal Chem 82:9476–9483
8. Hommerson P, Khan AM, de Jong GJ, Somsen GW (2011) Ionization techniques in capillary electrophoresis-mass spectrometry: Principles, design, and application, *Mass Spectrom*. Rev 30:1096–1120
9. Ramautar R, Busnel J-M, Deelder AM, Mayboroda OA (2012) Enhancing the Coverage of the Urinary Metabolome by Sheathless Capillary Electrophoresis-Mass Spectrometry. Anal Chem 84:885–892
10. Haselberg R, de Jong GJ, Somsen GW (2007) Capillary electrophoresis-mass spectrometry

for the analysis of intact proteins. J Chromatogr A 1159:81–109

11. Maxwell EJ, Zhong X, Zhang H, Zeijl NV, Chen DDY (2010) Decoupling CE and ESI for a more robust interface with MS. Electrophoresis 31:1130–1137
12. Li F-A, Huang J-L, Her G-R (2008) Chip-CE/MS using a flat low-sheath-flow interface. Electrophoresis 29:4938–4943
13. Moini M (2001) Design and Performance of a Universal Sheathless Capillary Electrophoresis to Mass Spectrometry Interface Using a Split-Flow Technique. Anal Chem 73:3497–3501
14. Moini M (2007) Simplifying CE-MS Operation. 2. Interfacing Low-Flow Separation Techniques to Mass Spectrometry Using a Porous Tip. Anal Chem 79:4241–4246
15. Schultz CL, Moini M (2003) The analysis of underivatized amino acids and their D/L enantiomers using sheathless CE-MS. Anal Chem 75:1508–1513
16. Enke CG (1997) A Predictive Model for Matrix and Analyte in Electrospray Ionization of Singly-Charge Ionic Analytes. Anal Chem 69:4885–4893
17. Sjoberg PJR, Bokman CF, Bylund D, Markides KE (2001) A Method for Determination of Ion Distribution within Electrosprayed Droplets. Anal Chem 73:23–28
18. Moini M, Schultz CL, Mahmood H (2003) CE/electrospray ionization-MS analysis of underivatized D/L-amino acids and several small neurotransmitters at attomole levels through the use of 18-crown-6-tetracarboxylic acid as a complexation reagent/background electrolyte. Anal Chem 75:6282–6287
19. Kminek G, Bada JL (2006) The effect of ionizing radiation on the preservation of amino acids on Mars, *Earth Planet.* Sci Lett 245:1–5
20. Moini M, Klauenberg K, Ballard M (2011) Dating Silk By Capillary Electrophoresis Mass Spectrometry. Anal Chem 83:7577–7581
21. Barbour Wood SL, Krause RA Jr, Kowalewski M, Wehmiller J, Simoes MG (2006) Aspartic acid racemization dating of Holocene brachiopods and bivalves from the Southern Brazilian shelf, South Atlantic. Quat Res 66:323–331
22. Cao Y, Wang B (2009) Biodegradation of Silk. Int J Mol Sci 10:1514–1524
23. Garza S, Chang S, Moini M (2007) Simplifying capillary electrophoresis–mass spectrometry operation: Eliminating capillary derivatization by using self-coating background electrolytes. J of Chromatogr A 1159:14–21
24. Hardenborg E, Zuberovic A, Ullsten S, Soderberg L, Heldin E, Markides KE (2003) Novel polyamine coating providing non-covalent deactivation and reversed electroosmotic flow of fused-silica capillaries for capillary electrophoresis. J Chromatogr A 1003:217–221
25. Horvath J, Dolnik V (2001) Polymer wall coatings for capillary electrophoresis. Electrophoresis 22:644–655
26. Righetti G (1996) Capillary Electrophoresis in Analytical Biotechnology. CRC Press, New York
27. Zhang J, Tran NT, Weber J, Slim C, Vivoy JL, Taverna M (2006) High-performance *electrophoresis* elimination of electro- endosmosis and solute adsorption. Electrophoresis 27:3086–3092
28. Katayama H, Ishihama Y, Asakawa N (1998) Stable cationic capillary coating with successive multiple ionic polymer layers for capillary electrophoresis. Anal Chem 70:5272–5277
29. Hsieh M, Chiu T, Lung W, Chang H (2006) Analysis of Nucleic Acids and Proteins by Capillary Electrophoresis and Microchip Capillary Electrophoresis using Polymers as Additives of the Background Electrolytes. Curr Anal Chem 2:17–33
30. Ullsten S, Zuberovic A, Wetterhall M, Hardenborg E, Markides KE, Bergquist J (2004) A polyamine coating for enhanced capillary electrophoresis-electrospray ionization-mass spectrometry of proteins and peptides. Electrophoresis 25:2090–2099
31. Li MX, Liu L, Wu J, Lubman DM (1997) Use of a Polybrene Capillary Coating in Capillary Electrophoresis for Rapid Analysis of Hemoglobin Variants with On-Line Detection via an Ion Trap Storage/Reflectron Time-of-Flight Mass Spectrometer. Anal Chem 69:2451–2456
32. Catai JR, Torano JS, De Jong GJ, Somsen GW (2006) Efficient and highly reproducible capillary electrophoresis–mass spectrometry of peptides using Polybrene-poly(vinylsulfonate)-coated capillaries. Electrophoresis 27:2091–2099
33. Jin XY, Kim J, Parus S, Zand R, Lubman DM (1999) On-Line Capillary Electrophoresis/μ Electrospray Ionization-Tandem Mass Spectrometry Using an Ion Trap Storage/Time of Flight Mass Spectrometer with Swift Technology. Anal Chem 71:3591–3597
34. Cao P, Moini M (1998) Capillary electrophoresis/electrospray ionization high mass accuracy time-of-flight mass spectrometry for protein identification using peptide mapping. Rapid Commun Mass Spectrom 12:864–870
35. Peng J, Elias JE, Thoreen CC, Licklider LJ, Gygi SPJ (2003) Evaluation of multidimensional chromatography coupled with tandem

mass spectrometry (LC/LC-MS/MS) for large-scale protein analysis: the yeast proteome. Proteome Res 2:43–50

36. Mawuenyega KG, Kaji H, Yamuchi Y, Shinkawa TJJ (2003) Large-scale identification of Caenorhabditis elegans proteins by multidimensional liquid chromatography-tandem mass spectrometry. Proteome Res 2:23–35

37. Kislinger T, Rahman K, Radulovic D, Cox B (2003) PRISM, a generic large scale proteomic investigation strategy for mammals. Mol Cell Proteomics 2:96–106

38. Jacobs JM, Mottaz HM, Yu LR, Anderson DJJ (2004) Multidimensional proteome analysis of human mammary epithelial cells. Proteome Res 3:68–75

39. Zubarev RA, Kelleher NL, McLafferty FW (1998) Electron Capture Dissociation of Multiply Charged Protein Cations. A Nonergodic Process. J Am Chem Soc 120:3265–3266

40. Syka JEP, Coon JJ, Schroeder MJ, Shabanowitz J, Hunt DF (2004) Peptide and protein sequence analysis by electron transfer dissociation mass spectrometry. Proc Natl Acad Sci USA 101:9528–9533

41. Aebersold R, Mann M (2003) Mass spectrometry-based proteomics. Nature 422:198–207

42. Gatlin CL, Eng JK, Cross ST, Detter JC, Yates JR 3rd (2000) Automated Identification of Amino Acid Sequence Variations in Proteins by HPLC/Microspray Tandem Mass Spectrometry. Anal Chem 72:757–763

43. Tipton JD, Tran JC, Catherman AD, Ahlf DR, Durbin KR, Kelleher NL (2011) Analysis of Intact Protein Isoforms by Mass Spectrometry. J Biol Chem 286:25451–25458

44. Voinov VG, Beckman JS, Deinzer ML, Barofsky DF (2009) Electron-capture dissociation (ECD), collision-induced dissociation (CID) and ECD/CID in a linear radio-frequency-free magnetic cell. Rapid Commun Mass Spectrom 23:3028–3030

45. Moini M, Huang H (2004) Application of capillary electrophoresis/electrospray ionization-mass spectrometry to subcellular proteomics of *Escherichia coli* ribosomal proteins. Electrophoresis 25:1981–1987

46. Arnold RJ, Reilly JP (1999) Observation of Escherichia coli ribosomal proteins and their posttranslational modifications by mass spectrometry. Anal Biochem 269:105–112

47. Hofstadler SA, Severs JC, Smith RD, Swanek FD, Ewing AG (1996) Analysis of single cells with capillary electrophoresis electrospray ionization Fourier transform ion cyclotron resonance mass spectrometry. Rapid Commun Mass Spectrom 10:919–922

48. Andersson B, Nayman PO, Strid L (1972) Amino acid sequence of human erythrocyte carbonic anhydrase B. Biochem Biophys Res Commun 48:670–677

49. Henderson LE, Henriksson D, Nyman PO (1973) Amino acid sequence of human erythrocyte carbonic anhydrase C. Biochem Biophys Res Commun 52:1388–1394, For a more recent CAII amino acid sequence, see the NCBI database at http://www.ncbi.nlm.nih.gov

50. Lindskog S (1997) Structure and mechanism of carbonic anhydrase. Pharmacol Ther 74:1–20

51. Aliakbar S, Brown PR (1996) Measurement of human erythrocyte CAI and CAII in adult, newborn, and fetal blood. Clin Biochem 29:157–164

52. Chiang W-L, Chu S-C, Lai J-C, Yang S-F, Chiou H-L, Hsieh Y-S (2001) Alternations in quantities and activities of erythrocyte cytosolic carbonic anhydrase isoenzymes in glucose-6-phosphate dehydrogenase-deficient individuals. Clin Chim Acta 314:195–201

53. Yoshida K (1996) Zinc Concentrations in Erythrocytes, Tohoku J. Exp Med 178:345–356

54. Nagai R, Kooh SW, Balfe JW, Fenton T, Halperin ML (1997) Renal tubular acidosis and osteopetrosis with carbonic anhydrase II deficiency: pathogenesis of impaired acidification. Pediatr Nephrol 11:633–636

55. Aramaki S, Yoshida I, Yoshino M, Kondo M, Sato Y, Noda K, Jo R, Okue A, Sai N, Yamashita F (1993) Carbonic anhydrase II deficiency in three unrelated Japanese patients. J Inherited Metab Dis 16:982–990

56. Moini M, Demars SM, Huang H (2002) Analysis of Carbonic Anhydrase in Human Red Blood Cells Using Capillary Electrophoresis/Electrospray Ionization-Mass Spectrometry. Anal Chem 74:3772–3776

57. Veenstra TD (1999) Electrospray ionization mass spectrometry: A promising new technique in the study of protein/DNA noncovalent complexes. Biochem Biophys Res Commun 257:1–5

58. Wang Y, Schubert M, Ingendoh A, Franzen J (2000) Analysis of non-covalent protein complexes up to 290 KDa using electrospray ionization and ion trap mass spectrometry. Rapid Commun Mass Spectrom 14:12–17

59. Light-Wahl KJ, Schwartz BL, Smith RD (1994) Observation of the Noncovalent Quaternary Associations of Proteins by Electrospray Ionization Mass Spectrometry. J Am Chem Soc 116:5271–5278

60. Boys BL, Konermann L (2007) Folding and Assembly of Hemoglobin Monitored by

Electrospray Mass Spectrometry Using an On-line Dialysis System. J Am Soc Mass Spectrom 18:8–16

61. Loo JA (2000) Electrospray Ionization Mass Spectrometry: A Technology for Studying Noncovalent Macromolecular Complexes Int. J Mass Spectrom 200:175–186
62. Loo JA (1997) Electrospray Ionization Mass Spectrometry: A Technology for Studying Noncovalent Macromolecular Complexes, Mass Spectrom. Rev 16:1–23
63. Smith RD, Bruce JE, Wu Q, Lei QP (1997) New Mass spectrometric methods for the study of noncovalent associations of biopolymers. Chem Soc Rev 26:191–202
64. Heck AJR, van den Huevel RHH (2004) Investigation of intact protein complexes by mass spectrometry, Mass Spectrom. Rev 23:368–389
65. Brenner-Weiss G, Kirschhofer F, Kuhl B, Nusser M, Obst U (2003) Analysis of non-covalent protein complexes by capillary electrophoresis–time-of-flight mass spectrometry, J. Chromatogr. A 1009:147–153
66. Martinovic S, Berger SJ, Pasa-Tolic L, Smith RD (2000) Separation and Detection of Intact Noncovalent Protein Complexes from Mixtures by On-Line Capillary Isoelectric Focusing-Mass Spectrometry. Anal Chem 72:5356–5360
67. Nguyen A, Moini M (2008) Analysis of Major Protein-Protein and Protein-Metal Complexes of Erythrocytes Directly from Cell Lysate Utilizing Capillary Electrophoresis Mass Spectrometry. Anal Chem 80:7169–7173
68. Garza S, Thomas PW, Fast W, Moini M (2010) Metal displacement and stoichiometry of protein-metal complexes under native conditions using capillary electrophoresis/mass spectrometry. Rapid Commun Mass Spectrom 24:2730–2734
69. Moini M, Martinez B (2010) Analysis of Protein Digests in about a Minute using a Handheld Ultrafast Capillary Electrophoresis Interfaced to MS using a Porous Tip. Proc Am Soc Mass Spectrom
70. Moini M, Jiang L, Bootwala S (2011) High-throughput analysis using gated multi-inlet mass spectrometry. Rapid Commun Mass Spectrom 25:789–794
71. Jiang L, Moini M (2000) Development of Multi-ESI-Sprayer, Multi-Atmospheric-Pressure-inlet Mass Spectrometry and Its Application to Accurate Mass Measurement Using Time-of-Flight Mass Spectrometry, Anal. Chem. 72, 2–24, and. Anal Chem 72:885
72. Jiang L, Moini M (2002) Mass spectrometer with multiple atmospheric pressure inlets (nozzles), University of Texas at Austin, US Patent 6465776 1B.

Chapter 9

Separation of Amino Acids by Capillary Electrophoresis with Light-Emitting Diode-Induced Fluorescence in the Presence of Electroosmotic Flow

Ming-Mu Hsieh and Po-Ling Chang

Abstract

In this chapter, we describe a method to identify amino acids (AA) by capillary electrophoresis in conjunction with light-emitting diode-induced fluorescence (LEDIF). First, amino acids labeled with naphthalene-2,3-dicarboxaldehyde (NDA) in the presence of cyanide are converted to highly fluorescent cyanobenz[*f*] isoindole (CBI) derivatives. Next, they are separated by gel electrophoresis in the presence of EOF. In the process, the CBI products were excited by a violet LED to produce green fluorescence. In addition to the optical setup of light-emitting diode-induced fluorescence, the preparation of poly(ethylene) oxide for amino acid separation is also described in this chapter.

Key words: Capillary electrophoresis, Light-emitting diode-induced fluorescence, Naphthalene-2,3-dicarboxaldehyde, Derivatization, Cyano[*f*]benzoisoindole, Poly(ethylene) oxide, Amino acid

1. Introduction

Capillary electrophoresis (CE) is one of the most popular separation techniques for the determination of biological analytes, offering advantages such as rapidity, high efficiency, and high resolving power, as well as only requiring minute amounts of samples and reagents. To ensure a high sensitivity, capillary electrophoresis is usually used in conjunction with laser-induced fluorescence (LIF) detection. Unfortunately, LIF detection depends on the use of bulky and expensive laser sources to generate sufficient excitation for analyte fluorescence, especially in UV-absorbing analytes. Light-emitting diodes (LEDs) are used to overcome this drawback due to their low cost, low power consumption, and availability in a wide spectral range (1–3). However, the light from LEDs is divergent

Nicola Volpi and Francesca Maccari (eds.), *Capillary Electrophoresis of Biomolecules: Methods and Protocols*, Methods in Molecular Biology, vol. 984, DOI 10.1007/978-1-62703-296-4_9, © Springer Science+Business Media, LLC 2013

and non-monochromatic, so it is difficult to focus light on small capillary columns. As a result, a relatively high fluorescence background usually occurs. To overcome such limitations, Chang et al. used a prototypical focusing system consisting of an objective lens constructed to allow light to be focused effectively onto a small capillary. CE-LEDIF has become a powerful technique to analyze small molecules. Many methods for separating small molecules were developed, especially for amino acids (4–7), amines (8), chiral amino acids (9, 10), and proteins (11). In this chapter, we describe the use of naphthalene-2,3-dicarboxaldehyde (NDA) as a derivatization reagent, followed by poly(ethylene oxide) (PEO)-assisted matrices for the separation of amino acids. In addition, a setup for a house-built violet LEDIF device for the detection of cyanobenz[*f*] isoindole (CBI) products is also described. This proposed method provides the following advantages: (a) the resolution and speed can be optimized by varying the capillary length, separation voltage, buffer composition, pH, and ionic strength; (b) the use of CE-LEDIF is relatively cheap compared to laser sources; (c) a small matrix effect was observed using PEO as the separation media; and (d) this method can be performed for large-volume sample stacking to improve the sensitivity.

2. Materials

2.1. Chemicals

1. Buffer solution for amino acid separation: 10 mM sodium tetraborate, pH 9.3, or 400 mM tris(hydroxymethyl)aminomethane (tris)-borate (TB), pH 9.0. Stored at 4°C.
2. Linear polymer: poly(ethylene) oxide, (PEO) (M_{avg}. 8,000,000 Da) (Sigma-Aldrich, USA).
3. Derivatization reagent: naphthalene-2,3-dicarboxaldehyde (NDA) (TOKYO CHEMICAL INDUSTRY, Tokyo, Japan).
4. Amino acids, acetonitrile (ACN), and methanol.

2.2. Equipment

1. Photomultiplier Tube (R3896, Hamamatsu Photonics, Hamamatsu, Japan).
2. 488 nm interference filter (Edmund Optics, NJ, USA).
3. Microscopic objective (10× and 40×) (Edmund Optics, NJ, USA).
4. Violet LED (InGaN, type no. LED513UV; Kwang-Hwa Electronic Material, Taichung, Taiwan) (see Note 1).
5. DV-E Viscometer (Brookfield Engineering Laboratories, Middleboro, MA, USA) (see Note 2).
6. Fused-silica capillary: 75 μm I.D., 365 μm O.D. (Polymicro Technologies; Phoenix, AZ, USA).

7. High-voltage power supply for electrophoresis (Gamma High Voltage Research Inc., FL, USA).
8. A laboratory-made power supply, which allowed the potentials to be adjusted up to 5.0 V, was used to drive the LED. In this study, the applied voltage was set at 3.9 V.

3. Methods

3.1. Preparation of Borate Buffer Solution, pH 9.3

1. Place 0.3814 g of sodium tetraborate into a 100 mL volumetric flask.
2. Add 80 mL of deionized water and swirl to dissolve the sodium tetraborate.
3. Add double deionized water to the 100 mL mark and invert the flask several times to ensure complete mixing.

3.2. Preparation of 400 mM TB Buffer Solution, pH 9.0

1. Dissolve 24.23 g of tris into 400 mL of double deionized water.
2. Adjust the solution with the appropriate amount of boric acid to pH 9.0.
3. Pour the solution into a 500 mL volumetric flask.
4. Add double deionized water to the 500 mL mark and invert the flask several times to ensure complete mixing.
5. The stock solution can be diluted to the desired concentration for the following experiments.

3.3. Preparation of PEO Buffer Solution

1. Pour 50 mL of buffer solution (TB or sodium tetraborate) into a 125 mL flask that contains an elliptical magnetic stirring bar (4×2 cm).
2. Gently pour the weighed polymer powder into the buffer solution and stir overnight (see Note 3).
3. The polymer solution was degassed by centrifugation (18,000×g for 10 min) or vacuum pump before electrophoresis.

3.4. Derivatization of Amino Acids with NDA (9)

1. The derivatization can be performed in 1.5 mL centrifuge tubes.
2. Equal volumes (50 μL) of 10 mM sodium tetraborate and 100 mM NaCN solutions were mixed with certain amounts of amino acids (see Note 4).
3. Add 200 μL acetonitrile to the reaction mixture and add the appropriate volume of double deionized water to bring the mixture to 450 μL.
4. Add 50 μL of 25 mM NDA (dissolved in methanol) to the mixture and mix by vortex.

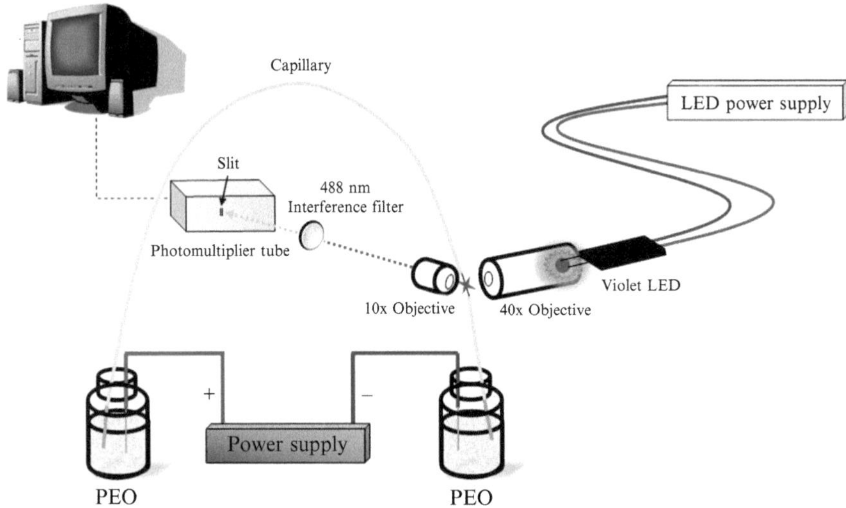

Fig. 1. Schematic illustration of the CE-LEDIF system used for the analysis of amino acid–cyanobenz[*f*]isoindole products.

5. The mixtures were allowed to react for 30 min at room temperature and remained stable for 3 days.

3.5. Setup the CE-LEDIF (Fig. 1)

1. A violet LED (InGaN; maximum output at 405 nm in the range 390–420 nm with a total output power of 0.4 mW) was used for excitation.
2. Insert the LEDs into the back of a 40× objective and secure them by sealing the space between the LEDs and the objective (Fig. 2a).
3. Move the LED-containing 40× objective to a capillary that contains buffer solution.
4. Move the focal point of light emitted by the 40× objective forward by adjusting the $x-y$ stage of the mounted capillary.
5. Use a 10× objective (numerical aperture: 0.25) to collect the fluorescence (Fig. 2b).
6. The angle between the two objectives should be about 120°.
7. Use an interference filter (488 nm) to block light scattered before the emitted light reaches the photomultiplier tube (PMT).
8. To ensure the fluorescence is correctly focused into slit of the PMT, the user can inject 1 mM fluorescein (dissolved in 10 mM sodium tetraborate, pH 9.3, or 100 mM TB, pH 9.0) into the capillary. The color of the focal point on the capillary should change from a violet to bright green fluorescence.
9. Carefully adjust the 10× objective until the fluorescence line reaches the square slit (~4 × 1 mm).

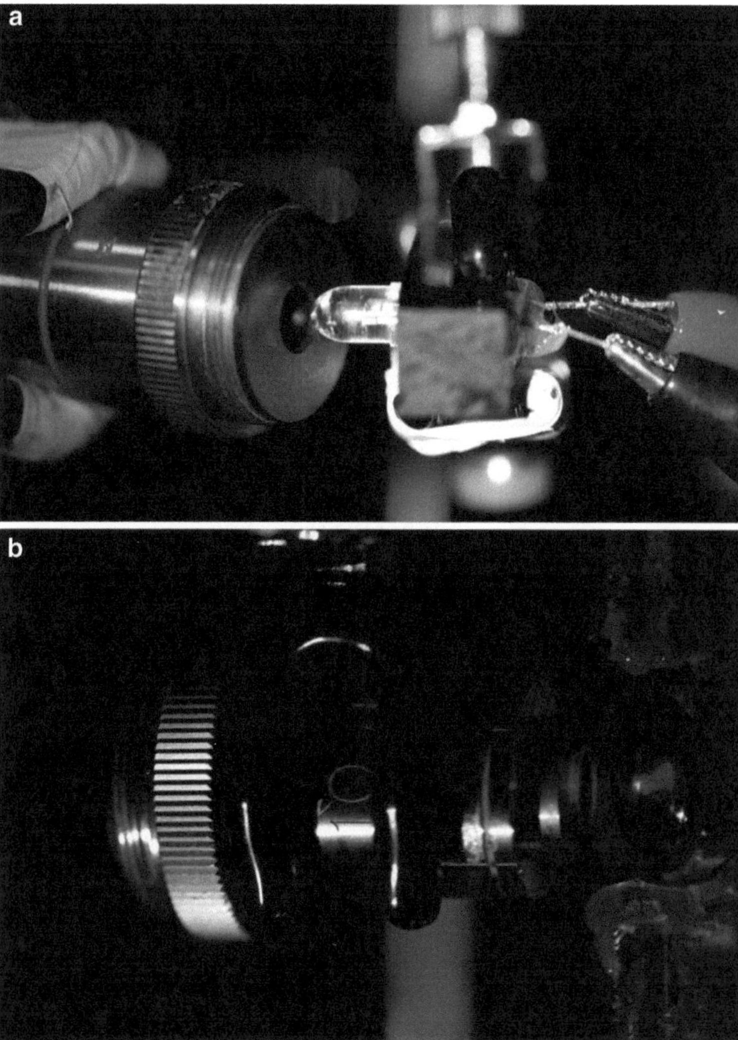

Fig. 2. Assemble the CE-LEDIF by microscopic objectives. (**a**) Insert the LEDs into the backside of a 40× objective. (**b**) Focus the excitation light onto the capillary and collect the scattered light and the fluorescence by a 10× objective. The angle between the two objectives is about 120°.

10. Carefully move the PMT along the fluorescence light path to find the focal plane (brightest fluorescence).
11. After finishing the light adjustment, remove the fluorescein from inside the capillary by buffer solution (sodium tetraborate or TB).

3.6. Electrophoresis Procedure

1. Prior to analysis, the 60 cm capillary (effective length: 50 cm) was treated overnight with 0.5 M NaOH solution.
2. Fill the capillary with 10 mM sodium tetraborate.

3. Inject the derivatized sample into the capillary by hydrodynamic injection at a height of 30 cm (the difference between the exit and entrance ends of the capillary).
4. After the sample was injected from the anode end, the two ends of the capillary were immersed in a PEO solution (prepared in 10 mM $Na_2B_4O_7$, pH 9.3 or 400 mM TB, pH 9.0).
5. Switch on the high-voltage power supply for separation (see Note 5).
6. When the electrophoresis is finished, push the polymer solution out from inside the capillary by connecting a high-pressure nitrogen gas (>200 psi) to the inlet end of the capillary.
7. After each run, the capillary was equilibrated with 0.5 M NaOH at 1 kV for 10 min (see Note 6).
8. Fill the capillary with the buffer solution for the next electrophoretic separation.

4. Notes

1. A violet LED (InGaN; maximum output at 405 nm over the range 390–420 nm) was used for excitation. A laboratory-made power supply, which allowed the potentials to be adjusted up to 5.0 V, was used to drive the LED. In this study, the applied voltage was set at 3.9 V.
2. Viscometer was employed to measure the viscosity of PEO solutions in a constant-temperature bath at 25.0 ± 0.2°C. All measurements were performed in triplicate.
3. PEO (0.1–1.5%) was gradually added to each of the prepared sodium tetraborate (pH 9.3) or TB buffer solutions (pH 9.0). During the addition of PEO, a magnetic stirring rod was used to produce a well-homogenized suspension. After the addition was completed, the solutions were stirred for at least 12 h. Prior to use in CE separation, the solutions were degassed with a vacuum system in an ultrasonic tank for 30 min (8).
4. The concentration of amino acids contained in the sample should not exceed the concentration of NDA (25 mM). For samples like ascites or serum, the initial volume of the sample should be reduced by a factor of 10 as a result of the large amount of amino acids in body fluid.
5. It is not necessary to fill the outlet vial with PEO as long as the ionic strength of the buffer solution contained by the inlet vial is maintained.
6. We note that injecting the sample and filling the capillary with PEO solution are not easily performed with pressure because

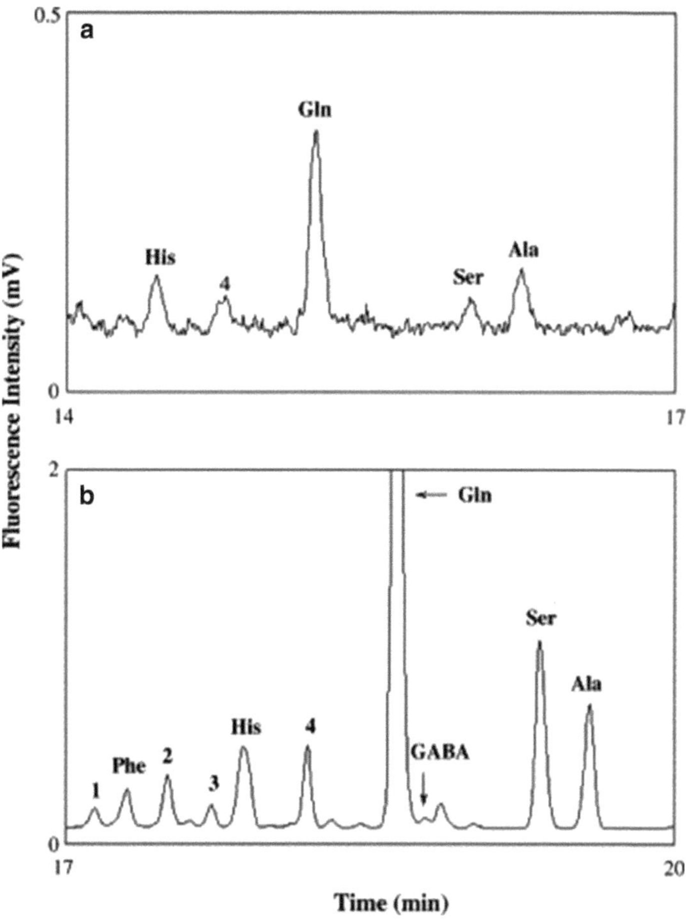

Fig. 3. Electropherograms of amino acid derivatives in a CSF sample in the presence of EOF using 2.0% PEO. (**a**) The sample was injected hydrodynamically at 30 cm height for 10 s; (**b**) hydrodynamic injection was conducted at 30 cm height for 180 s. Peaks assigned as numbers 1–4 separately correspond to four different unidentified analytes (reproduced from ref. 6 with permission from Elsevier).

the PEO solution is a very viscous polymer. Because the electroosmotic flow (EOF) was greater than the electrophoretic mobilities of the amino acid–NDA derivatives, the CBI products can be easily brought into the detection window by EOF.

7. The purpose to filled with 10 mM $Na_2B_4O_7$ buffer was to remove PEO molecules and to generate the bulk EOF. This process for the capillary cleaning was quite successful at generating constant bulk EOF mobility; the relative standard deviation (RSD) from five consecutive runs was below 2.0% (Fig. 3) (6).

8. To increase the reproducibility of quantitation, the derivatized sample was stored at −20°C to maintain the stability of the CBI products. The sample remains in a liquid phase at −20°C because the sample contains 40% ACN (Fig. 4) (12).

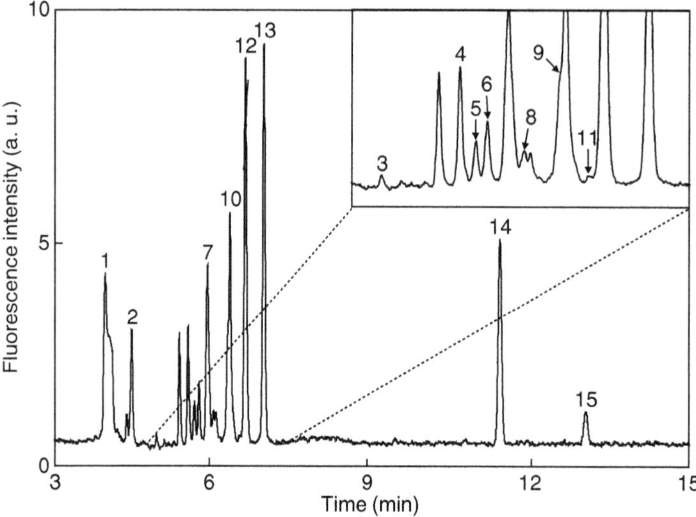

Fig. 4. Electropherogram of one representative ascites specimen that was analyzed by CE with LEDIF detection. Peak identities: (1) arginine, (2) lysine, (3) tryptophan, (4) phenylalanine, (5) Leu, (6) Ile, (7) Val, (8) methionine, (9) glutamine, (10) asparagine, (11) serine, (12) alanine, (13) glycine, (14) glutamic acid, and (15) aspartic acid. The PEO solutions that were prepared in 10 mM sodium tetraborate (pH 9.3). The capillary having inner diameter of 75 μm and outer diameter of 365 μm was 60 cm long (effective length: 50 cm). The sample was introduced into capillary by hydrodynamic injection in 5 s (30 cm height difference between the two ends). The separation was conducted at 20 kV (<25 μA) and at 25°C (reproduced from reference 12 with permission from Wiley-VCH Verlag GmbH & Co. KGaA).

9. The sodium tetraborate is useful for general amino acid separation for whatever solution is used as the free solution or preparation of PEO. However, if the sample is a body fluid, such as cerebrospinal fluid, ascites, or serum, the 400 mM TB (pH 9.0) is suggested for the preparation of the PEO solution to acquire better resolution for amino acids (6).
10. If the amount of amino acid contained in the sample is high enough to detect, direct reduction of the sample load at NDA derivatization may also decrease the effect of protein and salt on separation efficiency (Fig. 4) (12).
11. For the detailed methodological descriptions used in the biological applications, please refer to references 6–8 and 11, 12.

Acknowledgments

The authors would like to thank the National Science Council of Taiwan, NSC 99-2113-M-017-04-MY2 (M.-M. Hsieh) and NSC100-2113-M-029-003-MY2 (P.-L. Chang) for providing the financial support for this work.

References

1. Liu H, Dasgupta PK, Zheng HJ (1993) High performance optical absorbance detectors based on low noise switched integrators. Talanta 40:1331–1338
2. Bruno AE, Maystre F, Krattiger B, Nussbaum P, Gassmann E (1994) The pigtailing approach to optical detection in capillary electrophoresis. TrAC Trends Anal Chem 13:190–198
3. Butler AG, Mills B, Hauser PC (1997) Capillary Electrophoresis Detector Using a Light Emitting Diode and Optical Fibres. Analyst 122:949–953
4. Chen S-J, Chen M-J, Chang H-T (2003) Light-emitting diode-based indirect fluorescence detection for simultaneous determination of anions and cations in capillary electrophoresis. J Chromatogr A 1017:215–224
5. Chiu T-C, Chang H-T (2007) Stacking and separation of fluorescent derivatives of amino acids by micellar electrokinetic chromatography in the presence of poly(ethylene oxide). J Chromatogr A 1146:118–124
6. Lu M-J, Chiu T-C, Chang P-L, Ho H-T, Chang H-T (2005) Determination of glycine, glutamine, glutamate, and γ-aminobutyric acid in cerebrospinal fluids by capillary electrophoresis with light-emitting diode-induced fluorescence detection. Anal Chim Acta 538:143–150
7. Chang P-L, Lee K-H, Hu C-C, Chang H-T (2007) CE with sequential light-emitting diode-induced fluorescence and electro-chemiluminescence detections for the determination of amino acids and alkaloids. Electrophoresis 28:1092–1099
8. Kao Y-Y, Liu K-T, Huang M-F, Chiu T-C, Chang H-T (2010) Analysis of amino acids and biogenic amines in breast cancer cells by capillary electrophoresis using polymer solutions containing sodium dodecyl sulfate. J Chromatogr A 1217:582–587
9. Lin KC, Hsieh MM, Chang CW, Lin EP, Wu TH (2010) Stacking and separation of aspartic acid enantiomers under discontinuous system by capillary electrophoresis with light-emitting diode-induced fluorescence detection. Talanta 82:1912–1918
10. Chiu T-C, Tu W-C, Chang H-T (2008) Stacking and separation of protein derivatives of naphthalene-2,3-dicarboxaldehyde by CE with light-emitting diode induced fluorescence detection. Electrophoresis 29:433–440
11. Chang P-L, Chiu T-C, Chang H-T (2006) Stacking, derivatization, and separation by capillary electrophoresis of amino acids from cerebrospinal fluids. Electrophoresis 27:1922–1931
12. Chang P-L, Chiu T-C, Wang T-E, Hu K-C, Tsai Y-H, Hu C-C, Bair M-J, Chang H-T (2011) Quantitation of branched-chain amino acids in ascites by capillary electrophoresis with light-emitting diode-induced fluorescence detection. Electrophoresis 32:1080–1083

Chapter 10

Quantification of Arginine and Dimethylated Arginines in Human Plasma by Field-Amplified Sample Injection Capillary Electrophoresis UV Detection

Angelo Zinellu, Salvatore Sotgia, Luca Deiana, and Ciriaco Carru

Abstract

We report a capillary electrophoresis method by field-amplified sample injection (FASI) for the separation and detection of free plasma arginine and dimethylated arginines. Protein elimination from plasma was performed by acetonitrile/ammonia precipitation. Supernatants were dried, reswollen in water, and directly injected on capillary without complex cleanup by solid-phase extraction and/or tedious sample derivatization procedures. Analytes were baseline-separated within 22 min by using 50 mmol/L Tris phosphate pH 2.3 as running buffer. Due to the stacking effects of electrokinetic injection, it is possible to perform a considerable on-line preconcentration of the analytes before running the electrophoresis. This procedure allows to reach a detection limit in real sample of 10 nmol/L for dimethylated arginines and 20 nmol/L for arginine. Recovery of plasma ADMA was 99–104 % and inter-day CV was less than 3 %.

Key words: Capillary electrophoresis, FASI, ADMA, SDMA arginine

1. Introduction

Asymmetric dimethylarginine (ADMA) is a posttranslationally modified form of arginine that is generated in all the cells during the process of protein turnover. Its biosynthesis is catalyzed by a family of enzymes named protein arginine *N*-methyltransferases (1), which, based on their end product, are classified into two groups, both of which first catalyze the formation of monomethylarginine (MMA) as an intermediate (1). Type I enzymes promote

Nicola Volpi and Francesca Maccari (eds.), *Capillary Electrophoresis of Biomolecules: Methods and Protocols*, Methods in Molecular Biology, vol. 984, DOI 10.1007/978-1-62703-296-4_10,

the formation of ADMA. Type II enzymes catalyze the formation of symmetric NG, N′G-dimethylated arginine (SDMA). Increased levels of ADMA may inhibit nitric oxide synthase (eNOS) activity resulting in nitric oxide (NO) availability drops. Nitric oxide plays an important role in physiological functions by inhibiting key processes in vascular diseases (1). Endothelial dysfunction as an early stage of atherosclerosis is a direct result of NO deficiency (2). NO, in fact, has been identified to participate in the regulation of vascular tone, neurotransmission, blood flow, and blood pressure (3–5). Measurement of arginine and its methyl derivatives in biological samples may increase our information on the role of these compounds in the pathophysiological mechanisms involved in atherosclerosis. However, the ADMA measurement is complicated by its low concentration in plasma and the difficulty in achieving separation of the two isomers of dimethylarginine. Currently, the available analytical methods for the determination of plasma concentrations of ADMA are based on RP-HPLC with fluorescence detection after sample cleanup by SPE and precolumn derivatization with *ortho*-phthaldialdehyde (OPA) reagent (6–11), but other derivatization reagents have been used as well (12–17). Methods based on capillary electrophoretic (17) and micellar electrokinetic (18) separations with LIF detection have also been reported but rarely employed. Mass spectrometric detection, either in combination with GC (19, 20), LC (21–25), or CE (26), has been increasingly used, but these methods are not ideal for routine clinical purposes because the procedures are time-consuming and the instrumentation is not always available in a routine clinical laboratory.

Therefore, a more practical way to increase assay sensitivity is to enrich the analyte concentration before the separation process. For achieving on-capillary preconcentration, various methods based on electrophoretic principles have been reported, such as field-amplified methods (27–29) or on-line isotachophoresis (30). Field-amplified sample staking methods are based on the conductivity differences between the sample solution and the background electrolyte (BGE) and are the most commonly used modes of stacking due to their simplicity and practicality. In 1979, Mikkers et al. (31) reported that sharper peaks were obtained when the sample was dissolved in water instead of the same separation buffer. Since then, many techniques have been developed, showing that sample stacking can provide up to 2,500-fold improvement in the detector response, thereby extending CE–UV applications in trace analysis. In this work, we provide a new capillary electrophoresis method in which the use of field-amplified on-line sample stacking enrichment allows to reduce both pre-analytical times and costs (sample cleanup by SPE and derivatization procedure has been avoided) and volume sample needed for analysis.

2. Materials

2.1. Equipment

1. Capillary electrophoresis system (Beckman MDQ) equipped with a diode array detector (Beckman, Palo Alto, CA, USA). The system was fitted with a 30 kV power supply with a current limit of 250 μA.
2. Vortex mixer (VELP SCIENTIFICA, Monza, ITA).
3. Hettich Mikro 200 Microcentrifuge (GMI Inc., Minneapolis, MN).
4. Jouan Speed Vac RC1010 (Abbott Laboratories, Abbott Park, Illinois).
5. Milli-Q system (Millipore, Bedford, USA).
6. inoLab pH meter (WTW, Weilheim, Germany).
7. Falc Magnetic Stirrer (Progen Scientific, London, England).
8. Branson Ultrasonic Cleaner 1510 (KELL-STROM, Wethersfield, CT).
9. Captair Toxicap 1200 (Erlab INC., Rowley, MA).
10. NewClassic MF balance (METTLER TOLEDO Inc., Polaris Parkway, OH).
11. Thermo Scientific Forma 8600 Ultra Low temperature Upright Freezer (Thermo Fisher Scientific, Waltham, MA).
12. EDTA glass tubes (Becton Dickinson, Franklin Lakes, NJ).
13. e-CAP uncoated capillary tubing 75 μm I.D., 375 μm O.D.(BECKMAN, Palo Alto, CA).
14. Millex-GP Syringe Driven Filter Unit 0.22 μm (Millipore, Bedford, USA).
15. 32 Karat™ Software Version 5.0 (Beckman, Palo Alto, CA).

2.2. Reagents

1. Arginine, asymmetric dimethylarginine dihydrochloride, symmetric dimethylarginine di-(*p*-hydroxyazobenzene-*p*′-sulfonate) salt, homoarginine hydrochloride, and Tris(hydroxymethyl)aminomethane (Tris) were obtained from Sigma (St Louis, USA).
2. Acetonitrile, ammonia, hydrochloric acid, and phosphoric acid were purchased from Carlo Erba Reagenti SpA (Rodano, Italy).

2.3. Solutions

All solutions are prepared using ultrapure water (prepared by purifying deionized water to attain a sensitivity of 18 MΩ cm at 25 °C) and analytical grade reagents. All reagents are prepared and stored at room temperature (unless indicated otherwise).

1. NaOH 0.5 mol/L: Weight 2 g NaOH in a 100 mL graduated cylinder and add water to a final volume of 100 mL.

2. HCl 0.1 mol/L: Transfer 833.3 μL of 37 % HCl in a 100 mL graduated cylinder and add water to a final volume of 100 mL.
3. Acetonitrile/ammonia 90/10 (v/v): Transfer 90 mL of acetonitrile in a 100 mL graduated cylinder and add 10 mL of 25 % NH_3 (see Note 1).
4. Phosphoric acid 1 mol/L: Transfer 3,333 mL of 85 % H_3PO_4 in a 50 mL graduated cylinder and add water to a final volume of 50 mL.
5. Tris 75 mmol/L: Weight 0,909 g of Tris in a 100 mL beaker and add 90 mL of water. Titrate the solution with 1 mol/L phosphoric acid until reach pH 2.30. Transfer the solution in a 100 mL graduated cylinder and add water to a final volume of 100 mL (see Note 2).
6. Standard mixture for calibration curve (top calibrator: arginine 1,600 μmol/L, ADMA 12 μmol/L, SDMA 12 μmol/L): Separately weight 10 mg of arginine, ADMA, and SDMA in a 1.5 mL eppendorf tube and add 1 mL of water. Thoroughly vortex-mix the solution until complete dissolution of standard. Transfer 278.7 μL of arginine, 3.33 μL of ADMA, and 9.23 μL of SDMA in a 10 mL graduated cylinder and add water to a final volume of 10 mL (see Note 3).
7. Internal standard (homoarginine 100 μmol/L): Weight 10 mg of homoarginine in a 10 mL graduated cylinder and add water to a final volume of 10 mL (44.5 mmol/L). Mix the solution until complete dissolution of standard (see Note 4).

3. Methods

3.1. Preparation of Calibration Curve

The standard calibration curves were prepared by spiking 0.1 mL of a plasma pool sample with increasing amounts of arginines (see Note 5).

1. Serially dilute the top calibrator three times with water to obtain the four calibrators to use to spike plasma:
   (a) Arginine 1,600 μmol/L, ADMA 12 μmol/L, SDMA 12 μmol/L
   (b) Arginine 800 μmol/L, ADMA 6 μmol/L, SDMA 6 μmol/L
   (c) Arginine 400 μmol/L, ADMA 3 μmol/L, SDMA 3 μmol/L
   (d) Arginine 200 μmol/L, ADMA 1.5 μmol/L, SDMA 1.5 μmol/L

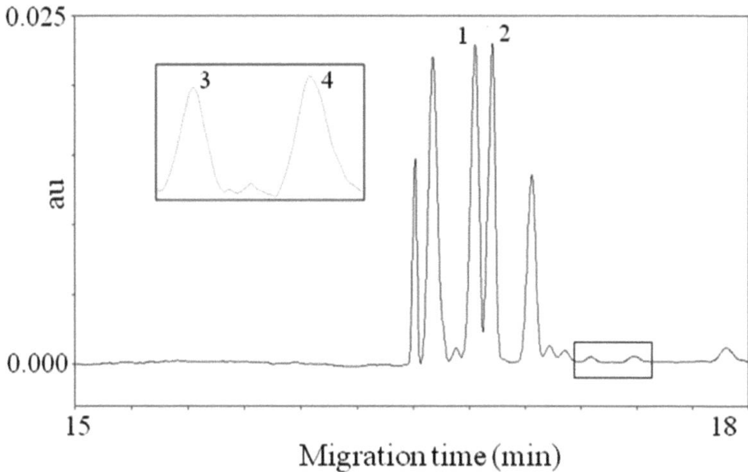

Fig. 1. Typical electropherogram for a plasma sample. 1 arginine; 2 homoarginine; 3 ADMA; 4 SDMA.

2. Mix 100 μL of plasma with 10 μL of each calibrator.
3. For blank, mix 100 μL of plasma with 10 μL of water.
4. Add 50 μL IS homoarginine (100 μmol/L).
5. Thoroughly vortex-mix the solution.
6. Add 300 μL of acetonitrile/ammonia 90:10 (see Note 6).
7. Thoroughly vortex-mix the solution.
8. Centrifuge at 3,000 × *g* for 5 min.
9. Transfer the supernatant in a 0.5 mL vial and dry it in a Speed Vac.
10. Add 200 μL of water to dried samples (see Note 7).
11. Put samples in a ultrasonic bath for 15 min (see Note 8).
12. Thoroughly vortex-mix the solution before injection on capillary electrophoresis.

3.2. Preparation of Biological Sample

1. Collect blood after an overnight fast by venipuncture into evacuated tubes containing EDTA and centrifuge immediately at 3,000 × *g* 5 min at 4 °C.
2. Mix 100 μL of the obtained plasma with 10 μL of water and proceed as in Subheading 3.1 steps 4–12. Figure 1 shows a typical electropherogram of a plasma sample.

3.3. Capillary Electrophoresis Method

3.3.1. Capillary Rinse

1. New capillaries are rinsed by 5 min of 0.1 mmol/L HCl, 5 min of 0.5 mmol/L NaOH, and 5 min of water. Rinse cycle is repeated for three times.
2. Between runs, capillary is rinsed for 1 min with 0.5 mmol/L NaOH, 1 min with 0.1 mmol/L water, and 1 min with 0.1 mmol/L HCl and equilibrated with run buffer for 1 min.

3.3.2. CE Method

1. Perform the analysis in an uncoated fused silica capillary, 75 μm ID and 57 cm length (50 cm to the detection window).
2. Inject 10 mm of water plug (3.5 KPa × 11 s) (see Note 9).
3. Inject sample by electrokinetic injection at 10 kV for 200s.
4. Inject a buffer plug of 4.5 mm (3.5 KPa × 5 s) (see Note 10).
5. Carry out the electrophoretic separation in a 75 mmol/L Tris buffer titrated with 1 mol/L phosphoric acid to the pH 2.30, 15 °C at normal polarity 15 kV for 20 min. Detection must be performed at 200 nm wavelength.

4. Notes

1. Solution must be prepared under safety cabinet and stored at 4 °C. It may be used for 2 weeks.
2. Solution could be stored at room temperature for 1 week.
3. Stock standard should be stored at −80 °C for 6 months.
4. Homoarginine is aliquoted and stored at −80 °C for 6 months. A single aliquot was thawed just prior to its use and diluted with water at a 100 μmol/L concentration. Transfer 22.47 μL of standard in a 10 mL graduated cylinder and add water to a final volume of 10 mL.
5. The problem with FASI procedure is that the peak intensity is strictly influenced by the sample matrix. This because a low-conductivity sample matrix is needed in order to obtain a field-enhanced sample injection. Peak signal is different on the basis of the matrix type. An aqueous matrix allows a higher gain in sensitivity if compared to saline matrix as 0.15 mol/L NaCl and 0.2 mol/L PBS or also compared with plasma matrix. These observations suggest that the calibration curves must be performed in the same matrix of samples. Note that the presence of internal standard would allow to prepare the calibration curve also in water. In this case, the injection times must be reduced to 10 s versus the 200 s of samples.
6. As reported in our previous study, the presence of ammonia is essential to obtain a complete recovery of analytes (32). When precipitation is carried out by 100 % ACN, the loss of arginine and dimethylarginines is of about 20% and 10 %, respectively. The recovery was not improved when protein elimination was performed by methanol or ethanol. Acidic precipitants like TCA or metaphosphoric acid were avoided since they increase conductivity of sample matrix reducing the on-line preconcentration and sensitivity. The sample recovery was improved by adding ammonia to ACN. As previously reported, when a

solution of acetonitrile/ammonia (90:10) was employed, the recovery was quantitative.

7. We found that the volume of water used to reswollen the samples after exsiccation may influence the signal peak. In particular, we determined that the most favorable volume of water to re-suspend dried samples was 200 μL. The modification of sample matrix with acetonitrile or trifluoroacetic acid, reported to be able to improve sensitivity in the FASI procedure, did not bring benefits in our case.
8. Ultrasonic bath allows to facilitate sample resuspension after dry procedure.
9. It has been reported that the introduction of a short plug of water before the sample electrokinetic injection can provide the proper electric field enhancement at the injection point, thus increasing sensitivity. We found that, by increasing the length of the water plug, there was a significant improvement in the assay sensitivity, and 10 mm resulted as the optimal length that allowed more than twofold increase in the peak height.
10. The use of buffer plug after electrokinetic injection allows to obtain a significant improvement in the assay reproducibility.

References

1. Bedford MT, Richard S (2005) Arginine methylation an emerging regulator of protein function. Mol Cell 18:263–272
2. Verma S, Anderson TJ (2002) Fundamentals of endothelial function for the clinical cardiologist. Circulation 105:546–549
3. Gryglewski RJ, Bottig RM, Vane JR (1998) Mediators produced by the endothelial cell. Hypertension 12:530–548
4. Moncada S, Higgs A (1993) The L-arginine-nitric oxide pathway. New Engl J Med 329:2002–2012
5. Furchgott LS, Zawadski JV (1980) The obligatory role of endothelial cells in the relaxation of arterial smooth muscle by acetylcholine. Nature 286:373–376
6. Pettersson A, Uggla L, Backman V (1997) Determination of dimethylated arginines in human plasma by high-performance liquid chromatography. J Chromatogr B 692:257–262
7. Tsikas D, Junker W, Frolich JC (1998) Determination of dimethylated arginines in human plasma by high-performance liquid chromatography. J Chromatogr B 705:174–176
8. Pi J, Kumagai Y, Sun G, Shimojo N (2000) Improved method for simultaneous determination of L-arginine and its mono- and dimethylated metabolites in biological samples by high-performance liquid chromatography. J Chromatogr B 742:199–203
9. Chen BM, Xia LW, Liang SX, Chen GH, Deng FL, Zhang WR, Tao LJ (2001) Simultaneous determination of L-arginine and dimethylarginines in human urine by high-performance liquid chromatography. Anal Chim Acta 444:223–227
10. Zhang WZ, Kaye DM (2004) Simultaneous determination of arginine and seven metabolites in plasma by reverse-phase liquid chromatography with a time-controlled ortho-phthaldialdehyde precolumn derivatization. Anal Biochem 326:87–92
11. Teerlink T (2005) Determination of the endogenous nitric oxide synthase inhibitor asymmetric dimethylarginine in biological samples by HPLC. Methods Mol Med 108:263–274
12. Park KS, Lee HW, Hong SY, Shin S, Kim S, Paik WK (1988) Determination of methylated amino acids in human serum by high-performance liquid chromatography. J Chromatogr 440:225–230
13. Marra M, Bonfigli AR, Testa R, Testa I, Gambini A, Coppa G (2003) High-performance liquid chromatographic assay of asymmetric

dimethylarginine, symmetric dimethylarginine, and arginine in human plasma by derivatization with naphthalene-2,3-dicarboxaldehyde. Anal Biochem 318:13–17

14. Heresztyn T, Worthley MI, Horowitz JD (2004) Determination of L-arginine and NG, NG- and NG, NG'-dimethyl-L-arginine in plasma by liquid chromatography as AccQFluor(TM) fluorescent derivatives. J Chromatogr B 805: 325–329
15. Nonaka S, Tsunoda M, Imai K, Funatsu T (2005) High-performance liquid chromatographic assay of NG-monomethyl- L-arginine, NG, NG-dimethyl-L-arginine, and NG, NG′-dimethyl-L-arginine using 4-fluoro-7-nitro-2,1,3-benzoxadiazole as a fluorescent reagent. J Chromatogr A 1066:41–45
16. Sotgia S, Zinellu A, Pinna GA, Deiana L, Carru C (2008) A new selective pre-column ninhydrin-based derivatization for a RP-HPLC determination of plasma asymmetric dimethyl-L-arginine (ADMA) by fluorescence detection. Amino Acids 34:677–682
17. Caussé E, Siri N, Arnal JF, Bayle C, Malatray P, Valdiguié P, Salvayre R, Couderc F (2000) Determination of asymmetrical dimethylarginine by capillary electrophoresis-laser-induced fluorescence. J Chromatogr B 741:77–83
18. Trapp G, Sydow K, Dulay MT, Chou T, Cooke JP, Zare RN (2004) Capillary electrophoretic and micellar electrokinetic separations of asymmetric dimethyl-L-arginine and structurally related amino acids: quantitation in human plasma. J Sep Sci 27:1483–1490
19. Tsikas D, Schubert B, Gutzki FM, Sandmann J, Frölich JC (2003) Quantitative determination of circulating and urinary asymmetric dimethylarginine (ADMA) in humans by gas chromatography-tandem mass spectrometry as methylester tri(N-pentafluoropropionyl) derivative. J Chromatogr B 798:87–99
20. Albsmeier J, Schwedhelm E, Schulze F, Kastner M, Böger RH (2004) Determination of NG, NG-dimethyl-L-arginine, an endogenous NO synthase inhibitor, by gas chromatography–mass spectrometry. J Chromatogr B 809:59–65
21. Vishwanathan K, Tackett RL, Stewart JT, Bartlett MG (2000) Determination of arginine and methylated arginines in human plasma by liquid chromatography-tandem mass spectrometry. J Chromatogr B 748:157–166
22. Martens-Lobenhoffer J, Bode-Böger SM (2003) Simultaneous detection of arginine, asymmetric dimethylarginine, symmetric dimethylarginine and citrulline in human plasma and urine applying liquid chromatography-mass spectrometry with very straightforward sample preparation. J Chromatogr B 798:231–239
23. Kirchherr H, Kühn-Velten WN (2005) HPLC-tandem mass spectrometric method for rapid quantification of dimethylarginines in human plasma. Clin Chem 51:249–252
24. Martens-Lobenhoffer J, Krug O, Bode-Böger SM (2004) Determination of arginine and asymmetric dimethylarginine (ADMA) in human plasma by liquid chromatography/mass spectrometry with the isotope dilution technique. J Mass Spectrom 39:1287–1294
25. Schwedhelm E, Tan-Andresen J, Maas R, Riederer U, Schulze F, Böger RH (2005) Liquid chromatography-tandem mass spectrometry method for the analysis of asymmetric dimethylarginine in human plasma. Clin Chem 51:1268–1271
26. Desiderio C, Rossetti DV, Messana I, Giardina B, Castagnola M (2010) Analysis of arginine and methylated metabolites in human plasma by field amplified sample injection capillary electrophoresis tandem mass spectrometry. Electrophoresis 31:1894–902
27. Quirino J, Terabe S (2000) Large volume sample stacking of positively chargeable analytes in capillary zone electrophoresis without polarity switching: use of low reversed electroosmotic flow induced by a cationic surfactant at acidic pH. Electrophoresis 21:355–359
28. Manetto G, Tagliaro F, Crivellente F, Pascall VL, Marigo M (2000) Field-amplified sample stacking capillary zone electrophoresis applied to the analysis of opiate drugs in hair. Electrophoresis 21:2891–2898
29. Macià A, Borrull F, Aguilar C, Calull M (2004) Application of capillary electrophoresis with different sample stacking strategies for the determination of a group of nonsteroidal anti-inflammatory drugs in the low microg x L(−1) concentration range. Electrophoresis 25:428–436
30. Raiz A, Kim B, Chung DS (2003) Capillary electrophoresis of trace metals in highly saline physiological sample matrices. Electrophoresis 24:2788–2795
31. Mikkers FEP, Everaerts FM, Verheggen TPEM (1979) Concentration distribution in free zone electrophoresis. J Chromatogr 169:11–20
32. Zinellu A, Sotgia S, Zinellu E, Pinna A, Carta F, Gaspa L, Deiana L, Carru C (2007) High-throughput CZE-UV determination of arginine and dimethylated arginines in human plasma. Electrophoresis 28:1942–1948

Chapter 11

Capillary Electrophoresis-Mass Spectrometry for Peptide Analysis: Target-Based Approaches and Proteomics/ Peptidomics Strategies

Carolina Simó, Alejandro Cifuentes, and Václav Kašička

Abstract

In this chapter, the potential of capillary electrophoresis-mass spectrometry (CE-MS) for peptide analysis is demonstrated by the presentation of two different strategies typically followed in analysis of these biomolecules by CE-MS. The first one is a target-based approach and it is used to detect a toxic oligopeptide in a complex matrix. Namely, CE-MS using an ion trap MS analyzer is applied to detect and quantify γ-glutamyl-*S*-ethenyl-cysteine (GEC) bioactive dipeptide in a legume plant. The second one is a shotgun-like methodology used for proteomic analysis. Particularly, CE-MS using a time-of-flight MS analyzer is employed to investigate the substantial equivalence between a genetically modified (GM) variety of soybean and its conventional isogenic counterpart. These generic methods have broad applications for the analysis of peptides in a large variety of matrices, including applications in the area of proteomics and peptidomics.

Key words: Peptides, Foodomics, Peptidomics, Proteomics, Target analysis, CE, CE-MS

1. Introduction

1.1. General

Capillary electrophoresis (CE) coupled with mass spectrometry (MS) represents an attractive analytical method due to the high separation efficiency, selectivity, sensitivity, speed of analysis, and structural information that CE-MS combination can provide (1–4). Thus, CE-MS using electrospray ionization (ESI) interface for the on-line coupling has given rise to a powerful hyphenated technique able to combine the advantages of both CE and MS instruments (5, 6). Moreover, CE-MS is a technique very well suited for peptide analysis as can be deduced from the large number of works already published on this topic (7–19). Analysis of peptides by

Nicola Volpi and Francesca Maccari (eds.), *Capillary Electrophoresis of Biomolecules: Methods and Protocols*, Methods in Molecular Biology, vol. 984, DOI 10.1007/978-1-62703-296-4_11, © Springer Science+Business Media, LLC 2013

CE-MS has become a key analytical step and its use includes applications in biological and biochemical research, clinical diagnosis, drug discovery, and pharmaceutical and food analysis.

Basically, there are two different approaches that can be followed for peptide analysis by CE-MS: a target-based method or a proteomics strategy (20, 21) (including the peptidomics approach in the latter strategy (11) since the fundamentals and analytical methodology are similar to those used for proteomics).

1.2. CE-MS of Peptides Following a Target-Based Approach

In target-based analysis, the analyte (or group of analytes) is known and the method is developed to adequately extract, separate, and characterize the peptidic target compound from the matrix. In some cases, quantitation can be included if standards are available. This method is straightforward; however, the information achieved on a total peptide composition of sample is rather limited.

In this work, CE on-line coupled via an ESI interface to an ion trap mass analyzer (CE-ESI-IT MS/MS) is applied following a target-based approach to the determination of γ-glutamyl-*S*-ethenyl-cysteine (GEC), a bioactive and unstable dipeptide present in *Vicia narbonensis* L. seeds. This compound is responsible for, among other negative effects, palatability reduction and grain toxicity. The method is applied to investigate the content of this compound in seeds of 21 genotypes of *V. narbonensis* L.

1.3. CE-MS of Peptides Following a Proteomic Strategy

Procedures used in proteomics are somehow different. Proteomics is defined as the comprehensive analysis of the whole protein content within an organism, tissue, cell, or organelle. Technically, this definition includes the analysis of all the peptides that can be obtained usually via enzymatic hydrolysis of proteins. This analysis is especially complex considering that proteome, in contrast to genome, is variable in time. Moreover, other additional limitations in a proteomic study are associated with the heterogeneity of analytes in terms of physicochemical properties and the huge differences in protein abundance. For example, a proteome can have a dynamic range of 7–12 orders of magnitude (22) and only a few orders can be analyzed simultaneously with the current proteomic platforms. Owing to this complexity, it is important to develop suitable methodologies of extraction, prefractionation, and/or enrichment of less abundant proteins, together with powerful separation procedures in order to allow their efficient identification by sensitive and selective MS analyses.

Thus, in a CE-MS proteomics approach, proteins are first extracted from the matrix and either (1) the total proteins are directly enzymatically hydrolyzed in a shotgun-like (bottom-up) approach (20) and the total peptide content is studied by CE-MS or (2) the proteins can be separated (usually by two-dimensional gel electrophoresis (23, 24)) and isolated and then each individual protein is enzymatically hydrolyzed and the resulting peptides are studied by CE-MS (top-down methodology). In both cases, analysis of peptides has shown to provide key information to characterize proteins and/or to understand the biological mechanisms within the organism, tissue, cell, or organelle under study.

The application shown in this work is on the use of CE on-line coupled to a time-of-flight mass analyzer (CE-ESI-TOF MS) in a shotgun-like proteomic study of peptides. The proteomic study is developed to corroborate (or not) the equivalence between a genetically modified (GM) transgenic soybean and its isogenic non-transgenic counterpart, in order to detect possible unintended effects. Thus, proteins are extracted from both types of soybeans and enzymatically hydrolyzed, and the resulting peptides are analyzed by CE-MS.

2. Materials and Instrumentation

2.1. GEC Peptide Identification and Quantification from Vicia narbonensis L. Seeds by CE-ESI-IT MS/MS

1. Deionized water for preparation of all solutions was obtained using a Milli-Q system (Millipore, Bedford, MA, USA).
2. Methanol–water (50:50 v/v).
3. Ethanol–water (70:30 v/v).
4. 0.1 M NaOH. To prepare 100 mL: 0.4 g of sodium hydroxide was dissolved in 95 mL deionized water. Deionized water was added to reach a final volume of 100 mL using a volumetric flask.
5. 50 mM SDS in 0.1 M NaOH. To prepare 100 mL: 1.44 g of sodium dodecyl sulfate was dissolved in 95 mL of 0.1 M NaOH solution. 0.1 M NaOH solution was then added to reach a final volume of 100 mL using a volumetric flask.
6. 0.1% v/v acetic acid in methanol–water (50:50 v/v).
7. 20 mM ammonium hydrogen carbonate, pH 7. To prepare 100 mL: 0.158 g of ammonium hydrogen carbonate was dissolved in 95 mL of deionized water. Formic acid was added to adjust the pH to 7. Deionized water was then added to reach a final volume of 100 mL using a volumetric flask.
8. *Vicia narbonensis* L. seeds were collected in different Spanish and Italian regions.
9. Purified GEC dipeptide was a gift from Dr. Max E. Tate from the University of Adelaide (Australia).
10. Bare fused silica capillary with outer polyimide coating from Composite Metals Service (Worcester, England), internal/external diameter (id/od) of 50/375 μm, and total (and MS detection) length of 87 cm.
11. Ultra Centrifugal mill ZM from Retsch (Newtown, PA, USA).

12. High-performance capillary electrophoresis (CE) system model P/ACE 5010 from Beckman (Fullerton, CA, USA).
13. Syringe pump from Cole Palmer (Vernon Hills, IL, USA).
14. Orthogonal electrospray interface model G1607A from Agilent Technologies (Palo Alto, CA, USA).
15. Ion trap mass spectrometer Esquire 2000 from Bruker Daltonics (Bremen, Germany).

2.2. CE-ESI-TOF MS Analysis of Protein Hydrolyzates from GM and non-GM Soybeans

1. Deionized water for preparation of all solutions was obtained using a Milli-Q system (Millipore, Bedford, MA, USA).
2. Acetonitrile (ACN)/water (80:20 v/v).
3. 2-propanol–water (50:50 v/v).
4. 0.1 M NaOH.
5. 0.5 M formic acid. To prepare 100 mL: 1.89 mL of formic acid was mixed with 95 mL deionized water in a volumetric flask. Deionized water was then added to reach a final volume of 100 mL.
6. 100 mM NH_4HCO_3. To prepare 100 mL: 0.79 g of ammonium hydrogen carbonate salt was dissolved in 95 mL deionized water. Deionized water was then added to reach a final volume of 100 mL using a volumetric flask.
7. 50 mM NH_4HCO_3. To prepare 100 mL: 0.39 g of ammonium hydrogen carbonate salt was dissolved in 95 mL deionized water. Deionized water was then added to reach a final volume of 100 mL using a volumetric flask.
8. 200 mM and 800 mM 1,4-dithiothreitol (DTT) in 100 mM NH_4HCO_3.
9. 1 M iodoacetamide (IAA) in 100 mM NH_4HCO_3.
10. Bovine serum albumin (BSA).
11. Bradford assay kit from Bio-Rad (Hercules, CA).
12. Trypsin solution (0.24 mg/mL) in 50 mM NH_4HCO_3.
13. 2% v/v Tuning Mix solution from Agilent (Palo Alto, CA, USA).
14. The soybean protein isolate (SPI) was from ICN (Aurora, OH, USA).
15. GM soybeans and non-GM soybeans used for the comparative peptide study were grown under the same conditions in a growth chamber.
16. Bare fused silica capillary with outer polyimide coating from Composite Metals Service (Worcester, England), id/od 50/375 μm, total (and MS detection) length 90 cm.
17. High-performance capillary electrophoresis (CE) system model P/ACE 5010 from Beckman (Fullerton, CA, USA).

18. Syringe pump from Cole Palmer (Vernon Hills, IL, USA).
19. Orthogonal electrospray interface model G1607A from Agilent Technologies (Palo Alto, CA, USA).
20. A time-of-flight micrOTOF MS instrument from Bruker Daltonics (Bremen, Germany).

3. Methods

3.1. GEC Peptide Identification and Quantification from Vicia Narbonensis L. Seeds by CE-ESI-IT MS/MS

3.1.1. Sample Preparation

1. *V. narbonensis* L. seeds were grounded with an Ultra Centrifugal mill. The obtained flour was stored at 4 °C in plastic bags until used.
2. GEC peptide extraction was carried out by the addition of 1 mL of ethanol–water (70:30 v/v) to 0.1 g of seed flour, followed by stirring for 45 min in darkness. Extracts were then centrifuged at 8,300 × *g* for 10 min at 5 °C, to remove all undesirable solid particles. Supernatant was directly analyzed by CE-ESI-IT MS/MS.

3.1.2. CE-ESI-IT MS/MS Analysis of GEC Peptide from Vicia narbonensis L. Seeds

1. All new capillaries were conditioned by flushing at 20 psi (1 psi = 6.895 kPa) with 0.1 M NaOH for 20 min, followed by water for 30 min. Background electrolyte (BGE) for CE analyses was 20 mM ammonium hydrogen carbonate at pH 7 (see Note 1). Before each analysis, the capillary was rinsed with 50 mM SDS in 0.1 M NaOH to clean the inner capillary surface (see Notes 2 and 3). After the cleaning procedure, conditioning was completed by flushing the capillary with water and BGE for 1.5 and 2 min, respectively.
2. Interface between CE and IT MS was carried out via a sheath-flow configuration, in which electrical contact at the electrospray needle tip was established via a flow of conductive sheath liquid, which consisted of methanol–water (50:50 v/v) containing 0.1% v/v acetic acid (see Note 4), and delivered at a flow rate of 3 μL/min by a syringe pump. The nebulizer/drying conditions were next: 11 psi nitrogen (see Note 5) and 8 L/min nitrogen, at 120 °C.
3. The mass spectrometer was operated in the positive ion mode. The spectrometer was scanned at 200–400 *m/z* range at 13,000 amu/s during separation.
4. Injections of sample were made at the anodic capillary end using N_2 pressure of 0.5 psi for 20 s. Separations were performed at +15 kV at a constant temperature of 25 °C. An example of the obtained electropherograms is shown in Fig. 1.
5. Sequential product ion fragmentation experiments (MS^2 and MS^3) were performed in order to confirm the structure of

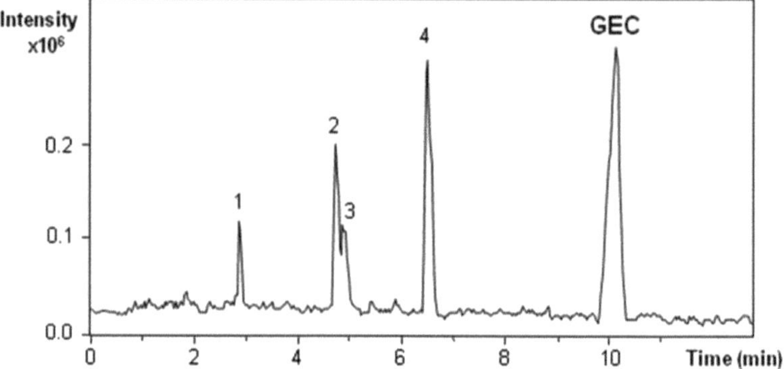

Fig. 1. CE-ESI-IT MS base peak electropherogram (BPE) of the *V. narbonensis* L. extract. CE-ESI-IT MS conditions: Bare fused silica capillary (total and MS detection length 87 cm, id/od 50/375 μm). BGE composition: 20 mM ammonium hydrogen carbonate, pH 7; separation voltage: +25 kV, capillary temperature: 25 °C, injection at 0.5 psi for 12 s. MS positive ion mode, sheath liquid composition: methanol–water (1:1, v/v) with 0.1% v/v acetic acid, at a flow rate of 3 μL/min, dry gas flow at 8 L/min. Temperature: 120 °C. Mass scan: 20–400 *m/z*. Peaks numbered from 1 to 4 indicate other extracted compounds from *V. narbonensis* L. seeds that were not identified with *m/z* values equal to 177.8, 174.0, 176.0, and 159.0, respectively.

GEC peptide. The parent ion 277.0 *m/z* obtained in the MS was chosen for fragmentation. The ions obtained in the MS^2 fragmentation spectra at 148.0, 156.0, and 122.0 *m/z* were selected for MS^3 fragmentation. Collision-induced-dissociation (CID) voltage amplitude for MS^2 and MS^3 fragmentation was 1 V. The obtained MS, MS^2, and MS^3 spectra are shown in Fig. 2.

6. To determine the limit of detection (LOD) and limit of quantification (LOQ), purified GEC peptide was injected at a concentration of 25 μg/mL. LOD (21 μg/mL) was calculated for a signal/noise (S/N) ratio equal to 3, and LOQ (71 μg/mL) was calculated for an S/N ratio equal to 10.
7. To quantify GEC peptide in plant seeds, purified GEC was dissolved in ethanol–water (70:30 v/v) at different known concentrations: 0.098, 0.245, 0.490, 0.980, 1.715, 2.450, and 4.900 mg/mL. To obtain the calibration curve, each concentration was injected in triplicate, and the migration time normalized area of the GEC peak (i.e., peak area divided by migration time) obtained in the electropherograms was plotted as a function of the concentration. The CE-ESI-IT MS procedure allows the quantification in the range studied with a determination coefficient (r^2) of 0.997.

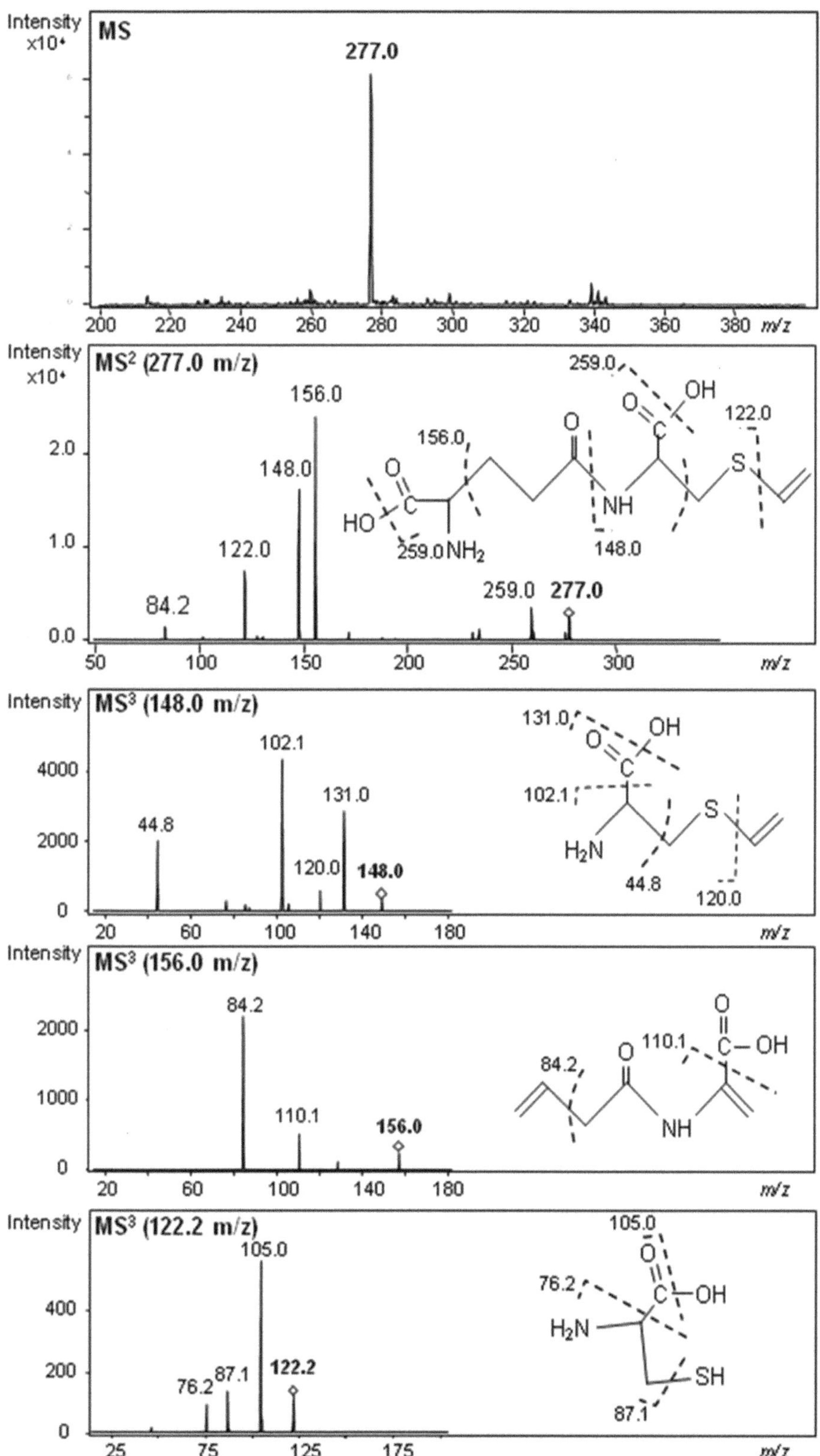

Fig. 2. MS spectra obtained for the last peak (10.1 min) in the electropherogram from Fig. 1. MS of the parent ion 277.0 *m/z*, MS² from the ion 277.0 *m/z*, and MS³ from the ions 148.0, 156.0, and 122.0 *m/z*.

3.2. CE-ESI-TOF MS Analysis of Protein Hydrolyzates from GM and non-GM Soybeans

3.2.1. Protein Extraction from Soybean and Total Protein Quantification

1. All beans were grounded with a domestic mill prior to extraction. The obtained material was stored at 4 °C in plastic bags until used.
2. Protein extracts were obtained by directly dissolving 150 mg of ground sample in 1.5 mL of ACN/water (80:20, v/v). After 10 min vortexing and 3 min sonication, extracts were centrifuged (9,000 × *g*) for 5 min. The supernatant fraction was collected and stored at −80 °C until used.
3. The total protein content was determined by the Bradford assay kit using a commercial dye reagent and using BSA as standard.
4. After protein quantification, all protein extracts were diluted to a concentration of 3 mg/mL with ACN/water (80:20, v/v) before enzymatic hydrolysis.

3.2.2. Enzymatic Hydrolysis

1. The digestion was performed under reductive alkylation conditions (see Note 6). To reduce disulfide bonds, 100 μL of the diluted protein extract was mixed with 5 μL of 200 mM DTT (in 100 mM NH_4HCO_3) and incubated at 90 °C for 5 min. After allowing the sample to cool at room temperature, alkylation of free thiol groups with 4 μL of 1 M IAA (in 100 mM NH_4HCO_3) for 45 min at room temperature was carried out. To neutralize the remaining IAA, 5 μL of 800 mM DTT (in 100 mM NH_4HCO_3) was added, and the solution was maintained at room temperature for 45 min.
2. 50 μL of freshly prepared bovine pancreas trypsin solution (in 50 mM NH_4HCO_3) was added at a 1:25, m/m substrate ratio, and the solution was incubated at 37 °C overnight under continuous shaking conditions in a thermomixer. The reaction was stopped by heating the mixture at 90 °C for 5 min. The suspension was then centrifuged at 14,000 × *g* for 5 min. The supernatant fraction was collected and stored at −20 °C prior to analysis by CE-ESI-TOF MS.

3.2.3. CE-ESI-TOF MS Analysis of Soybean Protein Hydrolyzates

1. All new capillaries were conditioned by flushing with 0.1 M NaOH for 20 min, followed by water for 30 min. BGE for CE was 0.5 M formic acid (see Note 7). Before each analysis, the capillary was rinsed with BGE for 4 min.
2. Interface between CE and TOF MS was carried out via a sheath-flow configuration, in which electrical contact at the electrospray needle tip was established via a flow of conductive sheath liquid, which consisted of 2-propanol–water (50:50 v/v) (see Note 8), and delivered at a flow rate of 3 μL/min by a syringe pump. The nebulizer and drying gas conditions were 0.4 bar N_2 and 4 L/min N_2, respectively, and maintaining the ESI chamber at 200 °C (see Note 9).
3. The mass spectrometer operated with the ESI source in the positive ion mode and the *m/z* range was 50–3,000. External

calibration of the TOF MS instrument (see Note 10) was performed by introducing a 2% v/v Tuning Mix solution from Agilent through the separation capillary (by applying a pressure of 20 psi) towards the TOF MS instrument.

4. CE-ESI-TOF MS injections were made at the anodic end using N_2 pressure of 0.5 psi for 20 s. Sample injections were made by duplicate. The electrophoretic separation was achieved using +25 kV as running voltage at a constant temperature of 25 °C. In Fig. 3, the base peak electropherogram (BPE) of the digested soybean protein isolate (SPI) extract and some extracted ion electropherogram (EIEs) are shown.
5. The number of mass spectra to be processed in a single electropherogram was huge; therefore, the only alternative was to carry out an automated interpretation. To do this, the abundance cutoff for automatic detection of peptides was set at 15% (see Note 11).
6. After the CE-ESI-TOF MS analysis of the non-GM and GM soybean (Fig. 4), exact masses of the peptides obtained from non-GM soybean were compared with those from the GM soybean. A reproducible set of 151 peptides was detected in the two samples. These results seem to corroborate the equivalence between GM soybean and its conventional counterpart for this particular sample.

4. Notes

1. When working with MS, the use of volatile sheath liquids and separation BGEs is needed because the presence of nonvolatile components decreases the MS sensitivity, increases the background noise, and, under extreme conditions, can clog the system.
2. When the rinse between injections was made with BGE, migration time of the GEC peptide was steadily increasing. This is typically produced by the adsorption of some of the compounds from seed extracts onto the inner capillary wall. In order to eliminate adsorbed compounds from the capillary wall, after each analysis, the capillary was rinsed with 50 mM SDS in 0.1 M NaOH to clean the inner surface.
3. In order to avoid any MS contamination with nonvolatile compounds, when the separation capillary is rinsed with 50 mM SDS in 0.1 M NaOH or 0.1 M NaOH, nebulizer pressure and MS voltages are switched off.
4. Different quantities of acetic and formic acid (0, 0.1, 0.5 and 1% v/v) to the sheath liquid, as well as methanol amount (0, 25, 50, 75 and 100% v/v) in the sheath liquid, were studied

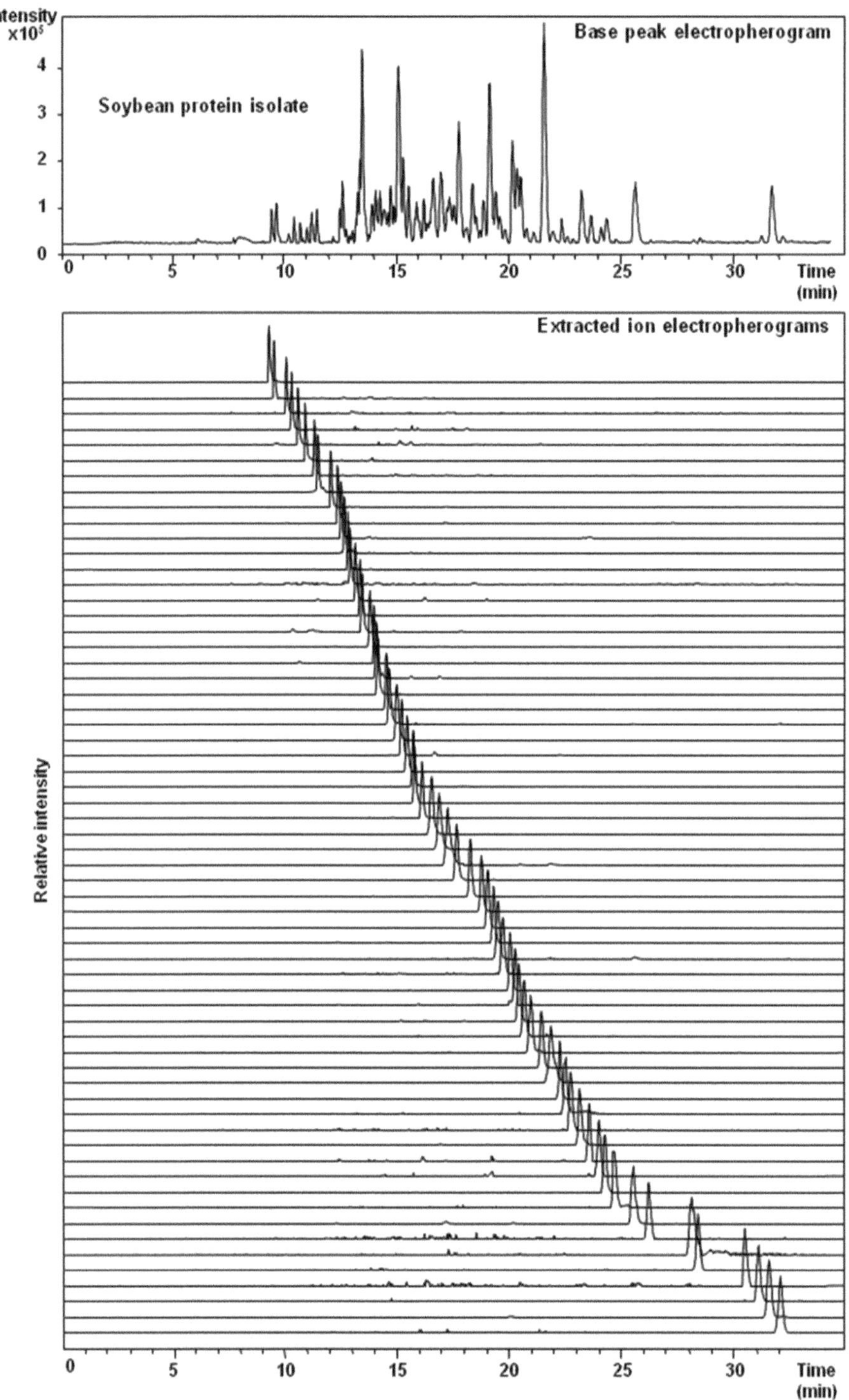

Fig. 3. CE-ESI-TOF MS BPE of the digested soybean protein isolate (SPI) and some representative extracted ion electropherograms (EIEs). CE-ESI-TOF MS conditions: Bare fused silica capillary (total and MS detection length 90 cm, id/od 50/375 μm). BGE composition: 0.5 M formic acid, separation voltage: +25 kV, capillary temperature: 25 °C, injection at 0.5 psi for 20 s. MS positive ion mode, sheath liquid composition: 2-propanol–water (50:50, v/v), at a flow rate of 3 μL/min, dry gas flow at 4 L/min. Temperature: 200 °C. Mass scan: 50–3,000 *m/z*.

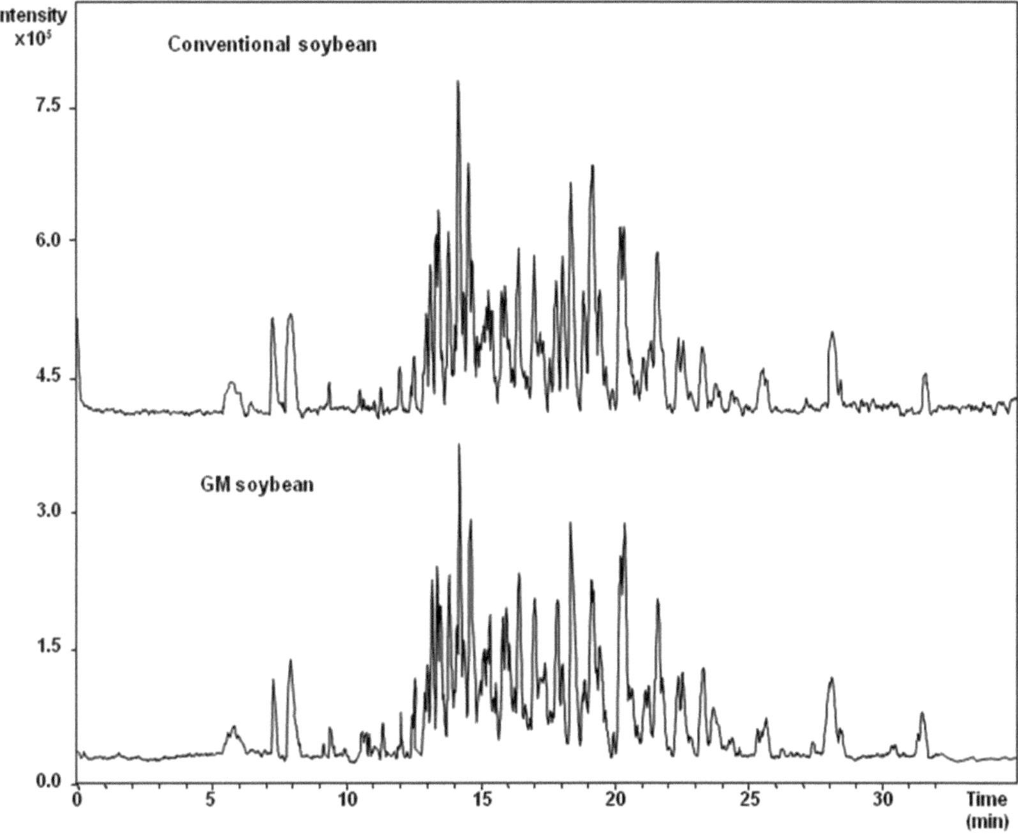

Fig. 4. CE-ESI-TOF MS BPE of the digested protein extract from conventional and GM soybean. CE-ESI-TOF MS conditions as in Fig. 3.

for GEC analysis. Best S/N ratio was obtained with a sheath liquid composed by methanol–water (50:50, v/v) and containing 0.1% acetic acid.

5. The effect of flow rate of the sheath liquid was studied over the range 1–6 μL/min observing a slight improvement on S/N when 3 μL/min was used for GEC analysis. A tendency towards better sensitivities was also observed when higher nebulizer pressures were used. In addition, lower migration times were obtained due to the suction effect produced by the nebulizer gas within the capillary column.

6. To carry out a suitable hydrolysis of the soybean proteins in the extract, it was necessary to take into account the presence of possible trypsin inhibitors in the sample. It was observed that heat treatment of the sample was not sufficient to completely inactivate trypsin inhibitors in the soybean protein extracts. A reduction/alkylation protocol was used before trypsin digestion. Although reduction and alkylation steps are generally used to favor the accessibility of the substrate in the

enzymatic proteolysis, this method can also be used to inactivate trypsin inhibitors because the reduction of disulfide bonds alters protein structure and reduces inhibitor activity (25).

7. BGEs containing acetic acid or formic acid at different concentrations (from 0.5 to 1.5 M) were studied in order to obtain best separations of the tryptic peptides in the shortest analysis time. Better results in terms of S/N ratio were obtained using formic acid-based BGEs. When increasing the concentration of formic acid, a gradual decreasing in the migration times was observed. On the other hand, a progressive decrease of the number of peaks was also observed. A BGE composed of 0.5 M formic acid was finally selected as a good compromise between number of detected peptides and analysis speed.
8. Different mixtures of ACN, methanol, and 2-propanol with water were tested as sheath liquid for the electrospray interface. It was observed that a 50:50 (v/v) 2-propanol–water mixture provided the best response in terms of sensitivity and signal stability of tryptic peptides. The addition of formic or acetic acid to this solution was not studied due to the high quantity of formic acid already present in the BGE.
9. The spray was shown to be stable using a sheath liquid flow rate between 2 and 6 μL/min, reaching the best S/N value at 3 μL/min during the CE-ESI-TOF MS analysis of the tryptic peptides. The influence of the nebulizer pressure (0.2–0.7 bar) was also studied observing that using a nebulizer pressure of 0.4 bar, the best results in terms of MS sensitivity were obtained.
10. External calibration of the TOF MS instrument was done prior to each analysis. Masses for the calibration of the TOF MS instrument were 322.0481, 622.0209, 922.0098, 1321.9842, 1521.9715, and 2121.9332 *m/z*.
11. Percent cutoff parameter indicates the minimum percentage abundance of an ion, relative to the most intense ion in the spectrum that will be considered for the deconvolution, determining which *m/z* signals from the mass spectra will be used to find the particular compounds.

Acknowledgments

This work was supported by AGL2008-05108-C03-01 (Ministerio de Educacion y Ciencia, Spain) and CSD2007-00063 FUN-C-FOOD projects (Programa CONSOLIDER, Ministerio de Educacion y Ciencia, Spain); by the Czech Science Foundation, grant no. 203/08/1428; and by the Academy of Sciences of the Czech Republic (ASCR), Research Project AV0Z40550506. Authors also thank 2008CZ0019 bilateral project (ASCR-CSIC) for the support.

References

1. Tomer KB (2001) Separations combined with mass spectrometry. Chem Rev 101:297–328
2. Klampfl CW, Buchberger W (2007) Coupling of capillary electroseparation techniques with mass spectrometric detection. Anal Bioanal Chem 388(3):533–536
3. Klampfl CW (2009) CE with MS detection: a rapidly developing hyphenated technique. Electrophoresis 30:S83–S91
4. Hommerson P, Khan AM, de Jong GJ, Somsen GW (2011) Ionization techniques in capillary electrophoresis-mass spectrometry: principles, design, and application. Mass Spectrom Rev 30:1096–1120
5. Olivares JA, Nguyen NT, Yonker CR, Smith RD (1987) Online mass-spectrometric detection for capillary zone electrophoresis. Anal Chem 59:1230–1232
6. Brocke A, Nicholson G, Bayer E (2001) Recent advances in capillary electrophoresis/electrospray-mass spectrometry. Electrophoresis 22:1251–1266
7. Schmitt-Kopplin P, Frommberger M (2003) Capillary electrophoresis-mass spectrometry: 15 years of developments and applications. Electrophoresis 24:3837–3867
8. Simó C, Cifuentes A (2003) Capillary electrophoresis-mass spectrometry of peptide from enzymatic protein hydrolysis. Simulation and optimization. Electrophoresis 24:834–842
9. Stutz H (2005) Advances in the analysis of proteins and peptides by capillary electrophoresis with matrix-assisted laser desorption/ionization and electrospray-mass spectrometry detection. Electrophoresis 26:1254–1290
10. Scriba GKE, Psurek A (2008) Separation of peptides by capillary electrophoresis. In: Schmitt-Kopplin P (ed) Capillary electrophoresis, Capillary electrophoresis, methods and protocols. Humana Press Inc., Totowa, NJ, pp 483–506
11. Herrero M, Ibáñez E, Cifuentes A (2008) Capillary electrophoresis-mass spectrometry for peptide analysis and peptidomics. Electrophoresis 29:2148–2160
12. Metzger J, Schanstra J, Mischak H (2009) Capillary electrophoresis-mass spectrometry in urinary proteome analysis: current applications and future developments. Anal Bioanal Chem 393:1431–1442
13. Neusüss C, Pelzing M (2009) Capillary zone electrophoresis-mass spectrometry for the characterization of isoforms of intact glycoproteins. Methods Mol Biol 492:201–213
14. Simó C, Dominguez-Vega E, Marina ML, Garcia MC, Dinelli G, Cifuentes A (2010) CE-TOF MS analysis of complex protein hydrolyzates from genetically modified soybeans—A tool for foodomics. Electrophoresis 31:1175–1183
15. Kašička V (2010) Recent advances in CE and CEC of peptides (2007–2009). Electrophoresis 31:122–146
16. Borges-Alvarez M, Benavente F, Gimenez E, Barbosa J, Sanz-Nebot V (2010) Assessment of capillary electrophoresis TOF MS for a confident identification of peptides. J Sep Sci 33:2489–2498
17. Haselberg R, Brinks V, Hawe A, de Jong GJ, Somsen GW (2011) Capillary electrophoresis-mass spectrometry using noncovalently coated capillaries for the analysis of biopharmaceuticals. Anal Bioanal Chem 400:295–303
18. Kašička V (2012) Recent developments in CE and CEC of peptides (2009–2011). Electrophoresis 33:48–73
19. Gimenez E, Ramos-Hernan R, Benavente F, Barbosa J, Sanz-Nebot V (2012) Analysis of recombinant human erythropoietin glycopeptides by capillary electrophoresis electrospray-time of flight-mass spectrometry. Anal Chim Acta 709:81–90
20. McDonald WH, Yates JR (2003) Shotgun proteomics: integrating technologies to answer biological questions. Curr Opinion Mol Therapeut 5:302–309
21. Ning ZB, Zhou H, Wang FJ, bu-Farha M, Figeys D (2011) Analytical Aspects of Proteomics: 2009–2010. Anal Chem 83:4407–4426
22. Righetti PG, Boschetti E, Lomas L, Citterio A (2006) Protein Equalizer (TM) Technology: the quest for a "democratic proteome". Proteomics 6:3980–3992
23. Görg A, Weiss W, Dunn MJ (2004) Current two-dimensional electrophoresis technology for proteomics. Proteomics 4:3665–3685
24. Westermeier R, Schickle H (2009) The current state of the art in high-resolution two-dimensional electrophoresis. Arch Physiol Biochem 115:279–285
25. Faris RJ, Wang H, Wang T (2008) Improving digestibility of soy flour by reducing disulfide bonds with thioredoxin. J Agric Food Chem 56:7146–7150

Chapter 12

High-Resolution Proteome/Peptidome Analysis of Body Fluids by Capillary Electrophoresis Coupled with MS

Amaya Albalat, Vasiliki Bitsika, Petra Zurbig, Justyna Siwy, and William Mullen

Abstract

The analysis of proteins and peptides in biological fluids is becoming more important as they are potential sources of diagnostic biomarkers of disease. The complexity of body fluids is such that no single technique can both identify and quantify all the constituents present. It therefore requires pre-concentration of the compounds of interest and hyphenated techniques like capillary electrophoresis (CE) or high-performance liquid chromatography (HPLC) coupled to a mass spectrometer (MS) to provide the separation, identification, and quantification of the proteins and peptides under investigation. Here we provide protocols for sample preparation of various body fluids, namely, urine, plasma, and cerebrospinal fluid, and how to obtain maximum sensitivity and selectivity of the small proteins and peptides under investigation by CE-MS.

Key words: Capillary electrophoresis, Mass spectrometry, Proteome/peptidome, Body fluids

1. Introduction

There has been an increasing need for highly selective and sensitive analytical methods for the analysis of the proteome/peptidome of body fluids. In particular the field of biomarker screening of body fluids for early detection of diseases has been a prime focus for these research tools. However, due to the large number of compounds present and their wide concentration range, no single technique has proven to be capable of both identifying and quantifying the full range of these compounds (1, 2). To overcome this problem a two-stage approach can be taken. More specifically, an upstream separation technique that can couple with a highly sensitive and selective downstream detector can be used. There are only two separation techniques that lend themselves to simple and routine on-line coupling with mass spectrometric (MS) detection systems. These are high-performance liquid chromatography

Nicola Volpi and Francesca Maccari (eds.), *Capillary Electrophoresis of Biomolecules: Methods and Protocols*, Methods in Molecular Biology, vol. 984, DOI 10.1007/978-1-62703-296-4_12,

(HPLC) and capillary electrophoresis (CE). This topic has been investigated previously with regard to off-line and on-line techniques (3). Although HPLC is more widely used in the biological analysis field, capillary electrophoresis can provide higher separation efficiencies within a shorter analysis time. In addition, it has a number of other advantages, such as the determination of mass range of proteins and peptides, and sample reproducibility. Moreover, it also does not suffer from memory or sample carryover effects as reported in HPLC analysis (3).

Capillary Electrophoresis—There are a number of separation techniques available in capillary electrophoresis, but the main one used in CE-MS is capillary zone electrophoresis (CZE). This technique separates compounds based on their different migration velocities which are set by the number of charges carried and the size of the molecule (4). The main electrolytes used in protein/peptide CE are mixtures of volatile solvents (acetonitrile or methanol) and acids (formic or acetic) which are ideal matches for the requirement for sensitive MS detection.

Mass Spectrometry—CE can be coupled to a range of different forms of mass analyzers. The choice of which type to use is dependant on the compounds being investigated. Quadrupole and ion trap systems are more widely available in laboratories, but they have slow scan speed and low resolution in full scan mode. Time-of-flight (TOF) systems have high resolution, adequate sensitivity, and the fast scan speed needed for protein and peptide detection in CE analysis.

Coupling Interface—The most frequently used technique for coupling CE with MS is electrospray ionization (ESI) (5). It has a number of factors that make it an ideal solution for joining CE to MS. It allows proteins and peptides in the liquid phase, as they exit the tip of the capillary, to be transferred directly into ions in the gas phase for passage into the MS. Furthermore, the mass range for measuring these compounds is greatly increased as ESI can form multiply charged species allowing the measurement of high molecular weight proteins. As all proteins and peptides that migrate through the CE system are already carrying a charge in the liquid phase, ESI is an ideal ionization technique for the analysis of these compounds.

One issue that should be addressed regards the electrical currents flowing in the CE and the MS systems. As the current in the CE is orders of magnitude greater than that in the electrospray interface there needs to be a separation of the two electrical circuits. The best solution to this problem is to have the ESI needle at ground potential and the voltage applied at the MS inlet, though this is not possible with all manufacturers' instruments (6).

There are a number of types of ESI interfaces but the most widely used system is the sheath-flow interface. It has a robust design and offers a greater degree of flexibility in the selection of the background electrolyte used for separation of the proteins (7). The supply of a sheath liquid allows the formation of a stable spray enhancing the robustness of the interface. The downside of the addition of a coaxial sheath flow is that it reduces the sensitivity when compared to a sheathless interface system. However, the aforementioned robustness more than compensates for this.

Body Fluids—Body fluids contain large amounts of information on the (patho)physiological state of an organism. Although this has been known for centuries, it is only over the last twenty years that the "omics" technologies have evolved sufficiently to attempt to make use of this information. The main body of work in this area has been conducted primarily in blood, cerebrospinal fluid, and urine.

Blood has been seen as the ideal source of biomarkers of disease as it is the transport mechanism for all molecules to and from all organs and tissues in the body. However, due to its complexity and issues regarding the wide concentration range these compounds exist in blood, a number of problems have been noted. It is not possible, with current technology, to measure the concentration of all compounds present without additional compound enrichment and/or depletion procedures (8). Furthermore, additional methodologies to investigate the proteome of blood just include specific sections, e.g., only glycosylated or phosphorylated proteins.

Cerebrospinal fluid (CSF) is not a readily accessible medium but its collection for clinical diagnosis for a range of neurological conditions can be advantageous. The level of proteins present in CSF is much lower than seen in plasma, so there are fewer problems with complexity and concentration range. As it is in contact with the central nervous system and the brain it should be an important source of biomarkers for neurological diseases (9).

Urine has some obvious advantages over the previous two body fluids since it can be easily collected in large quantities using noninvasive procedures and without the need of medically trained personnel. However, from an analytical point of view, the major advantages are with regard to its stability and lack of complexity, compared to blood. By the time urine can be collected it has been stored in the bladder for a number of hours and any proteolytic processes may have run to completion by the time of collection.

Urine samples show variability due to diet and exercise. To reduce this and produce consistent proteomic data, it is recommended to collect the second urine of the day (10). It is possible to compensate for concentration variability in urine using urinary peptides, which are generally present in human urine and can work as housekeeping peptides (11). Taking all these factors into consideration, urine appears to be an excellent choice as a source for proteomics/biomarkers research. Biomarkers from urine reflect on the (patho)physiological condition not only of organs in direct contact with it, such as kidney and bladder, but also of the vascular system.

In this chapter, we will describe in detail the CE-MS approach for the analysis of urine and CSF samples aiming at proteomic profiling.

2. Materials

2.1. Components for Sample Preparation

1. Disposable de-salting PD10-columns (GE Healthcare Biosciences AB, Bedford, USA).
2. Disposable de-salting Illustra NAP-5 (GE Healthcare Biosciences AB, Bedford, USA).
3. Centristart ultrafilter tubes 20 and 10 kD, 0.62 cm^2 tubes (Sartorius Stedim Limited, UK).
4. BCA protein quantification kit (Uptima, Interchim, Montluçon, France).
5. Lyophilization Christ Speed-Vac RVC 2-18/Alpha 1–2 (Christ, Osterode am Harz, Germany).

2.2. Components for CE-MS Analysis

1. P/ACE MDQ capillary electrophoresis system (Beckman Coulter, Crea, CA, USA) (see Note 1).
2. MicrOTOF (time-of-flight) mass spectrometer (Bruker Daltonics, Bremen, Germany).
3. Electro-ionization sprayer (Agilent Technologies, Santa Clara, USA) (see Note 2).
4. Capillary 90 cm, 50 μm I.D. fused silica capillary non-coated (New Objective, Woburn, USA).
5. Cole-Parmer single-syringe infusion pump (Cole-Parmer Instrument Company, USA).
6. Hamilton syringe 2.5 ml (Hamilton Company, USA).

2.3. Software for Data Processing and Statistical Analysis

1. Software for peak detection: MosaiquesVisu (Biomosaiques, Hannover, Germany).
2. Commercial statistical packages: SAS (www.sas.com) and MedCal (www.medcalc.be).
3. Multitest R-package available at www.bioconductor.org.
4. SVM-based program for multiparametric data classification: MosaCluster (Biomosaiques, Hannover, Germany).

3. Methods

3.1. Collection of Urine, Cerebrospinal Fluid, and Plasma

Urine samples (minimum 2 ml) should be collected from midstream spontaneously voided morning urine (collected between 7 and 10 a.m.). Ideally the collected urine should be from the second urine of the morning as variability is lowest at this time. Samples should be collected in clean urine collectors.

CSF (minimum 1 ml) is obtained by lumbar puncture in a sitting position according to standard protocols (12). Samples are placed into polypropylene tubes.

Blood samples are taken via vein puncture and collected into heparin-coated tubes.

3.2. Pretreatment and Storage of Urine and Cerebrospinal Fluid and Blood

When possible, urine samples should be centrifuged at 1,000 × *g* for 10 min before freezing to remove cell debris and casts. This centrifugation step is not absolutely needed but recommended. If centrifugation is not possible samples should be well mixed before aliquoting and samples with turbid appearance should be noted as these samples will produce a lower quality of proteomic analysis and should be discarded. In centrifuged samples, supernatants should be aliquoted appropriately and then stored preferably at −80°C or at −20°C. The use of preservatives is not recommended unless freezing time is expected to be longer than 3 h. In this instance, urine samples should be stored at 4°C or on ice with the addition of 10 mM NaN_3 (or 200 mM boric acid) in order to delay microbial growth.

CSF should be centrifuged immediately after collection at 16,000 × *g* for 10 min at 4°C, aliquoted into polypropylene test tubes, frozen within 30–40 min after puncture, and stored at −80°C until used.

Blood should be centrifuged immediately after collection at 16,000 × *g* for 15 min at 4°C, aliquoted into polypropylene test tubes, and immediately frozen and stored at −80°C until used.

3.3. Buffer and Stock Solutions for the Sample Preparation

3.3.1. Stock Solutions

Do not use the stock solutions for longer than three months (see Note 3).

1. Ammonium hydroxide solution (25 %) ultrapure (Merck, Nottingam, UK).
2. Phenylmethylsulfonyl fluoride or PMSF: Prepare a saturated solution in ethanol.
3. Sodium chloride (NaCl): Prepare a solution 5 M in deionized water.
4. Sodium dodecyl sulfate (SDS): Prepare a 2 % solution in deionized water.
5. Urea solution 8 M (Sigma-Aldrich Company Ltd, Dorset, UK).

3.3.2. Buffers

Buffers for sample preparation should be freshly prepared on the day of analysis.

1. Urea buffer for the analysis of urine (2 M urea, 100 mM NaCl, 0.0125 % NH_4OH, and 0.01 % SDS): Add around 20 ml of deionized water to a 50 ml graduated glass cylinder. Add the following solutions: 12.5 ml of 8 M urea solution, 1 ml of 5 M NaCl solution, 25 μl of 25 % ammonia solution, and 250 μl of 2 % SDS solution. Mix, check pH (pH: 10–12), and make up to 50 ml with deionized water.
2. Urea buffer for the analysis of CSF (2 M urea, 100 mM NaCl, 0.0125 % NH_4OH; note that no SDS is added): Add around 20 ml of deionized water to a 50 ml graduated glass cylinder. Add the following solutions: 12.5 ml of 8 M urea solution, 1 ml of 5 M NaCl solution, and 25 μl of 25 % ammonia solution. Mix, check pH (pH: 10–12), and make up to 50 ml with deionized water.
3. NH_4-buffer: Add 400 μl of ammonia solution (25 %) into 1,000 ml of deionized water. Mix the solution by inversion and check the pH value. The pH should be between 10.5 and 11.5; if it is too low add a few microliter of ammonia solution (25 %) to adjust the pH of the solution.

3.4. Sample Preparation Urine and CSF

1. Urine and CSF samples are defrosted at room temperature. From time to time samples have to be mixed gently by inversion.
2. While samples are thawing, mix them with PMSF solution in a ratio of 1/1,000 (v/v) (1 μl PMSF solution/1,000 μl of urine sample). Mix samples by inversion.
3. Remove internal filters of the 20 kD centrisart tubes and store them in a dry and clean place.
4. Pipette 0.7 ml of urea buffer in the centrisart tubes and add 0.7 ml of PMSF-treated urine or CSF; mix solutions gently by pipetting.
5. Place the filter unit back in each centrisart tube and centrifuge at 3,400 × *g* at 4°C until 1.1 ml of filtrate is obtained.
6. Maximum storage time of defrosted PMSF-treated samples at room temperature is 1 h. Therefore, remaining samples have to be placed back in the freezer.
7. While samples are being centrifuged, PD10-columns must be opened as specified by the manufacturer and each column should be placed on 50 ml collection tubes with pierced lids.
8. Equilibrate PD10-columns by adding 25 ml of NH_4-buffer. After this step, columns should be kept filled with a few ml of buffer and sealed with their lids until samples are ready.

9. Once urine or CSF samples have been centrifuged, add 1.1 ml of the centrisart filtrate in each dry PD10-column and wait until sample is inside the column.
10. Add 1.9 ml of NH_4-buffer into the column.
11. Once columns are dry, place PD10-columns on 15 ml collection tubes and elute by adding 2 ml of NH_4-buffer.
12. Eluted samples have to be aliquoted: Two aliquots of 0.9 ml each for CE-MS analysis and one aliquot of 0.2 ml for protein concentration determination. Protein concentration is measured by BCA assay (see Note 4). Aliquoted samples are frozen at −80°C and then lyophilized.
13. Lyophilized urine samples can be stored at −20°C for several years.
14. Aliquots for CE-MS analysis are re-suspended in deionized water to a final concentration of 2 μg/μl shortly before analysis.

3.5. Sample Preparation of Plasma

1. Plasma samples are defrosted at room temperature. From time to time samples have to be mixed gently by inversion.
2. Lipids are removed from the plasma by adding a 0.7 ml of *n*-butanol/*iso*propyl ether (40:60) to an equal volume of plasma. The solution is mixed by vortexing and then centrifuged at 10,000 × *g* at 4°C for 10 min. The yellow plasma phase is at the bottom of the tube and is removed for further treatment (13).
3. Remove internal filters of 10 kD MWCO centrisart tubes and store them in a dry and clean place. Note: 20 kD MWCO filters do not completely remove highly abundant proteins.
4. Pipette 0.5 ml of urea buffer in the centrisart tubes and add 0.5 ml of the delipidated plasma and 1 ml of deionized water.
5. Place the filter unit back in each centrisart tube and centrifuge at 3,400 × *g* at 4°C until 1.4 ml of filtrate is obtained.
6. While samples are being centrifuged, PD10-columns must be opened as specified by the manufacturer and each column should be placed on 50 ml collection tubes with pierced lids.
7. Equilibrate PD10-columns by adding 25 ml of NH_4-buffer. After this step, columns should be kept filled with a few ml of buffer and sealed with their lids until samples are ready.
8. Once plasma samples have been centrifuged, add 1.1 ml of the centrisart filtrate in each dry PD10-column and wait until sample is inside the column.
9. Add 1.9 ml of NH_4-buffer into the column.
10. Once columns are dry, place PD10-columns on 15 ml collection tubes and elute by adding 2 ml of NH_4-buffer.

11. Eluted samples have to be aliquoted: Two aliquots of 0.9 ml each for CE-MS analysis and one aliquot of 0.2 ml for protein concentration determination. Protein concentration is measured by BCA assay (see Note 4). Aliquoted samples are frozen at −80°C and then lyophilized.

3.6. Sample Preparation of Urine or CSF if Sample Volume Is Restricted

1. Urine and CSF samples are defrosted at room temperature. From time to time samples have to be mixed gently by inversion.
2. While samples are thawing, mix them with PMSF solution in a ratio of 1/1,000 (v/v) (1 μl PMSF solution/1,000 μl of urine sample). Mix samples by inversion.
3. Remove internal filters of centrisart tubes and store them in a dry and clean place.
4. Pipette 0.15 ml of urea buffer in the centrisart tubes and add 0.15 ml of PMSF-treated urine or CSF; mix solutions gently by pipetting.
5. Place the filter unit back in each centrisart tube and centrifuge at 3,400 × *g* at 4°C until 0.2 ml of filtrate is obtained.
6. While samples are centrifuging Illustra NAP-5 columns must be opened as specified by the manufacturer and each column should be placed on 50 ml collection tubes with pierced lids.
7. Equilibrate Illustra NAP-5 columns by adding 15 ml of NH_4-buffer. After this step, columns should be kept filled with a few ml of buffer and sealed with their lids until urine samples are ready.
8. Once urine or CSF samples have been centrifuged, add 0.2 ml of the centrisart filtrate in each dry Illustra NAP-5 column and wait until sample is inside the column.
9. Add 0.3 ml of NH_4-buffer into the column.
10. Once columns are dry, place Illustra NAP-5 columns on 15 ml collection tubes and elute by adding 0.7 ml of NH_4-buffer.
11. Eluted samples are frozen at −80°C and lyophilized.
12. Lyophilized urine samples can be stored at −20°C for several years.
13. Aliquots for CE-MS analysis are re-suspended in 10 μl deionized water shortly before analysis.

3.7. Solutions and Buffers for CE-MS Analysis

Buffers for CE-MS analysis should be freshly prepared every week.

1. Sodium hydroxide (NaOH): Prepare a solution 1 M in deionized water.
2. Ammonium hydroxide buffer: Add 3.76 ml of ammonium solution (25 %) and adjust volume to 50 ml with deionized water.

3. Running buffer: Add 10 ml of acetonitrile and 472 µl of formic acid and adjust volume to 50 ml with deionized water.
4. Sheath-flow liquid: Add 15 ml of 2-propanol and 200 µl of formic acid and adjust volume to 50 ml with deionized water.
5. Standard protein/peptide solution (0.5 pmol/µL) for the calibration of CE-MS analysis. This solution contains lysozyme (14,303 Da, L4919, Sigma-Aldrich, Dorset, UK), ribonuclease (13,681 Da, R5500, Sigma-Aldrich, Dorset, UK), aprotinin (6,513 Da, A1153, Sigma-Aldrich, Dorset, UK), and 4 synthetic peptides: ELMTGELPYSHINNRDQIIFMVGR (2,832 Da), TGSLPYSHIGSRDQIIFMVGR (2,333 Da), GIVLYELMTGELPYSHIN (2,048 Da), and REVQSKIGY GRQIIS (1,733 Da).

3.8. CE-MS Analysis

1. Install the capillary carefully into the CE. Condition the capillary first with NaOH (pressure 50 psi) for 10 min. After this wash the capillary with NH_4OH solution (50 psi) for 10 and 20 min with running buffer (50 psi).
2. Do not connect the capillary to the MS until after the conditioning procedure has been completed so that NaOH does not reach the MS detector.
3. Each time the capillary is changed the MS should be recalibrated using the tune mix.
4. The Beckmann CE must be set to run in "reverse mode" when connected to the MS system using the external detector adapter (EDA) (see Note 5).
5. Prior to each injection the capillary is rinsed with running buffer (pressure applied 50 psi) for 2 min. The sample (see Note 6) is then injected at a pressure of 2.0 psi for 99 s resulting in a loading volume of 290 nl of sample volume. Separation is carried out with +25 kV at the injection side for 30 min with a capillary set temperature of 35°C. Then, in addition to the +25 kV, pressure is applied (0.1 psi for 1 min, 0.2 psi for 1 min, 0.3 psi for 1 min, 0.4 psi for 1 min, and 0.5 psi for the following 30 min) (see Note 7).
6. Sheath liquid is applied coaxially at a running speed of 0.02 ml/h.
7. After each run the capillary is rinsed with deionized water (50 psi) for 1 min followed by a washing step with NH_4OH solution (50 psi) for 3 min, followed by a flushing step with deionized water (50 psi for 3 min).
8. ESI sprayer is grounded to achieve electric potential zero and the electrospray interface potential is set between –4,000 and –5,000 V.

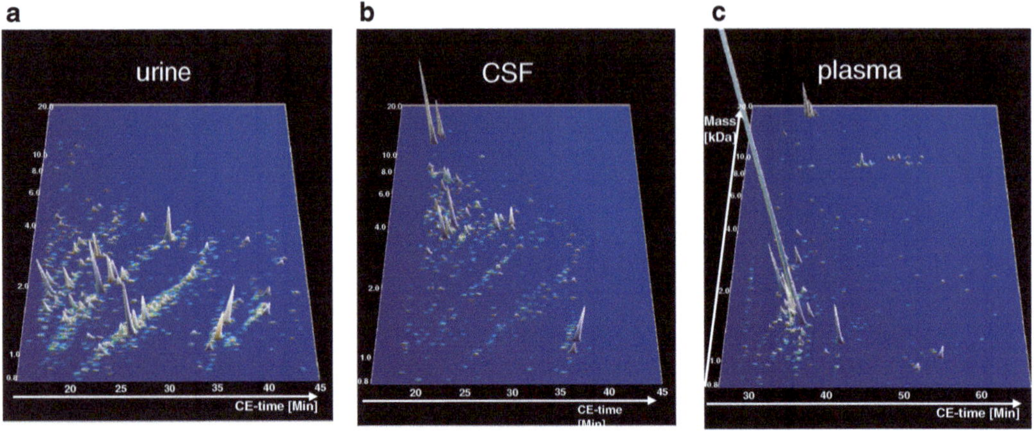

Fig. 1. Deconvoluted 3-D plots of the relevant signals after CE-time normalization and mass calibration of a sample of (**a**) urine, (**b**) cerebrospinal fluid, and (**c**) plasma.

9. MS spectra are accumulated every 3 s, over a mass-to-charge range from 400 to 3,000 m/z for about 60 min, depending on the analysis requirements.
10. Mass calibration of the MS is performed on a weekly basis using the standard protein/peptide solution (0.5 pmol/μl) for CE-MS analysis.

3.9. Data Processing

1. Mass spectral ion peaks are deconvoluted into single masses using MosaiquesVisu software (14) (www.proteomiques.com). Only signals observed in a minimum of 2 consecutive spectra with a signal-to-noise ratio of at least 4 are considered. Signals with a calculated charge of +1 are automatically excluded to minimize interference with matrix compounds or drugs. MosaiquesVisu employs a probabilistic clustering algorithm and uses both isotopic distributions and as conjugated masses for charge-state determination of the entities.
2. TOF-MS data are calibrated utilizing 80 accurately characterized reference masses (mass deviation <0.5 ppm) by FT-ICR-MS applying linear regression.
3. Furthermore, as migration time and ion counts show significant variability between samples mainly due to different salt and peptide concentrations, CE migration time and ion signal intensity are normalized. For normalization, MS signal intensities are normalized relative to a number of internal peptides with small relative standard deviation. For calibration, linear regression is performed (11, 15). The resulting peak list contains the molecular mass (Da) and normalized CE migration time (min) for each feature. Normalized signal intensity can be used as a measure of relative abundance (Fig. 1).

4. Data sets are accepted only if the following quality control criteria are met: A minimum of 950 chromatographic features (mean number of features minus one standard deviation) must be detected with a minimal MS resolution of 8,000 (required resolution to resolve ion signals with $z=6$) in a minimal migration time interval (the time window, in which separated signals can be detected) of 10 min. After calibration, the mean deviation of migration time (compared to reference standards) must be below 0.35 min.

4. Notes

1. It is necessary to acquire an EDA (external detector adapter) kit for the CE. In addition the CE-hardware must be switched to the EDA mode to prevent damage to the power supply.
2. The ESI sprayer tip should be cleaned once a week with methanol in an ultrasonic bath. Once a month, it is advised to disassemble the ESI sprayer and clean the parts separately with methanol in an ultrasonic bath. It is important that the inner steel capillary is placed centrally and aligned with the tip of the ESI sprayer.
3. Prepare all solutions using ultrapure water or deionized water and analytical grade reagents.
4. The BC assay is a colorimetric assay. It involves the reduction of Cu^{++} to Cu^{+} by the peptide bonds of proteins. The BC assay chelates Cu^{+} ions with very high specificity to form a stable, water-soluble purple-colored complex. Absorbance is directly proportional to the protein concentration and is measured at 562 nm.

   A standard curve with BSA is prepared using 1.5 mL Eppendorf micro test tubes. Protein concentration for the calibration curve should range from 1 to 0 mg BSA/ml.

   Aliquoted samples for protein determination are thoroughly resuspended in 50 μL water.

   The required amount of BC assay reagent is prepared by adding 1 part of reagent B (blue solution) to 50 parts of reagent A (clear solution). 0.2 mL of the BC assay reagent is added both to the resuspended samples and to the six standard curve samples. All samples must be thoroughly mixed by vortexing.

   Incubation is performed either for two hours at room temperature or in a heating cabinet at 37°C for 30 min.

   Afterwards the samples are measured at 562 nm. The conversion from absorbance to the protein concentration in the samples is calculated by using the calibration curve constructed from standards of known concentrations.

5. A description of how to create a method is described in the 32 Karat™ Software User's Guide.

   There are some basic settings that all methods have in common:

   - Current max. 20 μA.
   - Cartridge temperature 35°C.
   - Sample storage temperature 5°C.
   - Trigger settings: Wait until cartridge coolant temperature is reached.
   - Analogue output scaling: Factor 1.
   - Inlet trays: Buffer—36 vials; sample—no tray.
   - Outlet trays: Buffer—36 vials; sample—48 vials.

   If running time of the peptides is too short, it is possible to run the samples at a lower cartridge temperature than 35°C.

6. Centrifuging the samples prior to the run for 20 min at 4°C helps to avoid blocking the capillary.
7. If the current of the CE falls towards zero during the analysis, the capillary may be blocked. This can occur if the concentration of peptides in the sample is too high or if air has entered the capillary. To check whether the capillary is in working order, wash the capillary with running buffer and turn on the voltage (CE software) at the same time. If current is available the capillary is in good condition. If there is no current the capillary is most likely broken and should be replaced.

References

1. Fliser D, Novak J, Thongboonkerd V, Argiles A, Jankowski V, Girolami M, Jankowski J, Mischak H (2007) Advances in urinary proteome analysis and biomarker discovery. J Am Soc Nephrol 18:1057–1071
2. Decramer S, de Gonzalez PA, Breuil B, Mischak H, Monsarrat B, Bascands JL, Schanstra JP (2008) Urine in clinical proteomics. Mol Cell Proteomics 7:1850–1862
3. Mischak H, Kolch W, Aivalotis M, Bouyssie D, Court M, Dihazi H, Dihazi GH, Franke J, Garin J, Gonzales de Peredo A, Iphöfer A, Jansch L, Lacroix C, Makridakis M, Masselon C, Metzger J, Monsarrat B, Mrug M, Norling M, Novak J, Pich A, Pitt A, Bongcam-Rudloff E, Siwy J, Suzuki H, Thongboonkerd V, Wang L, Zoidakis J, Zurbig P, Schanstra J, Vlahou A (2010) Comprehensive human urine standards for comparability and standardization in clinical proteome analysis. Proteomics Clin Appl 4:464–478
4. Zürbig P, Renfrow MB, Schiffer E, Novak J, Walden M, Wittke S, Just I, Pelzing M, Neususs C, Theodorescu D, Root C, Ross M, Mischak H (2006) Biomarker discovery by CE-MS enables sequence analysis via MS/MS with platform-independent separation. Electrophoresis 27:2111–2125
5. Tomer KB (2001) Separations Combined with Mass Spectrometry. Chem Rev 101:297–328
6. Neususs C, Pelzing M, Macht M (2002) A robust approach for the analysis of peptides in the low femtomole range by capillary electrophoresis-tandem mass spectrometry. Electrophoresis 23:3149–3159
7. Schmitt-Kopplin P, Frommberger M (2003) Capillary electrophoresis-mass spectrometry:

15 years of developments and applications. Electrophoresis 24:3837–3867

8. Tu C, Rudnick PA, Martinez MY, Cheek KL, Stein SE, Slebos RJ, Liebler DC (2010) Depletion of abundant plasma proteins and limitations of plasma proteomics. J Proteome Res 9:4982–4991
9. Roche S, Gabelle A, Lehmann S (2009) Clinical proteomics of the cerebrospinal fluid: towards the discovery of new biomarkers. Proteomics Clin Appl 2:428–436
10. Weissinger EM, Wittke S, Kaiser T, Haller H, Bartel S, Krebs R, Golovko I, Rupprecht HD, Haubitz M, Hecker H, Mischak H, Fliser D (2004) Proteomic patterns established with capillary electrophoresis and mass spectrometry for diagnostic purposes. Kidney Int 65:2426–2434
11. Jantos-Siwy J, Schiffer E, Brand K, Schumann G, Rossing K, Delles C, Mischak H, Metzger J (2009) Quantitative Urinary Proteome Analysis for Biomarker Evaluation in Chronic Kidney Disease. J Proteome Res 8:268–281
12. Lewczuk P, Kornhuber J, Wiltfang J (2006) The German competence net dementias: standard operating procedures for the neurochemical dementia diagnostics. J Neural Transm 113: 1075–1080
13. Cham BE, Knowles BR (1976) A solvent system for delipidation of plasma or serum without protein precipitation *J.* Lipid Res 17: 176–181
14. Neuhoff N, Kaiser T, Wittke S, Krebs R, Pitt A, Burchard A, Sundmacher A, Schlegelberger B, Kolch W, Mischak H (2004) Mass spectrometry for the detection of differentially expressed proteins: a comparison of surface-enhanced laser desorption/ionization and capillary electrophoresis/mass spectrometry. Rapid Commun Mass Spectrom 18:149–156
15. Theodorescu D, Wittke S, Ross MM, Walden M, Conaway M, Just I, Mischak H, Frierson HF (2006) Discovery and validation of new protein biomarkers for urothelial cancer: a prospective analysis. Lancet Oncol 7:230–240

Chapter 13

Contribution of CE to the Analysis of Protein or Peptide Biomarkers

Kiarach Mesbah, Romain Verpillot, François de l'Escaille, Jean Bernard Falmagne, and Myriam Taverna

Abstract

Biomarker analysis is pivotal for disease diagnosis and one important class of biomarkers is constituted by proteins and peptides. This review focuses on protein and peptide analyses from biological fluids performed by capillary electrophoresis. The various strategies that have been reported to prevent difficulties due to the handling of real samples are described. Innovative techniques to overcome the complexity of the sample, to prevent the adsorption of the analytes on the inner capillary wall, and to increase the sensibility of the analysis are summarized and illustrated by different applications. To fully illustrate the contribution of CE to the analysis of biomarkers from human sample, two detailed protocols are given: the analysis from CSF of five amyloid peptide, biomarkers of the Alzheimer disease, and the analysis of sialoforms of transferrin from human serum.

Key words: Capillary electrophoresis, Biomarkers, Coating, Preconcentration, Biological fluids

1. Introduction

With the extensive work performed in the field of biomarker discovery, the area of diagnosis methods and tools has witnessed a considerable development in the past decade and represents currently a pivotal role in the medical and pharmaceutical environment. According to the definition given by the working group formed by the National Institutes of Health (1), a biomarker is "a characteristic that is objectively measured and evaluated as an indicator of normal biologic processes, pathogenic processes, or pharmacologic responses to a therapeutic intervention". In this context biomarkers may be anatomic, physiologic, biochemical, and molecular biological parameters. The most commonly recognized molecular biomarkers fall into different molecule classes among which are the proteins and peptides, nucleic acids (DNA, RNA),

Nicola Volpi and Francesca Maccari (eds.), *Capillary Electrophoresis of Biomolecules: Methods and Protocols*, Methods in Molecular Biology, vol. 984, DOI 10.1007/978-1-62703-296-4_13, © Springer Science+Business Media, LLC 2013

single-nucleotide polymorphism (SNP), and dissolved ions, gases, and metabolites. One of the best-known metabolite is probably glucose, enabling the diagnosis and the monitoring of diabetes mellitus.

These biochemical biomarkers can be detected in tissues or biological fluids such as blood, cerebrospinal fluid (CSF), urine, saliva, and tears. The selection of the biological fluid for a diagnosis purpose is mainly dictated by the physiopathological process of the disease. Indeed the concentration of the biomarkers in the selected body sample should reflect as much as possible the disease state of the patient. But other criteria are the accessibility of the fluids, their complexity, and the concentration of the target analyzed inside. In this chapter, we will consider specifically protein or peptide-based biomarkers which are common targets that require particular attention for the diagnosis method development. We will also focus on the different strategies already reported for their analysis by capillary electrophoresis (CE) in the context of diagnostic test development performed on body fluids.

CE offers many advantages for the detection or quantitation of biomarkers such as high efficiency and resolution, small sample volume required, and possibility to achieve high sensitivity when coupled with laser-induced fluorescence detection (LIF) or mass spectrometry (MS). When dealing with biomarker analysis by CE, we encounter several difficulties linked first to the complexity of the sample matrix and second to their low concentration. Blood is, for instance, a biological fluid that contains many proteins and at concentrations that may be very different. This raises problems not only of interfering substances that deteriorates the separation but also of possible interaction between the targeted analytes and the plasma proteins altering the assay accuracy. This complexity of biological fluids, generally combined to the very low abundance of the investigated biomarkers with regard to other molecules or cells present in that matrix, is one of the main issues we have to deal with. In addition, proteins and peptides tend to interact with the surface of the silica capillary. In order to maintain high resolution and sensitivity, different methods to avoid this kind of interaction must be envisaged in addition to the initial CE separation system.

Once parameters, allowing the analysis of the standard biomarkers by capillary electrophoresis, have been defined and optimized, the amount of biomarkers in the considered biological sample has to be compared to the method sensitivity. In most cases, the very low level of the target in the biological fluid does not allow for relevant detection in the real sample. Complementary techniques have to be applied to increase the sensitivity which can fall into two categories: those relying on preconcentration or enrichment steps and those targeting rather the detection sensitivity of the method.

1.1. State of the Art on CE Strategies to Analyze Biomarkers in Biological Samples

1.1.1. Strategies to Address Complexity of Matrix

Biological materials that contain biomarkers consist of tissues or aqueous fluids which contain many potential interfering substances that may have similar behavior during the analysis process. In most cases, a sample preparation step has to be performed before the CE analysis in order to quantify the analytes of interest in the complex biological matrix. The sample preparation step may be needed not only to eliminate the non-interesting compounds but also to increase the biomarker concentration.

In this part, our aim is to give an overview of innovative methods that are able to prepare samples for CE analysis in order to focus on biomarkers present in biological samples and to eliminate other highly concentrated components. Strategies consisting in increasing the detectability of biomarkers will be discussed in the third part of this chapter.

Body fluids commonly used for protein analysis are plasma, serum, cerebrospinal fluid, and urine, but the noninvasive approach of using saliva or tear fluid is very attractive. Many biochemical constituents present in blood are also reflected in tears, although at lower concentrations (2). This matrix has, from the analytical point of view, a strong advantage due to the relative lack of impurities compared to other biological fluids. Tears could be compared to an aqueous sample and a CE-UV separation of two proteins, lysozyme and lactoferrin (two biomarkers of Sjögren's syndrome and dry eyes disease), was achieved without any sample treatment (3, 4). The separation buffer was acidic and was combined to a cationic coating to prevent nonspecific adsorption (4). However, protein analysis from more complex matrices usually needs sample cleanup. Many analytical tools have been proposed for the pretreatment method to extract, isolate, fractionate, filtrate, or concentrate the biological target from these matrices. Methods for eliminating interfering substances can be divided in at least four classes: first one regrouping the technique based on the affinity for a solid phase, such as particle phases and monoliths; second composed by immunoaffinity supports with antibodies; third consisting of membrane-based techniques such as dialysis or filtration; and last one consisting of the liquid–liquid-based extraction techniques such as centrifugation or precipitation. Whatever the method selected, an extensive knowledge on the physicochemical characteristics of the biomarkers is essential to select the best strategy. The main information needed is the size or molecular mass, isoelectric point, hydrophobicity, solubility, and binding affinity to antibody or other ligands.

When protein or peptide extraction is needed from a biological matrix, the characteristic of the sample is critical. For instance, high-viscosity samples should be either diluted or at least homogenized. Samples that contain particles need microfiltration. High salt concentration samples can be handled by various off-line techniques such as fast-protein liquid chromatography (5), desalting columns, ultrafiltration (6), or dialysis (7) to be compatible with the subsequent CE analysis. Extraction and precipitation are considered to be harsh sample preparation techniques and therefore the functionality of the target protein or peptide is often altered. In addition they cannot easily be implemented to the capillary separation and remain off-line sample pretreatments.

Solid extraction phases are mainly composed of silica-based and polymer sorbents with reverse-phase, ion-exchange stationary phase, monolith sorbent, magnetic beads. Many commercial solid phases are now available. One major advantage of these supports is their possible on-line integration to the fused silica capillary inlet end to lead to an integrated two-step analysis of complex matrix. Immunoaffinity capillary electrophoresis (IACE) has been first reported in 1990 by Guzman et al. (8, 9). Since, many applications to biomarkers in different biological fluids have been published (10, 11). Most of the time immunoaffinity methods are developed using the antigen–antibody interaction, one of the most specific molecular recognition processes. Other kind of ligands may be used as proteins, aptamers, phages, or peptides. The immune-affinity ligands are immobilized on beads, polymeric matrix, gels, or directly onto the inner surface of the channel wall.

Solid-phase extraction (SPE) microcartridges packed inside the capillary, containing successively different sorbents, have been tested for the analysis of opioid peptides such as Met-enkephalin, endomorphin, or dynorphin in human plasma. The analysis was based on the combination of different techniques: first a precipitation with acetonitrile followed by a centrifugation and a filtration of the serum, prior to the CE determination of peptides. In this work a C18 stationary phase was placed into the capillary inlet, permitting an on-line preconcentration of the peptides. The preliminary sample pretreatments were found necessary to avoid the saturation of this microcartridge. The results were quite impressive with LOD of 100 ng/ml for the analyzed neuropeptides in biological fluids using UV detection (12). A lab-made support has also been described for the removal of interfering components and preconcentration of low-abundant proteins. A butyl methacrylate-*co*-ethylene dimethacrylate (BuMA-*co*-EDMA) monolith was employed for on-line cleanup of proteins followed by capillary electrophoresis separation. The monolith was synthesized, over a 1 cm section, by UV-initiated polymerization at the inlet end of a 75 μm diameter fused silica capillary that had been previously coated with poly(vinyl) alcohol. For the analysis of standard proteins and polypeptide biomarkers (lysozyme and trypsinogen A for acute pancreatitis and cystic fibrosis), this system provided reproducible migration times and peak areas with RSD less than 5%. The BuMA-*co*-EDMA monolithic preconcentrator CE was coupled to a protein G monolithic column via an Upchurch zero dead volume union for on-line removal of IgG. Although not applied to real samples, this system could lead to many applications on serum samples (13).

Sample preparation of urine for the discovery of polypeptide biomarkers was carried out by the centrifugation of the sample, then an off-line purification using Pharmacia C2-column, to remove urea, electrolytes, salts, and other interfering components, thus to decrease matrix effects and enrich the target. After lyophilization to eliminate the elution solution, the sample was rediluted in water prior to the analysis. This method allowed a CE-MS profiling of different proteins in order to compare healthy people and patients with diseases to discover new biomarkers (14).

In a recent work, a purification of alpha 1-acid glycoprotein (AGP) and its isoforms from human serum for CZE analysis using an immune-affinity column has been demonstrated. Different existing methods for AGP purification were compared. The house-made anti-AGP column showed the best results. The natural serum samples were first precipitated in acidic conditions (near AGP pI) to eliminate the less water-soluble proteins. The solution was then injected in the immuno-chromatographic column, and fractions were collected after elution with glycine–HCl pH 2.2 buffer. This quite simple method permitted to obtain a purified AGP sample from human serum for a CE-UV analysis (15). A different approach which still relies on Ab–Ag complex is the noncompetitive assay which has been successfully applied to CE with electrochemical detection for the simultaneous separation of three tumor biomarkers, PSA, CEA, and hCG, and for the quantification of alpha-fetoprotein (AFP), a serum biomarker of hepatocellular carcinoma. The method consisted in incubating the human serum sample containing the target antigen with an excess of labeled Ab. The Ag–Ab complex was then separated from the non-complexed Ab. A calibration method relating the free species and the amperometric signal permitted an evaluation of the quantity of the different proteins. This system was able to detect PSA, hCG, and CEA at, respectively, 19.74, 21.54, and 14.30 ng/ml in human serum sample with a recovery of more than 90%. The same system used for the detection of AFP led to a linear range and a detection limit (S/N = 3), respectively, from 1.5 to 66.6 ng/mL and 0.48 ng/mL (16) (Fig. 1).

Determination of neuropeptides in urine was carried out by IACE. The system contained Fab fragments derived from a polyclonal antibody against the peptide. The fragments were immobilized on glass beads. The cruciform microreactor was composed of two inlets and two outlets. Four frit structures allowed the path of fluid but confined the beads. The reactor was adapted to the capillary inlet and placed in the CE instrument (Fig. 2). Samples and buffers were transported through the transport tube by mechanical pressure using a syringe, or controlled vacuum, as showed in Fig. 2a. The sample was injected directly in the immobilized affinity

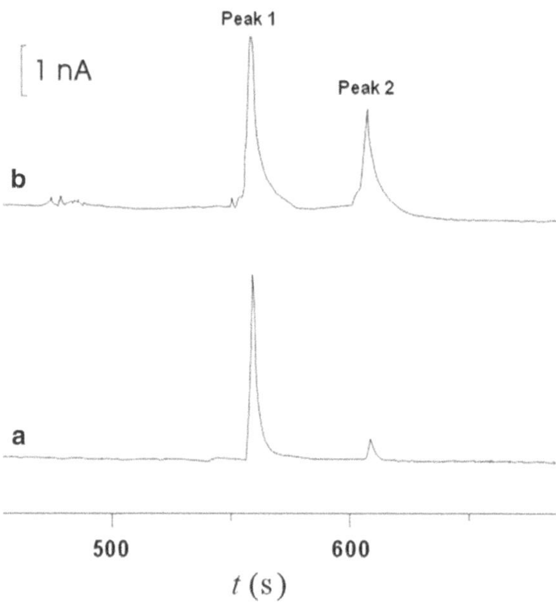

Fig. 1. Illustration of CE-based enzyme immunoassay represented by typical electropherograms of alpha-fetoprotein for normal (**a**) and hepatitis-infected (**b**) human serum samples. Peaks 1 and 2 correspond to anti-AFP–HRP and AFP–anti-AFP–HRP, respectively. CE conditions: BGE, 1.0 mM H_2O_2 and 10 mM Britton–Robinson buffer (BR, pH 5.0); separation voltage, 15 kV (Adapted from ref. 16).

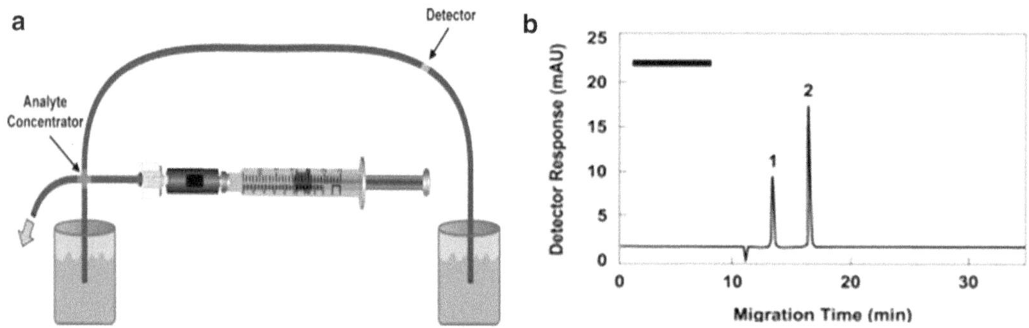

Fig. 2. Schematic representation of a typical homemade IACE module, containing the analyte concentrator–microreactor device near the inlet of the capillary, adapted to a CE system (**a**). Application to the analysis of two neuropeptides extracted from a urine (**b**). Peaks: Angiotensin II (1) and neurotensin (2). CE conditions: BGE, 50 mM sodium tetraborate buffer, pH 8.5; UV detection at 214 nm; 25°C; 26 kV (17).

ligand module, the peptide was retained, and after a few washes and conditioning with appropriate buffers, the peptide was eluted with a small plug of 0.3 M glycine–HCl buffer, pH 3.4, or neutral pH buffers containing acetonitrile or other organic solvents. The CE separation was performed with 50 mM sodium tetraborate as running buffer, at pH 8.5. A representative electropherogram is shown in Fig. 2b. The urine sample used in this experiment was diluted 1:1 v/v with 50 mM sodium tetraborate buffer, pH 9.0

and spiked with angiotensin II and neurotensin at 5 ng/mL each. The concentration limit of detection, under the experimental conditions used for angiotensin II and neurotensin, was 0.5 ng/mL (17). Another example of immune-based diagnosis is the method developed by Analis, for the quantification of transferrins in human serum by removing interfering substances via affinity column purification. This technique will be fully developed in the protocol section.

The immuno-based CE separation technique is widely used in clinical laboratories; readers are advised to read an interesting review detailing many applications in this domain (18).

1.1.2. Strategies to Address Biomarkers Adsorption to the Capillary

The capillary coating takes an important place in the success of biomarker analysis by CE. Indeed the high propensity of proteins to be adsorbed on the fused silica capillary surface can easily prevent the electrophoretic migration of low-abundant biomarkers and so their detection. It can also deteriorate the resolution or lead to irreproducible separation. The adsorption process is mainly due to electrostatic interactions between silica and oppositely charged areas of the protein but it can also involve at different stages, hydrogen bondings, dipole–dipole interactions, van der Waals forces, and hydrophobic effects (19–23). A cooperative effect of the adsorption between protein molecules already adsorbed and new analyzed ones has been also described by Righetti et al. (24). The amino acid sequence, conformation, structural stability, charge, and size of the protein determine its adsorption tendency (25). The extent of protein adsorption can also be related to the protein environment and medium. For instance, on hydrophilic surfaces, adsorption is higher when pH is close to the protein pI. High ionic strengths reduce adsorption both at pH higher and lower than protein pI (26).

Strategies consisting of carefully optimizing the rinsing steps or the separation electrolyte are often not enough to eliminate durably this kind of interaction. To prevent protein adsorption, the best strategy consists in surface modification of the fused silica wall of the capillary with either dynamic or static coatings. Coatings also contribute to tune the electroosmotic flow in order to obtain efficient and high-resolution separation of proteins or peptides. For dynamic modification, the compounds which interact with the silica surface are added to the background electrolyte (BGE). The main compounds used for dynamic modification of capillaries are monoamines, alkylamines, or polymers (neutral or cationic mainly) and detergents.

Static coatings can be physically either adsorbed for a nonpermanent or covalently bounded for a permanent coating. In general, higher efficiencies are achieved using static adsorbed polymer coatings. The preparation of many covalent coatings is time consuming and these coated capillaries often exhibit low stability at extreme pH. That is why recent researches have mainly focused on physically adsorbed coatings. Physically adsorbed neutral or positively charged polymers have many advantages such as simplicity, rapidity of coating formation, and possibility of regeneration. Detailed reviews are devoted to capillary coating and summarize the different polymers that can be applied to this purpose. Among these reviews the one of Lucy et al. (27) and that of Stutz (28) are very comprehensive, giving the different strategies that can be employed.

More recently, successive multiple ionic polymer layer (SMIL) coatings have been introduced as a physically adhered innovative wall coating. In SMIL, layers of counter-charged polyelectrolytes (PE) are alternately attached onto the capillary surface by successive rinsing steps with appropriate polyanionic and polycationic polymers. Adsorption and interaction between PE layers occur via Coulomb interactions and ion–dipole bonds (29).

Sialoforms of transferrin (Tf) or carbohydrate-deficient Tf (CDT) both in standard solutions and serum samples have been successfully separated by CE using the 1,4-diaminobutane (DAB) as a dynamic coating (30, 31). For example, DAB may be used not only for reducing protein adsorption but also to modify the EOF in order to perform high-resolution separation. Thormann's group also compared two amine modifiers, DAB and spermine, as dynamic coatings for analyzing CDT in human serum. Having 3 mM DAB or 0.02 mM spermine in a borate-based running buffer at pH 8.3 provided data of remarkable similarity as well as similar percent area of disialo-Tf related to the tetrasialo-Tf. However, DAB was clearly superior to spermine in terms of method reproducibility (31).

Grafted copolymer hydroxyethylcellulose-graft-polydimethylacrylamide, a novel multifunctional separation medium, has recently been evaluated as a suitable coating suppressing EOF and allowing separation of proteins. Appropriate proportion of each polymer was the key to obtain an efficient coating in terms of separation. The disadvantages of each polymer used alone were inhibited by this grafting. This innovative approach has been successfully applied to the analysis of human saliva, showing a high separation efficiency, good stability, and excellent repeatability for the separation of lysozyme by CE-UV (32).

Sassi A. et al. (33) reported a sheathless CE-ESI-MS system that could be used to analyze human serum samples. This method used a covalently linked hydrophilic, positively charged coating on the inner surface of the fused silica capillary for the separation of protein and peptide biomarker from biological fluids. The electroosmotic flow (EOF) was found to vary by less than 5% from batch to batch. This coating, compatible with ESI-MS, allowed detection of low concentration of different species in the approximate range of 10–100 nM in serum, and this for approximately 500 serum components. The same group (34) developed well-defined

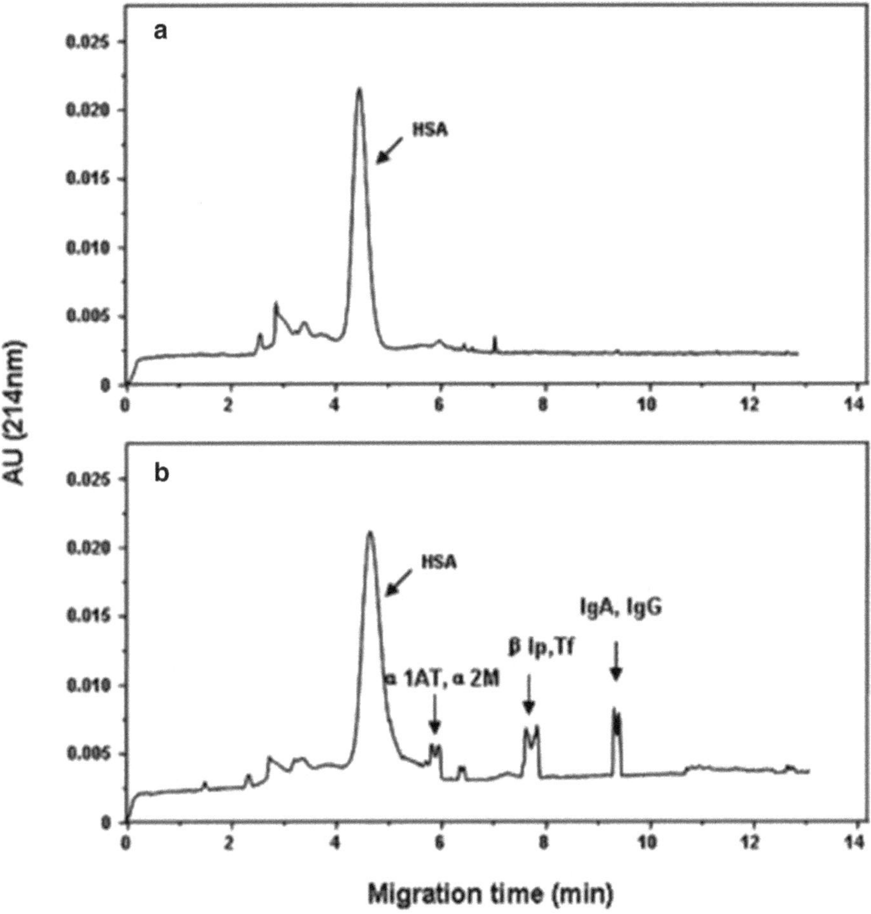

Fig. 3. Electropherograms illustrating the interest of an innovative coating for plasma analysis. Separation was performed on (**a**) a bare fused silica capillary and (**b**) a PEO113-*b*-P4VP294-coated capillary. CE condition: Separation buffer, 19 mM NaOH–$Na_2B_4O_7$ at pH 9.7; 500 V/cm; 40 cm capillary (30 cm to the detector); temperature 25°C; UV detection (34).

di-block copolymers of PEO-b-P4VP with different molecular weight of the P4VP block synthesized by atom transfer radical polymerization. In addition to its separation performance, this copolymer has a really promising potential for real sample application and diagnosis purposes. This potential was showed by comparing this coating to fused silica capillaries for the analysis of human serum. While the non-coated capillary led to one main peak attributed to HSA, the coated capillary allowed detection of many proteins such as transferrin or IgG. Furthermore, RSD for migration time of HSA was extremely low (less than 1%) (Fig. 3).

A very fast, simple, and stable coating, based on a new polycationic polymer, poly-LA 313, showed high tolerance against complex biological samples (plasma and CSF). With RSD of EOF less than 3%, poly-LA 313 appeared as a valuable coating for clinical analysis. This cationic polymer has been successfully used to perform

CE-ESI-MS of HSA in body fluids which confirmed its potential for biomarkers (35).

A SMIL based on polybrene and sodium dextran sulfate alternation led to high-durability coated capillary with time-efficient preparation and showed an excellent compatibility with MS. In addition, this coating procedure provided a stable EOF over a wide pH range 2–13. A strong reliability and chemical stability have been demonstrated for this SMIL (36).

Plenty of strategies are available to decrease nonspecific adsorption and to tune EOF for a higher resolution separation. The real challenge is to choose the most adapted one. The target is to render the method reproducible while keeping in mind that the target protein biomarker may not significantly adsorb to the capillary wall (being either coated or uncoated) but that the plasma proteins present in the sample may do, which will affect the overall reproducibility of the results.

1.1.3. Strategies to Increase Detection Sensitivity

For many diseases, early diagnosis is required to fight efficiently against the pathology progression. At these early stages, most often the biomarkers are not present at high concentrations in body fluids. It becomes therefore a real challenge to detect and to quantify them in CSF, blood, or urine. Sensitivity limitation encountered with CE analyses can be overcome through different methods.

Apart from the extraction of the biomarkers from the sample, as described in paragraph 2, one solution is to decrease the limit of detection by using more sensitive detections than UV one. This generally is performed by labeling the biomarkers with fluorescent dyes or by attaching covalently electrochemical moieties to them in order to use either laser-induced fluorescence detector or electrochemistry ones. However, only fluorescence, LIF, UV, and DAD are currently commercially available for CE. Mass spectrometry coupled to CE was extensively applied to protein and peptide analysis and biomarker discovery over the last 5 years and can not only add complementary information to CE but also lead in some cases to increased sensitivity (37). However, the MS needs specific conditions in terms of buffer, coating, voltage, and sample matrices. One major and widely used technique to increase the sensitivity of CE is the laser-induced fluorescence detection mode. An exhaustive review focuses on analyses of biological samples by CE–LIF (38). This mode permits lower limit of detection compared to UV detection but presents its own limitations mainly due to the optimization of the labeling method. Indeed, the derivatization of proteins and peptides, which can be done off-line or on-line the CE separation and can be applied to biological samples, has to be optimized in order to label quantitatively the target biomarker in a reproducible manner. In addition, derivatization may give rise to many species if the dye can potentially be attached to different sites in the protein. In fact, most of fluorescent dyes react directly with either amino acids containing primary or secondary amino groups or thiol functions and less often to carboxyl groups. As peptides or proteins may have several groups available for labeling, the same biomarker derivatization will result in several tagged forms with different degrees of derivatization in addition to the side products generated by the reaction itself.

Derivatization of biomarkers present in a biological sample is most of the time quite problematic. The main problem is to develop a separation method with sufficient resolution to separate the peaks of interest from the potential biological interfering substances. Plenty of fluorescent dyes are commercially available; their properties and their reactivity toward the biomarkers must be well studied to obtain the optimal performances. One option to circumvent the direct derivatization of the analyte, keeping the high sensitivity afforded by electroluminescence, fluorescence, or chimioluminescence methods, is to combine CE separation and immuno- or enzyme bioassays. Different strategies have been developed and applied to biological samples such as competitive immunoassays, noncompetitive immunoassays, or enzyme assays. All immunoassay methods are detailed in a recent review covering the analysis of biological samples by CE–LIF (38).

Another strategy to increase sensitivity is to preconcentrate the sample for a more sensitive detection using stacking methods which exploit electrokinetic processes performed in CE and permit the compression of the injected sample plug into a narrow band after the injection, leading to a high concentration zone. Most of stacking processes are based on differences between the sample medium and the BGE. Several comprehensive reviews summarize the advances of preconcentration techniques based on stacking in CE. Among these techniques we can find different categories: field-amplified sample stacking (FASS), field-amplified sample injection (FASI), isotachophoresis (ITP), large-volume sample stacking (LVSS), dynamic pH junction, and sweeping. Among the electrokinetically based methods, FASS, LVSS, and FASI are the most common and are based on the difference in conductivity between the sample and the BGE. The main limitation of those techniques is the high conductivity of biological samples which are not compatible with these approaches. However combined to an adapted sample preparation, some applications from real matrices are reported (39, 40). Isotachophoresis is particularly useful to preconcentrate biological samples which have a high salt content. This technique is based on the difference of conductivity of two buffers: one containing the leading ion and the second the terminating ion. The sample is injected between the two buffers. The electrophoretic mobility of the analyte must be higher than that of the terminating ion and lower than the leading ion. This concentration technique leads to assembling specific molecules into small focused zones. Concentration factors approaching 500-fold have been already reported for proteins (41). Dynamic pH junction is also well adapted to stack samples containing peptides or proteins because the concentration mechanism is based on changes in ionization states of the analytes or electrophoretic velocities between the sample zone and the separation buffer using pH variations.

Several applications applied to biological samples and relying on a CE-based competitive immunoassay have been already reported. A fast, selective, and reproducible analysis of prion protein in blood, a biomarker for spongiform encephalopathy, has been reported by Yang et al. (42). A sample pretreatment of the sheep blood was needed to perform the detection of prion protein. This included centrifugation, cells lysis, liquid–liquid extraction, and solid-phase extraction among other techniques. Two steps were needed. First a labeled derivative of the prion protein was incubated with its specific antibody, and then the resulting free-labeled protein and immuno-complex were separated by CE. In the next step, labeled derivative protein and antibody were added to the blood sample containing the prion protein. The competition between labeled and non-labeled biomarker was observed by the decrease of the peak corresponding to the immuno-complex. This decrease of peak area was correlated to the concentration of prion protein in the blood sample. The separation was achieved at pH 8.8 using TAPS (25 mM IS) as running buffer containing 0.6% of carboxymethyl-β-cyclodextrin to suppress the adsorption which allowed a rapid and high-performance separation. The detection limit of the assay was estimated to be about 80 ng/mL.

In another work, the methionine enkephalin (ME) (a potential biomarker of liver disease, cirrhosis, or cancer) was quantified by a similar CE-based competitive immunoassay. ME concentration was measured in human plasma using a specific antibody and a labeled ME. With the optimized separation conditions, it was possible to separate the antibody bound to ME and free fluorescein conjugated ME by a capillary electrophoresis–laser-induced fluorescence (CE–LIF) analysis using an uncoated fused-silica capillary. The assay specificity, selectivity, and accuracy were excellent with RSD less than 1% for migration time of the two species. The LOD was not reported but a comparison of ME levels in the plasma of healthy and cancer patients was possible with detection level in the nanomolar range (43).

A fast and efficient separation of phosphorylated and non-phosphorylated forms of ERK has been developed based on an enzyme assay (45). The principle was based on the incubation of an excess of labeled substrate and the target enzyme. After the required time, the product was separated from the non-reacted substrate. The peak area corresponding to the product of the reaction was correlated to the amount of enzyme present in the sample.

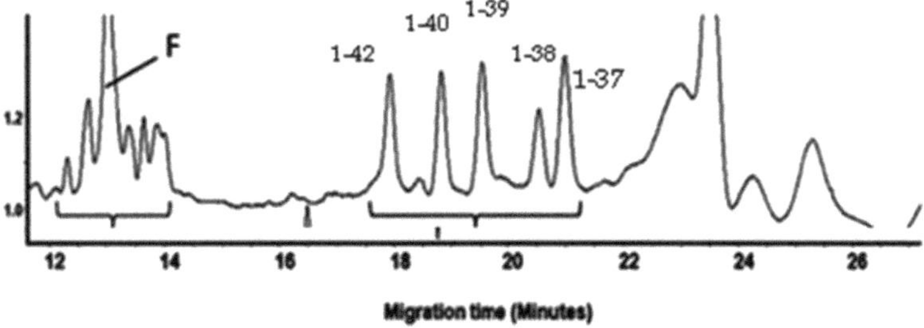

Fig. 4. CE–LIF of amyloid peptides, which are potential biomarkers of Alzheimer disease, after their labeling with Fluoprobe 488. CE conditions: BGE, borate buffer pH 9; IS 40 mM with 3.25 mM of DAB; detection: LIF (λ exc 488 nm) (Adapted from ref. 44).

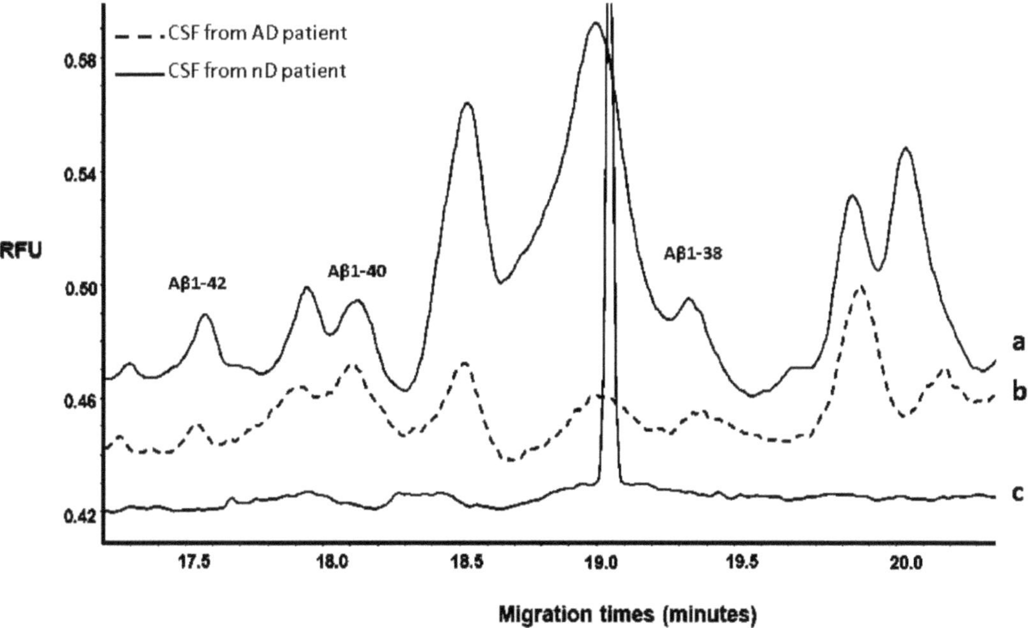

Fig. 5. Application of CE–LIF condition to CSF samples for the detection of Alzheimer biomarkers. Detection of Aβ peptides 1–42, 1–40, and 1–38 by CE–LIF in CSF from a non-demented subject (nD) (**a**) and one patient suffering AD. (**b**) Blank of immunocapture (**c**). CE conditions as in Fig. 5 (44).

In a recent study different fluorophores for the labeling of 5 beta-amyloid peptides that are potential biomarkers of Alzheimer disease have been compared in order to analyze them simultaneously by CE–LIF. The Fluoprobe 488 NHS gave the best results, and LOD close to 35 nM could be obtained using a LIF detector at 488 nm (Fig. 4). The method was applied to the separation and quantification of three of these peptides from CSF samples. To achieve the sensitivity required to detect these peptides in CSF, an off-line immuno-capture was performed prior to the CE analysis (44) (Fig. 5).

An application of LVSS was showed by Chen et al. (46) for CE–LIF detection of bradykinin-related peptides, such as bradykinin (BK), kallidin, and neurokinin A. Detection limit at the pmol/L level was obtained for BK-related peptides. In this case, in order to achieve the sample stacking without sacrificing the high resolution of CE, the sample matrix had to be removed. To this purpose, a negative voltage combined to a back pressure was applied after the sample injection, to keep the analytes into the capillary while removing the sample matrix. The method was applied to the detection of bradykinin and kallidin in human saliva and the detection of neurokinin in CSF. The analytes were successfully detected at the nanomolar level. In another work, combined with CE–LIF, tITP allowed the detection of different bradykinin fragments at a picomolar concentration. The tITP–CZE method was applied to assay the resulting peptides from BK in real human saliva and plasma samples with satisfactory results. RSD obtained for both saliva and human plasma was less than 5% and peptides were detected at concentration around 100 picomolar (41).

2. Materials

2.1. Analysis of Amyloid-β Peptides in Cerebrospinal Fluid Samples by Capillary Electrophoresis Coupled with LIF Detection (45)

2.1.1. Chemicals, Reagents, and Samples

Buffers: (see Notes 1 and 2)

- RIPA: Prepare 50 mM tris(hydroxymethyl)aminomethane (Tris)–HCl, pH 7.2 including 150 mM NaCl, 1% Triton X-100, 0.1% SDS, 1 mM sodium orthovanadate (Na_3VO_4), 1 mM phenylmethanesulfonyl fluoride (PMSF), 1 mM ethylenediaminetetraacetic acid (EDTA), 1 mM protease inhibitor cocktail (Roche, Basel, Switzerland).
- Prepare 80 mM borate in ddH_2O and adjust the pH at 10.5 using NaOH.
- Prepare 50 mM Tris in ddH_2O and adjust the pH at 7.4 using HCl.
- Running buffer: Prepare 40 mM borate buffer and adjust the pH at 9 using NaOH, then add 3.25 mM of DAB.
- Dimethyl sulfoxide (DMSO) 99.9% purity for analysis and diaminobutane chloride (DAB) were obtained from Sigma (St. Louis, MO, United States).
- The Fluoprobe 488 NHS was obtained from Interchim (Montlucon, France).
- Magnetic microparticles Dynabeads M-280 and monoclonal antibody anti-Aβ 6E10 were obtained respectively from Invitrogen (Grand Island, NY) and from Covance (Emeryville, CA).
- The CSF samples analyzed in this study were obtained from patients attending the outpatient memory clinic in Ulm (see Note 3).

Apparatus and Material

- Uncoated capillaries were obtained from Phymep (Paris, France).
- Beckman Coulter PA 800 ProteomLab equipped with LIF detector.
- Data acquisition and instrument control were carried out using Karat 7.0 software.
- Deionized water used in all experiments was purified using a Direct-Q 3 UV purification system (Millipore, Milford, MA, USA).
- pH of buffer solutions was adjusted with inoLab WTW series pH 730 pH meter.
- Buffer ionic strength (IS) calculations were performed using the Phoebus program (Analis, Suarlée, Belgium).
- Magnetic separator.

2.2. Preparation of Affinity Column for Purification of Transferrin from Serum

2.2.1. Buffer and Solutions

- Coupling buffer: 0.1 M carbonate buffer 0.5 M NaCl.

  In a beaker, weight 3.02 g of $NaHCO_3$, 0.4 g of Na_2CO_3, and 5.84 g of NaCl and add bi-distilled water up to 200 ml.
- Stopping buffer: 0.5 M ethanolamine and 0.5 M NaCl.

  In a beaker, weight 12.22 g of ethanolamine and 5.84 g of NaCl and add bi-distilled water up to 150 ml. Then adjust the pH with HCl 32% until pH 8 is obtained. Add bi-distilled water until 200 ml is reached (see Note 4).
- Dilution buffer: 10 mM Tris buffer 0.15 M NaCl.

  In a beaker, weight 0.268 g of Tris/HCl, 0.036 g of Tris, and 1.763 g of NaCl and add bi-distilled water up to 200 ml.
- Stripping buffer: 0.10 mM glycine–HCl buffer pH 2.5.

  In a beaker, weight 7.51 g of glycine and add bi-distilled water up to 150 ml. Then adjust the pH with HCl 32% until pH 2.5 is obtained. Add bi-distilled water until 200 ml is reached (see Note 4).
- Neutralization buffer: 1 M Tris/HCl buffer pH 8.

  In a beaker, weight 24.23 g of Tris and add bi-distilled water up to 150 ml. Then adjust the pH with HCl 32% until pH 8 is obtained. Add bi-distilled water until 200 ml is reached (see Notes 4 and 5).
- Column storage buffer: 10 mM Tris buffer 0.15 M NaCl 0.1% NaN_3 (TBS-Az).

  In a beaker, weight 0.877 g of Tris/HCl, 0.533 g of Tris, 1.763 g of NaCl, 0.2 g of sodium azide and add bi-distilled water up to 200 ml (see Notes 6 and 7).
- PBS buffer: Phosphate buffer saline 0.1% azide.

In a beaker, weight 8 g of NaCl, 0.149 g of KCl, 1.424 g of $Na_2HPO_4.H_2O$, 0.272 g KH_2PO_4, 1 g NaN_3 and add bi-distilled water up to 1 L.

- HCl cold solution: 1 mM HCl kept at 4°C.

  In a beaker, weight 2.3 g of concentrated HCl 32% and add bi-distilled water up to 200 ml (0.1 M solution). This solution is further diluted 100 times (1 ml into 100 ml) and kept at 4°C.

2.2.2. Biochemicals and Reagents

- NHS-activated Sepharose 4 Fast flow PN = 17-0906-01 (GE Healthcare).

  NHS-activated Sepharose 4 Fast flow is supplied as suspension in 100% isopropanol (in this protocol 3 ml of suspension was used).
- Polyclonal Rabbit Anti-Human Transferrin PN = A0061 (DakoCytomation).

  The concentration is 15 g/l.
- Albumin 20% in phosphate buffer saline 0.1% azide.

  Mix 1 g of albumin bovine serum fraction V (BSA PN = A4503) with 4 ml of PBS and store at 4°C.
- Apo-transferrin in dilution buffer.

  Mix 3 g of apo-transferrin (PN = T2252) with 5 ml of dilution buffer and store at 4°C.
- CEofix™ NTMP (Analis, Suarlée, Belgium) (see insert to have the running conditions).
- CEofix™ CDT (Analis, Suarlée, Belgium) (see insert to have the running conditions).

3. Methods

3.1. Analysis of Amyloid-β Peptides in Cerebrospinal Fluid Samples by Capillary Electrophoresis Coupled with LIF Detection (45)

3.1.1. Dissolution and Storage of Fluorescent Dyes

- Dissolution of the Fluoprobe 488 NHS powder in DMSO at concentration of 10 mg/ml by adding 1 mg of fluorescent dye powder in 100 μL of DMSO.
- Prepare 5–10 mL aliquots of the fluorescent dye and keep them in the dark and at −20°C for a few months.

3.1.2. Preparation of CSF Sample

- Collection and analyses of the CSF sample were approved by the Ethics Committee in Ulm. All samples were anonymized before transmission to our laboratory (see Note 8).

- CSF was collected, aliquoted within 2 h, and stored at –80°C.
- CSF could be directly used for the labeling reaction (see Note 9).

3.1.3. Preparation of Magnetic Beads

- The beads were supplied as a suspension containing 6 or 7 E+08 beads/ml approx. (10 mg/ml) in phosphate-buffered saline (PBS) pH 7.4 with 0.1% bovine serum albumin (BSA) and 0.02% sodium azide (NaN_3).
- 1 mg of the beads are washed two times with PBS pH 7.4 containing 0.1% bovine serum albumin (BSA) and using a magnetic separator to separate the beads from the solution.
- Add 100 μL of monoclonal antibody anti-Aβ (1 mg/mL) to 1 mg of magnetic microparticles M280 Dynabeads (10 mg/mL, 6.7 × E+08 beads/mL).
- Incubate the mixture overnight at 4°C.
- Add 0.02% w/V sodium azide (NaN_3) (see Note 10).
- The activated beads can be stored at 4°C for few months.

3.1.4. Immunoprecipitation

- Mix 250 μL of CSF with 250 μL of borate buffer pH 10.5, 80 mM.
- Add 10 μL of Fluoprobe to this diluted CSF (see Note 9).
- The pH value of the sample mixture is checked to be 10.5.
- This solution is mixed and incubated for 5 min in the dark.
- Add 500 μL of RIPA buffer and 25 μL of magnetic microparticles coated with the monoclonal antibody 6E10 to the sample mixture.
- Sample is incubated under rotation for 15 h at 4°C.
- After mixing, the sample is washed first with RIPA buffer; then with 50 mM Tris/HCl, pH 7.4; and last with deionized water using a magnetic separator to separate the beads from the solution for each rinsing step.
- For the elution step, the beads are isolated with the magnetic separator and 400 μL of 0.16% NH_4OH is added to the beads (see Note 11).
- The eluted Aβ peptides are fully lyophilized in polypropylene tubes.
- The sample is reconcentrated in 2 μL of 0.16% NH_4OH prior to CZE–LIF analysis.

3.1.5. Capillary Electrophoresis Conditions

- Pretreat new capillaries by applying pressures of 138 kPa to the capillary inlet and using the following sequence: 0.1 M NaOH for 5 min, 1 M NaOH for 5 min, and then water for 5 min.
- Carry out five successive cycles to stabilize the silica surface: One cycle entails one rinsing step with the running buffer

(10 min), then voltage application at 30 kV (for 15 min), followed by one rinse with water (10 min) and 0.1 M NaOH (5 min).

- The in-between rinsing run is carried out by pumping deionized water through the capillary for 5 min, DMSO/water (50:50) for 3 min and then water 3 min using reversed flow, and finally the running buffer for 10 min (see Note 12).
- The sample is introduced into the capillary by hydrodynamic injection for 5 s and at 3.4 kPa. The capillary is thermostated at 25°C.
- The samples are maintained at 10°C by the storage sample module of the PA 800.
- The peptides are detected using LIF detection (a 3.5 mW argon-ion laser having an excitation at 488 nm; emission was collected through a 520 nm band-pass filter).
- The separations are carried out at 30 kV with positive polarity at the inlet and using the running buffer composed of borate buffer pH 9.0, 40 mM, containing 3.25 mM of DAB.
- The running electrolyte is renewed after every run and prepared daily.

3.2. Preparation of Affinity Column for Purification of Transferrin from Serum

3.2.1. Coupling Protocol

- Prepare a solution of the antibody: 1 g of polyclonal rabbit anti-human transferrin is diluted into 2 g of coupling buffer $NaHCO_3/Na_2CO_3/NaCl$. Take 100μL of the solution and analyse it with the CEofix™ NTMP to see the anti-human transferrin concentration.
- Wash NHS-activated Sepharose 4 Fast flow with 10–15 medium volumes of cold 1 mM HCl immediately before use to wash out the isopropanol (see Note 13).
- Take the NHS-activated Sepharose 4 Fast flow washed and let the coupling reaction occur with the solution of polyclonal rabbit anti-human transferrin in the 0.1 M carbonate buffer 0.5 M NaCl.
- Close the vials and let mix for 1–2 h (see Note 14).
- After the coupling is completed, rinse the gel suspension with the stopping buffer 0.5 M ethanolamine/0.5 M NaCl to wash out the uncoupled antibodies and to inactivate all residual activated ester (see Note 15).
- After all the uncoupled antibodies are washed with column storage buffer, put the anti-transferrin-tagged Sepharose in a spin column.

3.2.2. Immuno-Capture of Transferrin Proteins

- Rinse the column three times with 500 μl of the dilution buffer.
- Place the end cap on the column tip. Immediately add 500 μl of the apo-transferrin solution in the column. Close the column with the top snap cap (see Note 16).

- Mix the beads and the sample completely by inverting and shaking the column.
- Place the sample on an end-to-end rotator and incubate it at room temperature for 15 min.
- Invert the column. Remove the end cap and place the column in a 2 ml centrifuge collection tube. Centrifuge for 30 s at 400×*g*. Collect the flow-through sample for further analyses.
- Repeat this operation ten times.

3.2.3. Rinsing of Nonspecifically Bound Proteins

- To remove proteins nonspecifically bound to microbeads, wash the beads with the dilution buffer, a total of four times.
- For each wash, always first insert the end cap, then add 500 μl of the dilution buffer, and close the column with the snap cap.
- Mix the beads and buffer completely by inverting and shaking the column.
- Place the sample on an end-to-end rotator and incubate it at room temperature for 3 min.
- Invert the column. Remove the end cap and place the column in a 2 ml centrifuge collection tube. Centrifuge for 30 s at 400×*g*. Collect the flow-through sample for further analyses.

3.2.4. Stripping of Bound Transferrin

- Strip off bound transferrin from beads using the stripping buffer, a total of three times.
- First add 50 μl of neutralization buffer (10× concentrated) in three different "2 ml centrifuge collection tube" (see Note 17).
- For each elution, always insert first the end cap, then add 500 μl of the stripping buffer, and close the column with the snap cap.
- Mix the beads and buffer completely by inverting and shaking the column.
- Place the sample on an end-to-end rotator and incubate it at room temperature for 3 min.
- Invert the column. Remove the end cap and place the column in a 2 ml centrifuge collection tube containing 50 μl of neutralization buffer (see above). Centrifuge for 30 s at 400×*g*. Collect the flow-through sample for further steps.
- After the centrifugation add 30 μl of 20% BSA in dilution buffer in the flow-through collected in the 2 ml centrifuge collection tube with the neutralization buffer from previous step.
- It is crucial to neutralize the beads immediately for column stability. A 1/10 dilution in water of neutralization buffer is immediately added to the beads (see Note 17).

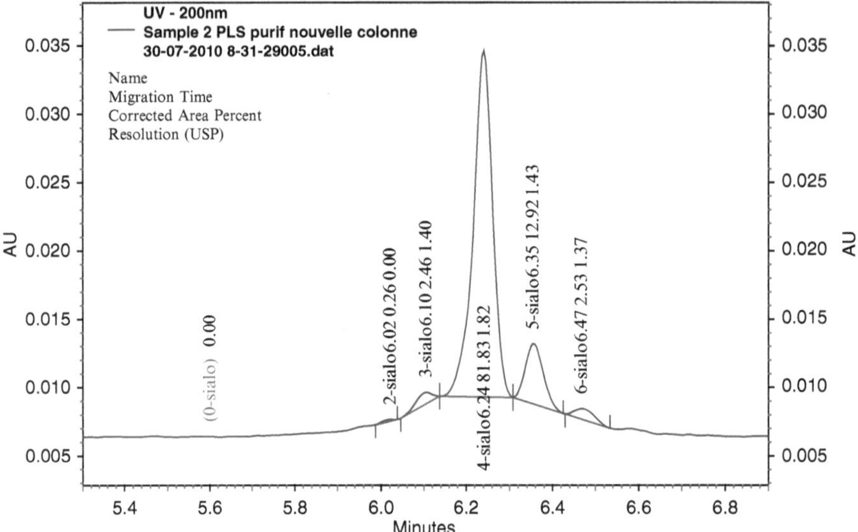

Fig. 6. CE analysis of apo-transferrin sialoforms from human serum sample after purification with a lab-made affinity column. CE conditions and purification step are described in protocol 2.

3.2.5. Spin Column Regeneration

- To regenerate the beads after stripping bound proteins, immediately neutralize beads with 0.5 ml of 1/10 diluted neutralization buffer.
- Mix the beads and buffer completely by inverting and shaking the column.
- Place the sample on an end-to-end rotator and incubate it at room temperature for 3 min.
- Invert the column. Remove the end cap and place the column in a 2 ml centrifuge collection tube. Centrifuge for 30 s at 400 × *g*.
- After the centrifugation, wash the column with the column storage buffer (TBS-Az) for a total of five times by elution.
- Let the beads be suspended in the column storage buffer after the fifth elution. Beads are ready for storage at 4°C.

3.2.6. Analyses of Bound Transferrin with the CEofix™ CDT Kit (Fig. 6)

- For CEofix™ CDT analyses, the stripped bounded transferrin has to be concentrated.
- First rinse the Vivaspin with 30 μl of 20% BSA in dilution buffer and 500 μl of the storage buffer 0.1% NaN_3 (see Note 18).
- Place the first collected flow-through sample with BSA from step 4 in the Vivaspin and centrifuge for 10 min at 10,000 × *g*.
- After the centrifugation, collect the flow-through sample for further analyses, then add the second collected flow-through sample with BSA from step 4 in the same Vivaspin, and centrifuge for 10 min at 10,000 × g.

- Repeat this step with the third collected flow-through from step4.
- After this step, rinse the three different "2 ml centrifuge collection tubes" with 500 μl of the column storage buffer 0.1%NaN_3, add into the 500 μl in the Vivaspin, and centrifuge for 10 min at 10,000×*g*.
- Repeat this washing a total of three times.
- After the concentration, the sample is ready to be used with the CEofix™ CDT kit.
- Dilute the sample in a solution made from 16 mg $NaHCO_3$ in 1 ml of Fe solution of the CEofix™ CDT kit. Add 50 μl of the sample in 150 μl of $NaHCO_3$/Fe solution (see Note 19).

4. Notes

1. The buffer concentrations are expressed in molarity.
2. In general the buffer volume needed for electrophoresis is very low, so the buffers can be prepared in 50 mL volumes. All the buffers should be filtered by 0.2 μm and kept in refrigerator before use.
3. Obtaining CSF from patients should be done by a specialist and processed in specialized laboratories (e.g., neurochemical labs).
4. Concentrated HCl can be used at first to narrow the gap from starting pH to the required pH. From then on it would be better to use a series of HCl with lower ionic strength to avoid a sudden drop in pH below the required pH (e.g., 6 N, 1 N, and 0.1 N).
5. Measuring pH on highly concentrated solution may not be accurate; we find better to record the value obtained after a tenfold or more dilution of the final buffer preparation.
6. Prepare all solution using ultrapure water (prepared by double distillation apparatus) and analytical grade reagents (see Note 20). Prepare and store all reagents at room temperature (unless indicated otherwise).
7. Another way to prepare this buffer is to dilute the 2 ml of 1 M Tris in a beaker with 1.763 g of NaCl, 0.2 g of sodium azide, and bi-distilled water up to 200 ml.
8. All individuals gave written informed consent to their participation in the study and underwent a clinical, neurological, and neuroradiological examination, as well as a short neuropsychological screening, including Mini-Mental State Examination to

investigate global cognitive functioning. If deterioration were suggested, then a detailed psychometric test battery covering executive functions, memory, constructional abilities, premorbid verbal intelligence, and depression was administered to assess more specifically for cognitive impairment.

9. Fluorescent labeling can also be done after immune-capture and on the eluted peptides.
10. NaN_3 is highly toxic; avoid inhalation or skin contact.
11. To avoid peptide aggregation, the addition of NH_4OH in the lyophilized samples is recommended by the suppliers.
12. We observed that Fluoprobe 488 is able to precipitate in the inlet end of the capillary. DMSO rinsing in reverse flow is applied to eliminate this issue.
13. An Ace Hirsch filter of 25 capacity and 10–20 μm porosity is used for easy and fast cleaning. The 1 mM HCl solution should be cold when operating.
14. We find that it is best to use screw cap vial to perform the coupling reaction. The suspension is then mixed gently on an end-to-end rotator.
15. Ethanolamine reacts with the NHS-activated ester to form a stable amide. It is essential to block all reactive sites to avoid nonspecific capture of the transferrin during the immunopurification step.
16. We find best to seal first the bottom of the column, then to add the transferrin in the binding buffer to the column, and finally to close the top slowly without pushing too much air into it. If there is too much pressure inside leakage can occur.
17. The neutralizing buffer is placed into the collection side of the diafiltration device to raise pH as quick as possible in order to lower the protein denaturation that can take place at low pH.
18. Without BSA present at this stage, the analysis of the transferrin with the CEofix™ CDT does not perform well. We explain this result by a facilitated refolding of the transferrin molecule in a medium close to a serum protein concentration.
19. The presence of carbonate ions helps to stabilize the binding of the second iron atom onto the transferrin.
20. The buffer concentrations are expressed in molarity. All buffers were prepared with deionized water and were filtered through a 0.22 μm membrane before use.

Acknowledgments

This review's writing was supported by the European Union's NADINE project under Contract No. 246513. François de l'Escaille and Jean-Bernard Falmagne are employees of Analis s.a. Belgium. Analis s.a. is producer of CEofix™ NTMP and CEofix™ CDT kits and editor of the PHoEBuS software.

References

1. Atkinson AJ, Wayne A et al (2001) Biomarkers and surrogate endpoints: preferred definitions and conceptual framework. Clin Pharmacol Ther 69:89–95
2. Chen R, Jin Z, Colón LA (1996) Analysis of tear fluid by CE/LIF: a noninvasive approach for glucose monitoring. J Capillary Electrophor 3:243–248
3. Tomosugi N et al (2005) Diagnostic potential of tear proteomic patterns in Sjögren's syndrome. J Proteome Res 4:820–825
4. Zhou L et al (2009) Identification of tear fluid biomarkers in dry eye syndrome using iTRAQ quantitative proteomics. J Proteome Res 8: 4889–4905
5. Van Eijk HMH et al (1999) Automated isolation of high-purity plasma albumin for isotope ratio measurements. J Chromatogr B 731: 199–205
6. Kakehi K et al (2001) Capillary electrophoresis of sialic acid-containing glycoprotein. Effect of the heterogeneity of carbohydrate chains on glycoform separation using an alpha 1-acid glycoprotein as a model. Anal Chem 73:2640–2647
7. Valcárcel M, Arce L, Ríos A (2001) Coupling continuous separation techniques to capillary electrophoresis. J Chromatogr A 924:3–30
8. Guzman NA, Trebilcock MA, Advis JP (1990) Paper presented at the First Annual Conference on Capillary Electrophoresis. Frederick, Maryland, October 15–16, Abstract no. 12.
9. Guzman NA, Trebilcock MA, Advis JP (1991) The Use of a Concentration Step to Collect Urinary Components Separated by Capillary Electrophoresis and Further Characterization of Collected Analytes by Mass Spectrometry. J Liq Chromatogr 14:997–1015
10. Dalluge J, Sander LC (1998) Precolumn Affinity Capillary Electrophoresis for the Identification of Clinically Relevant Proteins in Human Serum: Application to Human Cardiac Troponin I. Anal Chem 70:5339–5343
11. Peoples MC, Karnes HT (2008) Microfluidic Capillary System for Immunoaffinity Separations of C-Reactive Protein in Human Serum and Cerebrospinal Fluid. Anal Chem 80:3853–3858
12. Yang YZ, Boysen RI, Hearn MTW (2006) Optimization of field-amplified sample injection for analysis of peptides by capillary electrophoresis–mass spectrometry. Anal Chem 78:4752–4758
13. Armenta J et al (2007) Coupled affinity-hydrophobic monolithic column for on-line removal of immunoglobulin G. preconcentrated of low abundance proteins and separation by capillary zone electrophoresis. J Chromatogr A 1148:115–122
14. Kaiser T et al (2004) Capillary electrophoresis coupled to mass spectrometer for automated and robust polypeptide determination in body fluids for clinical use. Electrophoresis 25:2044–2055
15. Ongay S et al (2010) Development of a fast and simple immunochromatographic method to purify alpha 1-acid glycoprotein from serum for analysis of its isoforms by capillary electrophoresis. Anal Chim Acta 663:206–212
16. Li XM, Zhang F, Zhang SS (2008) Capillary electrophoresis enzyme immunoassay for alpha-fetoprotein and thyroxine in human serum with electrochemical detection. J Sep Sci 31:336–340
17. Guzman NA et al (2003) Improved solid-phase microextraction device for use in on-line immunoaffinity capillary electrophoresis. Electrophoresis 24:3718–3727
18. Amundsen LK, Sirén H (2007) Immunoaffinity CE in clinical analysis of body fluids and tissues. Electrophoresis 28:99–113
19. Van der Veen M, Norde W, Stuart MC (2004) Electrostatic interactions in protein adsorption probed by comparing lysozyme and succinylated lysozyme. Colloids Surf B Biointerfaces 35:33–40
20. Gray JJ et al (2004) the interaction of proteins with solid surfaces. Curr. Opin. Struct. Biol 14:110–115

21. Malmsten M et al (1998) Formation of Adsorbed Protein Layers. J Colloid Interface Sci 207:186–199
22. Ding HM et al (2005) Silica nanotubes for lysozyme immobilization. J Colloid Interface Sci 290:102–106
23. Hamrníkova I et al (1999) Binding of proline- and hydroxyproline-containing peptides and proteins to the capillary wall. J Chromatogr A 838:167–177
24. Verzola B, Gelfi C, Righetti PG (2000) Protein adsorption to the bare silica wall in capillary electrophoresis quantitative study on the chemical composition of the background electrolyte for minimising the phenomenon. J Chromatogr A 868:85–99
25. Nakanishi K, Sakiyama T, Imamura K (2001) On the adsorption of proteins on solid surfaces, a common but very complicated phenomenon. J BiosciBioeng 91:233–244
26. Essa H et al (2007) Influence of pH and ionic strength on the adsorption, leaching and activity of myoglobin immobilized onto ordered mesoporoussilicates. J Mol Catal Enzym 49: 61–68
27. Lucy CA, MacDonald AM, Gulcev MD (2008) Non-covalent capillary coatings for protein separations in capillary electrophoresis. J Chromatogr A 1184:81–105
28. Stutz H (2009) Protein attachment onto silica surfaces–a survey of molecular fundamentals, resulting effects and novel preventive strategies in CE. Electrophoresis 30:2032–2061
29. Weinbauer M, Stutz H (2010) Successive multiple ionic polymer layer coated capillaries in the separation of proteins - Recombinant allergen variants as a case study. Electrophoresis 31:1805–1812
30. Giordano BC et al (2000) Dynamically-coated capillaries allow for capillary electrophoretic resolution of transferrin sialoforms *via* direct analysis of human serum. J Chromatogr B 742:79–89
31. Lanz C et al (2002) Evaluation and optimization of capillary zone electrophoresis with different dynamic capillary coatings for the determination of carbohydrate-deficient transferrin in human serum. J Chromatogr A 979:43–57
32. Yang R, Liu Y, Wang Y (2009) Hydroxyethylcellulose-graft-poly (4-vinylpyridine) as a novel, adsorbed coating for protein separation by capillary electrophoresis. Electrophoresis 30:2321–2327
33. Sassi AP et al (2005) An automated, sheathless capillary electrophoresis-mass spectrometry platform for discovery of biomarkers in human serum. Electrophoresis 26:1500–1512
34. Liu H et al (2008) A well-defined diblock copolymer of poly-(ethylene oxide)-block-poly (4-vinylpyridine) for separation of basic proteins by capillary zone electrophoresis. Electrophoresis 29:2812–2819
35. Puerta A et al (2006) Novel adsorptive polyamine coating for enhanced capillary electrophoresis of basic proteins and peptides. J Chromatogr B 838:113–121
36. Isemura T, Kitagawa F, Otsuka K (2009) Separation of complex mixtures of fluorobenzoic acids by capillary electrophoresis. J Sep Sci 32:381–387
37. Mischak H, Schanstra JP (2011) CE-MS in biomarker discovery, validation, and clinical application. Proteomics Clin Appl 5:9–23
38. Szökő E, Tábi T (2010) Analysis of biological samples by capillary electrophoresis with laser induced fluorescence detection. J Pharm Biomed Anal 53:1180–1192
39. Breadmore MC et al (2011) Recent advances in enhancing the sensitivity of electrophoresis and electrochromatography in capillaries and microchips (2008–2010). Electrophoresis 32: 127–148
40. Mala Z et al (2011) Contemporary sample stacking in analytical electrophoresis. Electrophoresis 32:116–126
41. Chen Y et al (2009) Assay of bradykinin metabolites in human body fluids by CE-LIF coupled with transient ITP preconcentration. Electrophoresis 30:2300–2306
42. Yang WC, Yeung ES, Schmerr MJ (2005) Detection of prion protein using a capillary electrophoresis-based competitive immunoassay with laser-induced fluorescence detection and cyclodextrin-aided separation. Electrophoresis 26:1751–1759
43. Babu S, Chung BC, Lho DS, Yoo YS (2006) Capillary electrophoretic competitive immunoassay with laser-induced fluorescence detection for methionine-enkephalin. JChromatogr 1111:133–138
44. Verpillot R et al (2011) Analysis of amyloid-β peptides in cerebrospinal fluid samples by capillary electrophoresis coupled with LIF detection. AnalChem 83:1696–1703
45. Tu J et al (2003) Application of multiplexed capillary electrophoresis with laser-induced fluorescence (MCE-LIF) detection for the rapid measurement of endogenous extracellular signal-regulated protein kinase (ERK) levels in cell extracts. J Chromatogr B 789:323–335
46. Chen Y, Xu L, Lin J, Chen G (2008) Assay of bradykinin-related peptides in human body fluids using capillary electrophoresis with laser-induced fluorescence detection. Electrophoresis 29:1302–1307

Chapter 14

Highly Charged Polyelectrolyte Coatings to Prevent Adsorption During Protein and Peptide Analysis in Capillary Electrophoresis

Reine Nehmé and Catherine Perrin

Abstract

Capillary electrophoresis (CE) is an interesting technique for protein and peptide analysis. However, one of the major problems concerns sample adsorption on the internal capillary wall. The use of non-covalent coatings using highly charged polyelectrolytes is an efficient, simple, and fast approach to reduce peptide and protein adsorption phenomena. We have studied in a systematic manner the effect of coating conditions on the stability and efficiency of multilayer coatings using poly(diallyldimethylammonium) chloride (PDADMAC) as polycation and polystyrene sulfonate (PSS) as polyanion. When optimal conditions defined in the protocols are used, very stable coatings are obtained and adsorption phenomena are eliminated. The coatings are stable over a large range of pH buffer (2–10) and in the presence of organic solvent. Hundreds of analyses can be performed without coating regeneration. Coated capillaries can be easily stored and reused.

Key words: Adsorption, Capillary electrophoresis, Capillary coating, Multilayer coating, Peptide, Protein, Polyelectrolyte, Electroosmotic flow

1. Introduction

Capillary electrophoresis (CE) is now a well-established technique for the analysis of biomolecules such as peptides and proteins. Very high peak efficiencies are obtained compared to HPLC since diffusion, which is mainly governed by molecular diffusion, is low for biomolecules such as peptides and proteins (the theoretical plate number is about 10^6 for proteins) (1). Furthermore, due to the absence of stationary phase, a great variety of analytical conditions can be used.

However, CE suffers from an inherent drawback that is the possible adsorption of peptides and proteins onto the silica surface of the capillaries used for analyses. Adsorption phenomena result

Nicola Volpi and Francesca Maccari (eds.), *Capillary Electrophoresis of Biomolecules: Methods and Protocols*, Methods in Molecular Biology, vol. 984, DOI 10.1007/978-1-62703-296-4_14, © Springer Science+Business Media, LLC 2013

into impaired analytical performances such as nonrepeatability of migration times and peak areas, and reduced efficiencies, which alters any quantitative analysis. Indeed, the inner surface of the capillaries used in CE is characterized by the presence of silanol groups that behave as a polyanion (2) with a continuous pKa distribution comprised between 2 and 9. Silanol groups will consequently ionize when in contact with an electrolyte of pH higher than 2 (3). Ionized silanols have a tenacious affinity for large organic biomolecules. Peptides and especially proteins are known to interact strongly with the silica surface by cooperative multipoint attachment due to the presence of hydrophobic patches and electrostatic (anionic and cationic) sites at their surface (4). Adsorption phenomena are particularly important for proteins with high values of molecular weight and pI.

On the other hand, the reproducibility of the electroosmotic flow (EOF) resulting from the global movement of the electrolyte under the influence of the electric field is of major importance to ensure good analytical performances in CE. The amplitude of the EOF is proportional to the ζ-potential, i.e., the electrical potential at the electrolyte/capillary interface. Towns and Regnier (5) showed that the uniformity of the ζ-potential is essential in capillary electrophoresis to obtain high separation efficiency. However, the ζ-potential may change when artifacts are present on the capillary surface (5–8) such as silicon–carbon bonds originating from the production process (9) or cracks (7, 9). In fact, cations are preferably adsorbed at the defect surface (10) which locally changes the ζ-potential and the EOF (9, 10). Defects also produce a recirculating disturbance movement where analytes may be trapped resulting in important dispersion (7). Furthermore, adsorption of sample components to the silica wall (2, 8, 9, 11–14) particularly alters the ζ-potential

Consequently, to obtain repeatable and high separation efficiencies for biomolecules in capillary electrophoresis, analyte–capillary interactions must be eliminated and the electroosmotic flow (EOF) appropriately controlled. Numerous approaches have been proposed (2, 4, 15–27) aiming to create columbic repulsion between the silica wall and the biomolecules and/or to mask the silanol groups. These strategies concern (a) the appropriate choice of analysis conditions, especially the background electrolyte (BGE) properties (pH and ionic strength) and (b) the chemical modification of the capillary inner wall with a coating agent (polymer, surfactant, etc.). This latter approach has shown to be the most effective and has a double interest. It reduces the many possible interactions between peptides/proteins and the capillary wall, by inducing electrostatic repulsion forces (15, 28) and/or by mechanical shielding of the silanol groups so that they will be no more accessible (29–31). It also allows tuning of the EOF by changing the charge (amount and nature) of the capillary surface and/or the nature of the BGE. This may concern the magnitude as well as the direction of the EOF. Controlling the EOF is an important issue in CE since it has a direct influence on the analysis time, resolution, repeatability, and separation efficiency (32–35).

The capillary coating agent can be covalently bound to the silanol groups (13, 36–40) but this approach usually requires multiple time-consuming steps. Coating agents can also be physically adsorbed to the capillary (see 3, 4, 17, 18, 23, 30, *and references therein)* via electrostatic, hydrogen, and hydrophobic interactions (23). Physical coatings present several advantages (2, 31, 41): (a) simplicity of the procedure, (b) low cost (no use of organic solvents), and (c) possibility for automation and regeneration of the coating (42).

Physical coatings can be obtained by either a dynamic or a static approach. With the dynamic approach (17, 23), the coating agent (surfactant (3), mono- and oligo-amines (43), polymer (44), etc.) is added to the BGE to prevent the bleeding of the coating. With the static approach, coating of the capillary is performed before the analysis run. Static coatings are stable during the electrophoretic runs with no need to add the coating agent in the BGE as it is the case for dynamic coatings (12, 45–49). Although very efficient for many applications, the presence of coating agent in the BGE in the dynamic approach can alter/limit selectivity, and may interfere with the sample causing protein denaturation (2, 4), and with analyte detection (e.g., spectrometric detection) (38, 50). For these reasons, the static approach is preferred.

For static coatings, the use of highly charged high-molecular-weight polymers (polyelectrolytes) has proved to provide very efficient coatings to limit peptide or protein adsorption. In this case, the coating procedure is very simple since it consists of rinsing the capillary with the appropriate coating agent. Highly charged polymers bind to silica via multisite electrostatic interactions. Therefore, polyelectrolytes limit the adsorption of charged molecules by electrostatic repulsion at a pH below their pI.

So-called monolayer coated capillaries are obtained using highly charged polycations such as polybrene (12, 34, 47, 49, 51–53), poly(diallyldimethylammonium) chloride (PDADMAC) (12, 54–58), chitosan (59, 60), polyethylenimine (PEI) (12, 48, 55, 61), and polyarginine (49). Under appropriate conditions, these polycations can be adsorbed via electrostatic bonds in an almost irreversible manner onto anionic silica (12).

To further improve the stability and the efficiency of this type of coatings in CE, Katayama et al. (52, 53) introduced a successive multiple ionic-polymer layer (SMIL) coating procedure, in which "a cationic polymer is sandwiched between an anionic polymer and the uncoated negative fused-silica capillary." In this approach, the capillary surface is covered with successive alternating layers of polycation and polyanion, i.e., polyelectrolyte multilayers (Fig. 1).

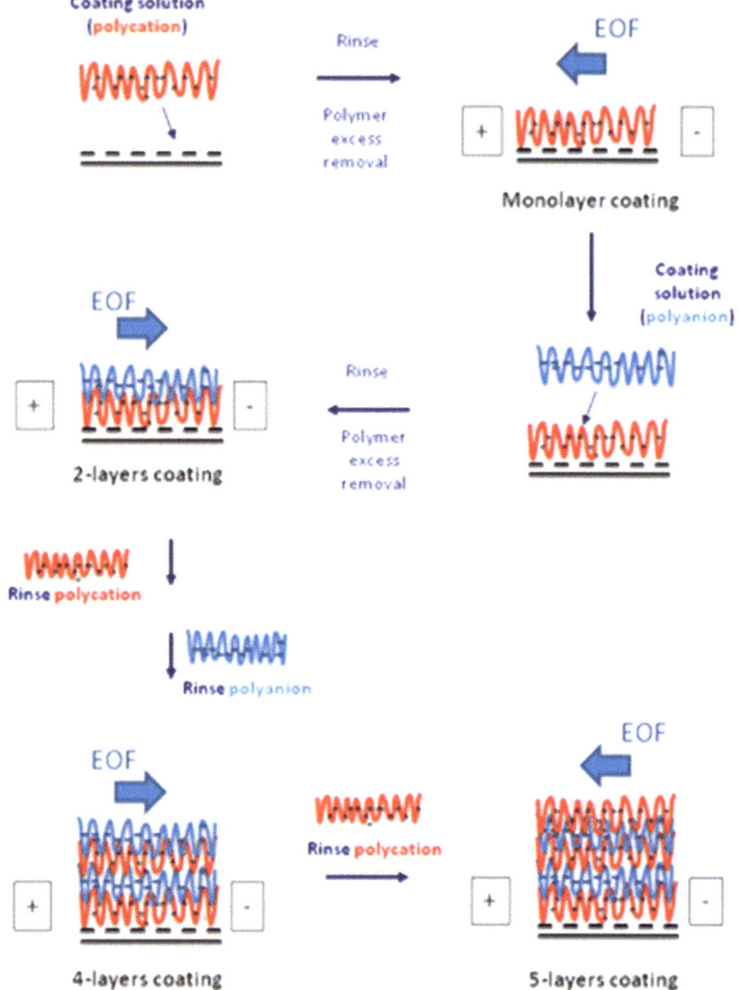

Fig. 1. General monolayer and multilayer coating procedure using highly charged polyelectrolytes.

With this approach either cationic or anionic coated capillaries can be obtained. For example, in order to obtain a negatively charged capillary surface, the first step is to rinse it with a polycation solution. As a result, the polycation adsorbs to the silica surface, compensates its charges, and then exposes an excess of positive charges. Any polymer not adhering to the surface is then removed by rinsing with a solution that does not contain polymer. The now cationic silica surface is subsequently exposed to a polyanion solution. Electrostatic interactions allow the adsorption of the polyanion layer on the polycation one previously adhered. The surface of the capillary is thus negatively charged. Hydrophobic interactions and hydrogen bonds can also occur between the two polyelectrolyte layers (62, 63). The process can be repeated several times by altering rinses with cationic and anionic polymers.

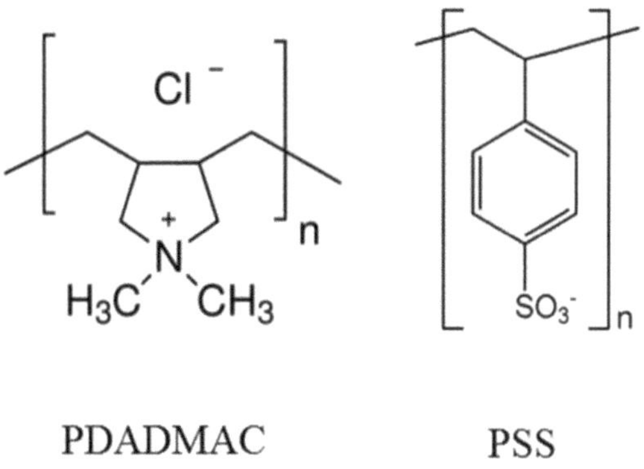

PDADMAC PSS

Fig. 2. Chemical structure of poly(diallyldimethylammonium) chloride salt (PDADMAC) and polystyrene sulfonate sodium salt (PSS).

Electroneutrality within a polyelectrolyte multilayer is always required. In the bulk of the multilayer, charges of oppositely charged polyelectrolyte layers quantitatively balance each other (intrinsic compensation). In contrast, small counterions from the coating solution balance excess polymer charge in the region near the multilayer surface (extrinsic compensation). In other terms, all excess charge resides at the multilayer surface (64–66) which will thus be reversed after each deposition step. The alternate rinse cycles with oppositely charged polymers result in the stepwise growth of the multilayer film. SMIL coatings have proven to be the most efficient to prevent peptide and protein adsorption (46, 52, 53). Their higher efficiency is mainly to be attributed to a better coverage of the silanol groups, leading to a more hydrophilic surface and therefore reduced protein adsorption. An important advantage of SMIL coatings is the possibility to use various coating agents which enables to introduce a characteristic selectivity based on interactions between the analytes and the coating. Among them, the combination of the polycation PDADMAC and the polyanion poly(sodium 4-styrenesulfonate) (PSS) have shown to produce very stable SMIL coatings. Indeed, the pair PDADMAC/PSS (Fig. 2) has a moderate hydrophobicity (67) which allows to improve the coating adherence to silica and to reduce hydrophobic interactions with proteins.

It is very important to note that the quality and stability of the SMIL coatings are largely influenced by the composition of the coating solution, especially if highly charged polyelectrolytes are used. We have studied (55–57) in a systematic manner the effect of the experimental coating conditions on the endurance, the performance, and the chemical stability (57) of these coatings for the analysis of peptides (55) and proteins (56) by CE. Cationic coatings (of one and five layer(s)) as well as anionic coatings (four layers) were evaluated. The ionic strength of the coating solution, the polyelectrolyte concentration, the number of coating layers, as well as the coating stabilization procedure have shown to have a great influence on the coating stability and performance in limiting peptide and protein adsorption. From these studies, we have developed coating protocols to produce stable cationic PDADMAC monolayer and 5-layer PDADMAC/PSS coatings as well as anionic 4-layer PDADMAC/PSS coatings. These coatings are extremely efficient in preventing peptide and protein adsorption. Indeed, various peptides and proteins have been successfully analyzed in CE by using these coatings in different analysis conditions (pH, ionic strength, etc.). For most applications, very good repeatabilities (less than 1 % on migration times and less than 5 % on peak areas) and great peak efficiencies (in the order of 700,000) were obtained.

Each coating protocol contains three steps: (1) activation of the capillary surface, (2) rinse of the capillary with PDADMAC or PSS, and (3) stabilization of the coating. These protocols are easy to conduct. Monolayer PDADMAC coatings can be used with BGE having a pH inferior to 7 and containing no organic solvent, while both 5-layer cationic and 4-layer anionic coatings can be used over a wide range of pH (2–10) and in the presence of various organic solvents. With these latter coatings, corrosive rinses (basic, acidic) can also be performed between analyses if necessary. Procedures to monitor coating stability before and during analyses as well as possible protein/peptide adsorption are also given. Using an appropriate protocol, coated capillaries can be stored for months without any modification of their performances. Hundreds of analyses can be performed in the same capillary without regeneration of the coating in most cases.

2. Materials

Prepare all solutions using ultrapure water and analytical grade reagents. Store the reagents and the solutions at 4 °C.

2.1. Silica Capillaries

1. Cut the fused silica capillary to the appropriate length depending on the application and instrumentation (see Note 1). 75, 50, or even 25 μm internal diameter capillaries are recommended (see Note 2).
2. Remove capillary polyimide (typically 5 mm) at both extremities of the capillary (see Note 3).
3. When UV or LIF (laser-induced fluorescence) detection is used, remove a small part of the polyimide coating (typically 2–3 mm) at the specified position depending on the instrumentation (see Note 3).
4. House the capillary in the appropriate cartridge depending on the instrumentation.

2.2. Polyelectrolyte Solutions

1. PDADMAC solution (High molecular weight M_r: 4.10⁵–5.10⁵) 20 % (w/w) in water (Sigma-Aldrich, Steinheim, Germany) (see Note 4).
2. PDADMAC solution 0.2 % (w/v): Weigh 50 mg PDADMAC, 12.1 mg Tris, and 43.4 mg NaCl (see Note 5). Add 0.5 mL HCl 0.1 N (see Note 6). Adjust to 5 mL with water. Dissolve for 5 min in an ultrasonic bath. Store at 4 °C (see Note 7).
3. PSS (average M_r~10^6) (Sigma-Aldrich, Steinheim, Germany) (see Note 4).
4. PSS solution 0.2 % (w/v): Weigh 10 mg PSS, 12.1 mg Tris, and 43.4 mg NaCl. Add 0.5 mL HCl 0.1 N (see Note 5). Adjust to 5 mL with water. Dissolve for 5 min in an ultrasonic bath. Store at 4 °C (see Note 7).

2.3. Other Solutions

1. EOF marker solution: 0.05 % (v/v) DMF (dimethylformamide) solution in water (see Note 8).
2. NaOH solutions: 1 M and 0.1 M in water.

3. Methods

All procedures are carried at 25 °C (see Note 9).

3.1. Conditioning of New Capillaries

New capillaries must be first conditioned with alkaline solutions to obtain a silica surface capable to bind homogeneously to polyelectrolytes.

1. Rinse capillary with NaOH 1 M. Rinse pressure and time should be adapted so that the capillary is rinsed about a hundred times (100 times the total capillary volume) (see Note 10).
2. Rinse capillary with NaOH 0.1.M. Rinse pressure and time are similar to those used in step 1.
3. Rinse capillary with water. Rinse pressure and time are similar to those used in step 1.

3.2. Monolayer Cationic PDADMAC Capillary Coating

Using the following procedure, the capillary will carry a high positive charge. This type of capillaries should therefore be used for the analysis of positively charged biomolecules at the pH of the BGE used for the analyses. pH of the BGE should be below 7 and presence of organic solvent in the BGE should be limited. The resulting EOF is directed to the anode (see Note 11 and Fig. 1).

1. Rinse capillary with PDADMAC solution 0.2 % (v/w). Rinse pressure and time should be adapted so that the capillary is rinsed about a hundred times.

2. Rinse capillary with water. Rinse pressure and time should be adapted so that the capillary is rinsed about 50 times.
3. Rinse the capillary with the BGE that is intended to be used for the analysis of the biomolecules. Rinse pressure and time are identical to those used in step 2.

3.3. 4-Layer Anionic PDADMAC/PSS Capillary Coating

Using this procedure, four successive layers of polyelectrolytes, alternatively PDADMAC and PSS, are deposited on the capillary surface. Consequently, the last deposited layer being PSS, the capillary will carry a high negative charge. This type of capillaries should therefore be used for the analysis of negatively charged biomolecules at the pH of the BGE used for the analyses. The resulting EOF is directed to the cathode (see Note 12 and Fig. 1).

1. Rinse capillary with PDADMAC solution 0.2 % (v/w). Rinse pressure and time should be adapted so that the capillary is rinsed about a 100 times.
2. Rinse capillary with water. Rinse pressure and time should be adapted so that the capillary is rinsed about 50 times.
3. Rinse capillary with PSS solution 0.2 % (v/w). Rinse pressure and time are identical to those used in step 1.
4. Rinse capillary with water. Rinse pressure and time are identical to rinse step 2.
5. Repeat rinse steps 1–4.
6. Rinse the capillary with the BGE that will be used for the analysis of the biomolecules. Rinse pressure and time are identical to step 2.

3.4. 5-Layer PDADMAC/PSS Capillary Coating

Using this procedure, five successive layers of polyelectrolytes, alternatively PDADMAC and PSS, are deposited on the capillary surface. Consequently, the last deposited layer being PDADMAC, the capillary will carry a high positive charge. This type of capillaries should therefore be used for the analysis of positively charged biomolecules at the pH of the BGE used for the analyses. The resulting EOF is directed to the anode (see Note 11 and Fig. 1).

Compared to monolayer PDADMAC coatings, 5-layer PDADMAC/PSS coating will present a greater thickness and therefore provide a better screening of the silanol groups responsible for biomolecule adsorption. This type of coating should be used for highly cationic or/and hydrophobic biomolecules.

1. Rinse capillary with PDADMAC solution 0.2 % (v/w). Rinse pressure and time should be adapted so that the capillary is rinsed about a 100 times.
2. Rinse capillary with water. Rinse pressure and time should be adapted so that the capillary is rinsed about 50 times.

3. Rinse capillary with PSS solution 0.2 % (v/w). Rinse pressure and time are identical to those used in step 1.
4. Rinse capillary with water. Rinse pressure and time are identical to those used during step 2.
5. Repeat rinse steps in the following order: 1, 2, 3, 4, 1, and 2.
6. Rinse the capillary with the BGE that will be used for the analysis of the biomolecules. Rinse pressure and time are identical to those used in step 2.

3.5. Stabilization of the Coatings

1. Rinse the capillary with the BGE that is intended to be used for the analysis of the biomolecules. Rinse pressure and time should be adapted so that the capillary is rinsed about 50 times.
2. Apply an electric field of approximately 320 V/cm for 10 min.
3. Let the capillary equilibrate with the BGE for 10 min (see Note 13).

3.6. Control of the Quality of the Coating Before Analysis

After and stabilization of the coating as shown above, we recommend to test the quality of the coating (in terms of adherence to the silica capillary).

1. Rinse the capillary with the BGE that is intended to be used for the analysis of the biomolecules. Rinse pressure and time should be adapted so that the capillary is rinsed about 50 times.
2. Inject EOF marker solution (see Note 14).
3. In the case of cationic monolayer PDADMAC or 5-layer PDADMAC/PSS coatings, apply a negative voltage adapted to your analytical conditions. For anionic 4-layer PDADMAC/PSS coatings, apply a positive voltage.
4. From the observed migration time of the EOF marker, calculate the EOF mobility as follows:

$$\mu_{EOF} = \frac{L.l}{V.t}$$

where L is the total capillary length (cm)

l is the effective length, i.e., capillary length to the detector (cm)

V is the applied voltage (V)

t is the migration time (s)

5. Repeat steps 1–4 at least 6 times ($n = 6$).
6. Evaluate the stability of the coating by calculating the RSD on μ_{EOF}:

$$RSD_{\mu_{EOF}} = \frac{s}{\overline{\mu}_{EOF}} 100$$

where s is the standard deviation of the measured EOF mobilities ($n = 6$)

$\bar{\mu}_{EOF}$ is the average EOF mobility

RSD should not be more than 3 %.

3.7. Separations Using Coated Capillaries with Highly Charged Polyelectrolytes

3.7.1. Cationic Monolayer or 5-Layer PDADMAC/PSS Coating

1. Rinse the capillary with the BGE that is intended to be used for the analysis of the biomolecules. Rinse pressure and time should be adapted so that the capillary is rinsed about 50 times.
2. Inject sample hydrodynamically (volume should not exceed 1 % of the total capillary volume).
3. Apply a negative voltage of about −300 V/cm with a rise time of 0.17 min.
4. UV detection is performed at 214 or 200 nm. An example of the separation of cationic protein at pH 2.5 and 7.0 as well as expected repeatability is given in Fig. 3.

3.7.2. Anionic 4-Layer PDADMAC/PSS Coating

1. Rinse the capillary with the BGE that is intended to be used for the analysis of the biomolecules. Rinse pressure and time should be adapted so that the capillary is rinsed about 50 times.
2. Inject sample hydrodynamically (volume should not exceed 1 % of the total capillary volume)
3. Apply a positive voltage of about + 300 V/cm with a rise time of 0.17 min.
4. UV detection is performed at 214 or 200 nm. An example of the separation of anionic protein at pH 9.3 as well as expected repeatability is given in Fig. 4.

3.8. Conservation of the Coated Capillaries

Coated capillaries can be easily stored taking care that they do not dehydrate. With appropriate storage conditions, coated capillaries can be reused for months without any deterioration of their performances (see thereafter coating regeneration after storage procedure).

1. Rinse capillary with water. Rinse pressure and time should be adapted so that the capillary is rinsed about 50 times.
2. Store capillary in water to prevent dehydration by dipping capillary ends into vials containing water.

3.9. Regeneration of the Coating After Storage

1. Rinse capillary with water. Rinse pressure and time should be adapted so that the capillary is rinsed about 50 times.
2. Rinse with the BGE that is intended to be used for the analyses. Rinse pressure and time should be adapted so that the capillary is rinsed about 100 times.
3. Apply an electric field of approximately 320 V/cm for 10 min.

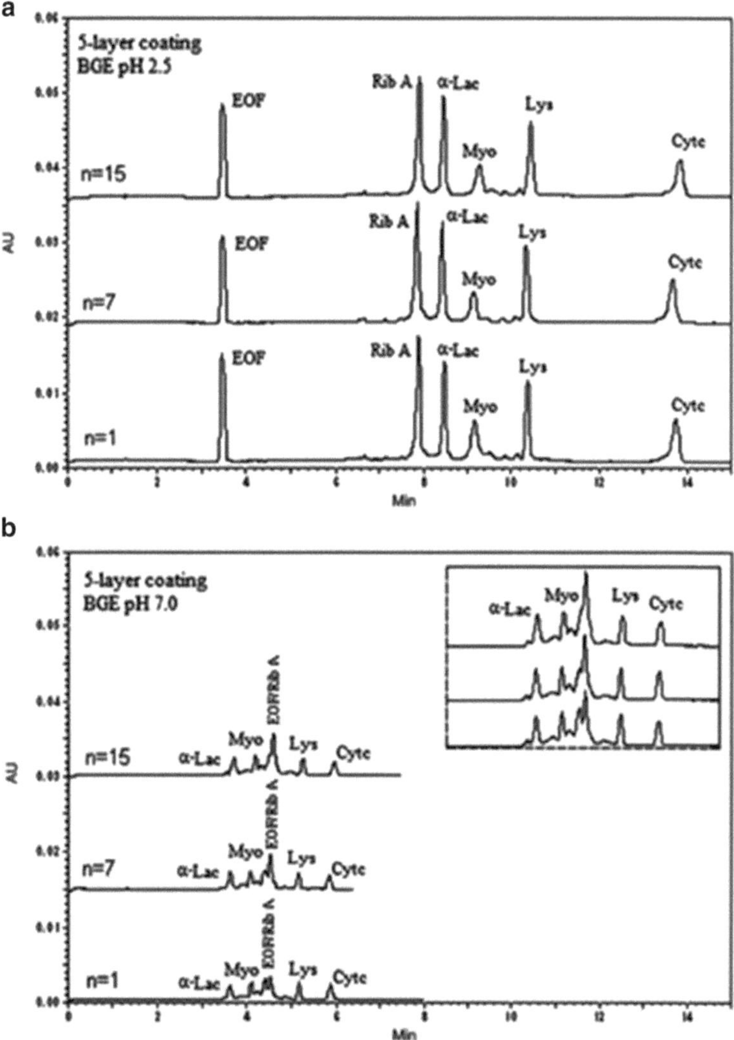

Fig. 3. Repeatability of protein analysis using 5-layer PDADMAC/PSS coated capillaries. Electropherograms obtained for the 1st, 7th and 15th injection (**a**) at pH 2.5 (100 mM TRIS-phosphate buffer) and (**b**) at pH 7.0 (100 mM phosphate buffer). Peak identification: Rib A (ribonuclease A); α-Lac (α-lactalbumin); Myo (myoglobin); Lys (lysine) and Cytc (cytochrome C). Experimental conditions: capillary: 31.2 cm total length, 50 μm i.d.—10 kV applied. UV detection 214 nm. 25 °C. Reprinted with permission from (57). Copyright 2011. Elsevier Science B.V.

4. Notes

1. A particular attention must be given to the cut of the capillary ends (using a ceramic cleaving stone) that must be perfectly flat. Any crack present at the capillary ends must be avoided since it presents a favorable site for protein adsorption.

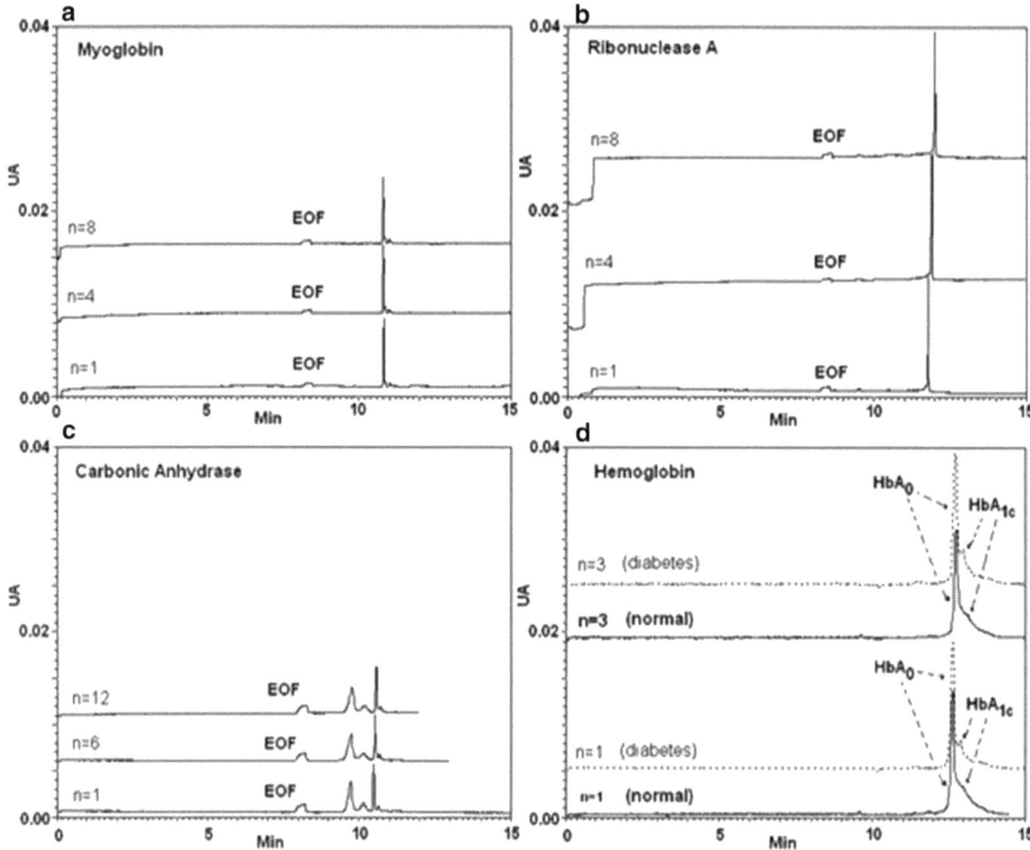

Fig. 4. Repeatability of protein analysis using 4-layer PDADMAC/PSS coated capillaries and 100 mM borate buffer pH 9.3. (**a**–**c**): Electropherograms obtained for (**a**) Myoglobin ; (**b**) Ribonuclease A and (**c**) Carbonic Anhydrase. (**d**): Electropherograms of hemoglobin (Hb) from normal and diabetic adult. HbA_0: normal adult Hb. HbA_{1c}: glycated variant (indicator of long-term diabetes). Capillary: 81.2 cm total length, 25 μm i.d. Experimental conditions: capillary: 81.2 cm total length, 50 μm i.d. +30 kV applied. UV detection 214 nm. 25 °C. Partially reprinted with permission from ref. 57. Copyright 2011. Elsevier Science B.V.

2. Using 25 μm internal diameter capillaries requires instrumentation able to provide pressure above 50 psi (3.5 bar), typically up to 100 psi (7 bar). Smaller internal diameter (typically 10 μm) capillaries are not recommended due to the viscosity of the polyelectrolytes. If such a small internal diameter is required, an external pump device must be used for the polyelectrolyte rinsing steps. For similar reasons, capillary length over 100 cm is not recommended.
3. A wide variety of tools can be used to remove polyimide such as a resistively heated hot wire device or butane lighter flame.
4. Molecular weight of both PDADMAC and PSS is of utmost importance for the stability of the capillary coatings. Lower molecular weights will lead to less stable coatings.

5. NaCl is added to obtain an ionic strength of the coating solution of about 1.5 M. This high ionic strength is necessary to obtain thick and stable coatings.
6. Adding HCl allows to adjust the pH to 8.3. When using different solution volume and/or HCl concentration, final pH must be controlled with a pH meter.
7. Polyelectrolyte solutions are stable for at least 1 week when stored at 4 °C.
8. Other neutral molecules such as mesityl oxide or methanol can be used as EOF marker.
9. Temperature between 15 and 37 °C can be used.
10. Pressure and rinse time combinations to be applied for appropriate rinse volumes can be easily calculated using the free software **CE Expert Lite** that can be downloaded at www.beckmancoulter.com.
11. Due to the high cationic charge of the last deposited layer (i.e., for monolayer PDADMAC or 5-layer PDADMAC/PSS coating), the EOF has a high mobility directed to the anode (i.e., reversed compared to EOF obtained with untreated fused silica capillaries). EOF exact mobility depends on the BGE used for analyses but the domain of variation is rather limited (since EOF magnitude for PDADMAC/PSS coated capillaries is almost independent of pH). For most applications, EOF mobility is usually greater than that of most positively charged proteins (peptides). Therefore, for most cases, all analytes will migrate towards the anode. This implies to use a negative separation voltage.
12. Due to the high anionic charge of the 4-layer PDADMAC/PSS coating, the EOF has a great mobility directed to the cathode (as in untreated fused silica capillary). EOF exact mobility depends on the BGE used for the analysis but the domain of variation is rather limited (since EOF magnitude for PDADMAC/PSS coated capillaries is almost independent of pH). However, for most applications, EOF mobility is usually greater than that of most negatively charged proteins (peptides). Therefore, for most cases, all analytes will migrate towards the cathode. This implies to use a positive separation voltage.
13. For capillary lengths greater than 50 cm, the stabilization time should be extended to 30 min.
14. The injected volume should be between 1 and 5 % of the total capillary volume. Adapt pressure and injection time to the characteristics of your capillary (i.e., capillary length and internal diameter).

References

1. Mosher R, Dewey D, Thormann W, Saville D, Bier M (1989) Computer simulation and experimental validation of the electrophoretic behavior of proteins. Anal Chem 61:362–366
2. Horvath J, Dolnik V (2001) Polymer wall coatings for capillary electrophoresis. Electrophoresis 22:644–655
3. Castelletti L, Verzola B, Gelfi C, Stoyanov A, Righetti PG (2000) Quantitative studies on the adsorption of proteins to the bare silica wall in capillary electrophoresis III: effects of adsorbed surfactants on quenching the interaction. J Chromatogr A 894:281–289
4. Rodriguez I, Li SFY (1999) Surface deactivation in protein and peptide analysis by capillary electrophoresis. Anal Chim Acta 383:1–26
5. Towns JK, Regnier FE (1992) Impact of Polycation Adsorption on Efficiency and Electroosmotically Driven Transport in Capillary Electrophoresis. Anal Chem 64:2473–2478
6. Potocek B, Gas B, Kenndler E, Stedry M (1995) Electroosmosis in capillary zone electrophoresis with non-uniform zeta potential. J Chromatogr A 709:51–62
7. Long D, Stone HA, Ajdari A (1999) Electroosmotic Flows Created by Surface Defects in Capillary Electrophoresis. J Colloid Interface Sci 212:338–349
8. Ghosal S (2004) Fluid mechanics of electroosmotic flow and its effect on band broadening in capillary electrophoresis. Electrophoresis 25: 214–228
9. Watzig H, Kaupp S, Graf M (2003) Inner surface properties of capillaries for electrophoresis. Trends Anal Chem 22:588–604
10. Kaupp S, Steffen R, Watzig H (1996) Characterization of inner surface and adsorption phenomena in fused-silica capillary electrophoresis capillaries. J Chromatogr A 744:93–101
11. Mohabbati S, Westerlund D (2006) Improved properties of the non-covalent coating with N, N-didodecyl-N, N, dimethylammonium bromide for the separation of basic proteins by capillary electrophoresis with acidic buffers in 25 μm capillaries. J Chromatogr A 1121:32–39
12. Cordova E, Gao J, Whitesides GM (1997) Noncovalent Polycationic Coatings for Capillaries in Capillary Electrophoresis of Proteins. Anal Chem 69:1370–1379
13. Hjertén S (1985) High-Performance Electrophoresis, Elimination of Electroosmosis and Solute Adsorption. J Chromatogr 347: 191–198
14. Bello MS, Capelli L, Righetti PG (1994) Dependence of the electroosmotic mobility on the applied electric field and its reproducibility in capillary electrophoresis. J Chromatogr A 684:311–322
15. El Rassi Z, Nashabeh W (1993) Capillary Zone Electrophoresis of Biopolymers with Hydrophilic Fused-Silica Capillaries. In: Guzman NA (ed) Capillary electrophoresis technology. CRC Press, New York, pp 383–434
16. Schwartz H, Pritchett T (1994) CE Techniques Applied to Proteins and Peptides. In: Schwartz H, Pritchett T (eds) Separation of proteins and peptides by capillary electrophoresis: application to analytical biotechnology. Beckman instruments, Inc., Fullerton, pp 1–66
17. Corradini D (1997) Buffer additives other than the surfactant sodium dodecyl sulfate for protein separations by capillary electrophoresis. J Chromatogr B 699:221–256
18. Doherty EAS, Meagher RJ, Albarghouthi MN, Barron AE (2003) Microchannel wall coatings for protein separations by capillary and chip electrophoresis. Electrophoresis 24:34–54
19. Landers JP (2008) Introduction to Capillary Electrophoresis in Capillary and Microchip Electrophoresis and Associated Microtechniques. In: Landers JP (ed) Handbook of Capillary and Microchip Electrophoresis and Associated Microtechniques. CRC Press, New York, pp 4–21
20. Dolnik V (1997) Capillary zone electrophoresis of proteins. Electrophoresis 18:2353–2361
21. Kasicka V (1999) Capillary electrophoresis of peptides. Electrophoresis 20:3084–3105
22. Dolnik V (1999) Recent developments in capillary zone electrophoresis of proteins. Electrophoresis 20:3106–3115
23. Righetti PG, Gelfi C, Verzola B, Castelletti L (2001) The state of the art of dynamic coatings. Electrophoresis 22:603–611
24. Dolnik V, Hutterer KM (2001) Capillary electrophoresis of proteins 1999–2001. Electrophoresis 22:4163–4178
25. Hutterer K, Dolník V (2003) Capillary electrophoresis of proteins 2001–2003. Electrophoresis 24:3998–4012
26. Dolnik H (2006) Capillary electrophoresis of proteins 2003–2005. Electrophoresis 27:126–141
27. Dolnik V (2008) Capillary electrophoresis of proteins 2005–2007. Electrophoresis 29: 143–156
28. Lauer HH, McManigill D (1986) Anal Chem 58:166–170

29. Cunico RL, Gooding KM, Wehr T (1998) Basic HPLC and CE of biomolecules. Bay Bioanalytical Laboratory, Inc., Richmond, California
30. Corradini D, Bevilacqua L, Nicoletti I (2005) Separation of basic proteins in bare fused-silica capillaries with diethylentriamine phosphate buffer as the background electrolyte solution. Chromatogr Suppl 62:S43–S50
31. Doherty EAS et al (2002) Critical factors for high-performance physically adsorbed (dynamic) polymeric wall coatings for capillary electrophoresis. Electrophoresis 23: 2766–2776
32. Khaledi MG (1998) High performance capillary electrophoresis: theory, techniques, and applications. Wiley Interscience Publications, New York Chichester
33. Wätzig H, Degenhardt M, Kunkel A (1998) Strategies for capillary electrophoresis: method development and validation for pharmaceutical and biological applications. Electrophoresis 19:2695–2752
34. Li MX, Liu L, Wu J-T, Lubman MD (1997) Use of a Polybrene Capillary Coating in Capillary Electrophoresis for Rapid Analysis of Hemoglobin Variants with On-Line Detection via an Ion Trap Storage/Reflectron Time-of-Flight Mass Spectrometer. Anal Chem 69: 2451–2456
35. Guryca V, Pacakova V, Tlust'akova M, Stulik K, Michalek J (2004) Topographical properties of polymer films deposited in capillaries for electrophoretic separations of large organic molecules. J Sep Sci 27:1121–1129
36. Figeys D, Aebersold R (1997) Capillary electrophoresis of peptides and proteins at neutral pH in capillaries covalently coated with polyethyleneimine. J Chromatogr B Biomed Sci Appl 695:163–168
37. Gilges M, Kleemiss MH, Schomburg G (1994) Capillary Zone Electrophoresis Separations of Basic and Acidic Proteins Using Poly(vinyl alcohol) Coatings in Fused Silica Capillaries. Anal Chem 66:2038–2046
38. Haselberg R, de Jong G, Somsen G (2007) Capillary electrophoresis - mass spectrometry for the analysis of intact proteins. J Chromatogr A 1159:81–109
39. Hjertén S (1967) Free zone electrophoresis. Chromatogr Rev 9:122–219
40. Liu Q, Lin F, Hartwick RA (1997) Free solution capillary electrophoretic separation of basic proteins and drugs using cationic polymer coated capillaries. J Liq Chrom Rel Technol 20:707–718
41. Chiari M, Cretich M, Stastna M, Radko SP, Chrambach A (2001) Rapid capillary coating by epoxy-poly-(dimethylacrylamide): performance in capillary zone electrophoresis of protein and polystyrene carboxylate. Electrophoresis 22: 656–659
42. Danger G, Ramonda M, Cottet H (2007) Control of the electroosmotic flow in capillary electrophoresis using polyelectrolytes of different charge densities. Electrophoresis 28:925–931
43. Verzola B, Gelfi C, Righetti PG (2000) Protein adsorption to the bare silica wall in capillary electrophoresis. Quantitative study on the chemical composition of the background electrolyte for minimising the phenomenon. J Chromatogr A 868:85–99
44. Verzola B, Gelfi C, Righetti PG (2000) Quantitative studies on the adsorption of proteins to the bare silica wall in capillary electrophoresis. II. Effects of adsorbed, neutral polymers on quenching the interaction. J Chromatogr A 874:293–303
45. Lin C-Y, Yu C-J, Chen Y-M, Chang H-C, Tseng W-L (2007) Simultaneous separation of anionic and cationic proteins by capillary electrophoresis using high concentration of poly(diallyldimethylammonium chloride) as an additive. J Chromatogr A 1165:219–225
46. Graul T, Schlenoff J (1999) Capillaries Modified by Polyelectrolyte Multilayers for Electrophoretic Separations. Anal Chem 71: 4007–4013
47. Pesek JJ, Matyska MT, Swedberg S, Udivar S (1999) Protein and peptide separations on high surface area capillaries. Electrophoresis 20:2343–2348
48. Erim FB, Cifuentes A, Poppe H, Kraak JC (1995) Performance of a physically adsorbed high-molecular-mass polyethyleneimine layer as coating for the separation of basic proteins and peptides by capillary electrophoresis. J Chromatogr A 708:356–361
49. Chiu RW, Jimenez JC, Monnig CA (1995) High Molecular Weight Polyarginine as a capillary coating for separation of cationic proteins by capillary electrophoresis. Anal Chim Acta 307:193–201
50. Righetti PG (2001) Capillary Electrophoretic Analysis of Proteins and Peptides of Bomedical and Pharmaceutical Intrest. Biopharm Drug Dispos 22:337–351
51. Roche ME, Anderson MA, Oda RP, Riggs LB, Strausbauch MA, Okazaki R, Wettstein PJ, Landers JP (1998) Capillary electrophoresis of insulin-like growth factors: enhanced ultraviolet detection using dynamically coated capillaries and on-line solid-phase extraction. Anal Biochem 258:87–95
52. Katayama H, Ishihama Y, Asakawa N (1998) Stable Capillary Coating with Successive

Multiple Ionic Polymer Layers for Capillary Electrophoresis. Anal Chem 70:5272–5277

53. Katayama H, Ishihama Y, Asakawa N (1998) Stable Capillary Coating with Successive Multiple Ionic Polymer Layers. Anal Chem 70:2254–2260
54. Nehmé R, Perrin C, Guerlavais V, Fehrentz J-A, Cottet H, Martinez J, Fabre H (2009) Use of coated capillaries for the electrophoretic separation of stereoisomers of a growth hormone secretagogue. Electrophoresis 30:3772–3779
55. Nehmé R, Perrin C, Cottet H, Blanchin MD, Fabre H (2008) Influence of polyelectrolyte coating conditions on capillary coating stability and separation efficiency in capillary electrophoresis. Electrophoresis 29:3013–3023
56. Nehmé R, Perrin C, Cottet H, Blanchin MD, Fabre H (2009) Influence of polyelectrolyte capillary coating conditions on protein analysis in capillary electrophoresis. Electrophoresis 30:1888–1898
57. Nehmé R, Perrin C, Cottet H, Blanchin MD, Fabre H (2011) Stability of capillaries coated with highly charged polyelectrolyte monolayers and multilayers under various analytical conditions-Application to protein analysis. J Chromatogr A 1218:3537–3544
58. Liu Q, Lin F, Hartwick RA (1997) Poly(diallyldimethylammonium chloride) as a cationic coating for capillary electrophoresis. J Chromatogr Sci 35:126–130
59. Huang X, Wang Q, Huang B (2006) Preparation and evaluation of stable coating for capillary electrophoresis using coupled chitosan as coated modifier. Talanta 69:463–468
60. Yao YJ, Li SFY (1994) Capillary zone electrophoresis of basic proteins with chitosan as a capillary modifier. J Chromatogr A 663:97–104
61. Nutku MS, Erim FB (1998) The Use of Cationic Polymer for the Separation of Inorganic Anions by Capillary Electrophoresis. J High Resol Chromatogr 21:505–508
62. Dubas S, Schlenoff J (1999) Factors Controlling the Growth of Polyelectrolyte Multilayers. Macromolecules 32:8153–8160
63. Dubas ST, Schlenoff JB (2001) Polyelectrolyte multilayers containing a weak polyacid: construction and deconstruction. Macromolecules 34:3736–3740
64. Schlenoff JB, Ly H, Li M (1998) Charge and Mass Balance in Polyelectrolyte Multilayers. J Am Chem Soc 120:7626–7634
65. Schlenoff JB, Dubas ST (2001) Mechanism of Polyelectrolyte multilayer growth: charge overcompensation and distribution. Macromolecules 34:592–598
66. Farhat T, Schlenoff JB (2001) Ion Transport and Equilibria in Polyelectrolyte Multilayers. Langmuir 17:1184–1192
67. Salloum DS, Schlenoff JB (2004) Protein adsorption modalities on polyelectrolyte multilayers. Biomacromolecules 5:1089–1096

Chapter 15

Capillary Electrophoresis with Laser-Induced Fluorescence Detection of Proteins from Two Types of Complex Sample Matrices: Food and Biological Fluids

Raul Garrido-Medina, Angel Puerta, Cristina Pelaez-Lorenzo, Zuly Rivera-Monroy, Andras Guttman, Jose Carlos Diez-Masa, and Mercedes de Frutos

Abstract

Sample preparation and laser-induced fluorescence detection are two key steps of the analytical methodology to determine by capillary electrophoresis low concentrations of proteins in complex sample matrices. In this chapter the options of performing both steps in different ways are shown by detailing the analysis of the allergen β-lactoglobulin in food products for infants and the analysis of the isoforms of alpha 1-acid glycoprotein, a potential biomarker, in serum and secretome.

Key words: Sample preparation, Capillary electrophoresis, LIF, Detection, Lactoglobulin, Alpha 1-acid glycoprotein, Orosomucoid, Milk, Isoform, Biomarker

1. Introduction

At present, there is no doubt that capillary electrophoresis (CE) is a very powerful technique for protein analysis (1). This method features high resolution for separating proteins in complex samples as well as to separate closely related isoforms of a given protein related to its micro-heterogeneities.

However, when analyzing complex sample matrices, there are two important factors that should be considered in order to succeed in adequate analysis of proteins by CE. The first factor is the

Nicola Volpi and Francesca Maccari (eds.), *Capillary Electrophoresis of Biomolecules: Methods and Protocols*, Methods in Molecular Biology, vol. 984, DOI 10.1007/978-1-62703-296-4_15,

large number of compounds which are present in a broad concentration range, making almost impossible to detect a single analyte in spite of the high-resolution CE can offer. The second factor to fight arises in those cases in which the proteins of interest are present in a very low concentration. In this instance, the frequently used UV–vis detection does not provide high-enough sensitivity.

Two key steps to solve these problems to make possible carrying out the CE analysis of the proteins of interest are sample preparation and detection. Both of them can significantly contribute to increase the selectivity and the sensitivity required for a successful assay. Sample preparation is aimed at eliminating possibly interfering compounds as much as possible for the CE analysis. This step can also provide pre-concentration of the analytes of interest in the purified sample. The chosen detection mode should be sensitive enough to analyze the compounds of interest in the purified sample and, additionally, it can provide selectivity for some compounds. Protein analyses by CE are most commonly performed using UV detection due to the almost universal character of this approach and its affordable price. However, the limit of detection provided by UV detection for proteins can only reach the micromolar range, i.e., it is not sensitive enough to analyze low-concentration proteins of interest, such as trace amounts of allergens in food products or low copy level biomarkers in biological fluids. One of the most sensitive detection modes for CE is laser-induced fluorescence (LIF) reaching detection limits as low as picomolar; however many proteins do not have native fluorescence; thus they need to be fluorescently tagged with a fluorescent or fluorogenic dye prior to analysis (2). It is important to note that labeling should be performed in such a way that it neither causes excessive band broadening nor multiple peaks for the species of interest as it would distort the CE separation.

The schematic representation of the procedures to perform sample preparation and consequent derivatization and CE-LIF analysis of different proteins in complex sample matrices is delineated in Fig. 1. The first case corresponds to the analysis of the allergenic protein β-lactoglobulin (βLG) in infant foods. The second case describes the analysis of the potential biomarker alpha 1-acid glycoprotein (AGP) isoforms in two biological fluids, namely serum and secretome.

Food allergy is an increasing problem worldwide and the immunologic reaction to milk proteins is considered as one of the most common. Bovine βLG, a protein with a molecular mass of about 18 kDa and a pI of 5.09 for the variant A and pI of 5.23 for the variant B, is considered as the main allergenic protein in cow's milk (3), even when present at trace levels. Currently the only

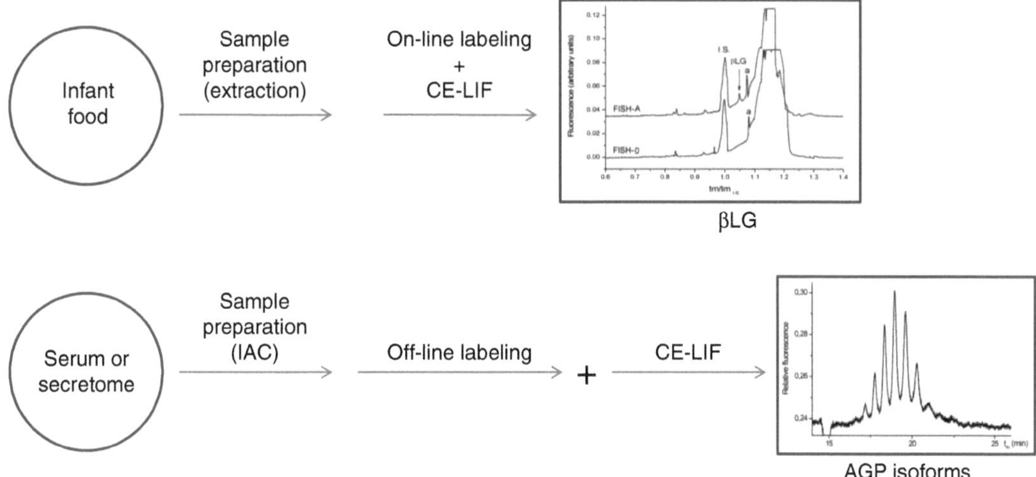

Fig. 1. Schematic representation of the two ways to perform sample preparation, fluorescent labeling and capillary electrophoresis with laser-induced fluorescence detection, for β-lactoglobulin (βLG) in baby foods and alpha 1-acid glycoprotein (AGP) isoforms in serum and secretome samples.

preventive measure for sensitized individuals is an avoidance diet. The recent regulations in the European Union (4) require that information about the presence of milk and its products in food, even in altered forms, should be provided to the consumer. Besides of the presence of milk proteins as food ingredients, in some instances milk proteins are not purposely added, i.e., their presence is the result of contamination during storage or production. Thus, analysis methods selective and sensitive enough for detecting trace levels of the allergen βLG in different food products and even during production processes are greatly needed.

Our group has previously reported that sample preparation methods have noticeable effects on the sensitivity of ELISA and CE analyses for βLG (5–7). The composition of solvents used to extract βLG from the samples was designed, taking into consideration the possibility of βLG being incorporated in casein micelles or to interact with other components of the food matrix when the food was submitted to heat treatment during processing. Thus, the recommended extraction solution contained a reducing agent (β-mercaptoethanol), guanidine hydrochloride to make specific regions of the polypeptide chain more accessible, a buffer to regulate pH (borate), and high salt concentration (sodium chloride) to enhance βLG solubility. In the present chapter it is detailed how to apply this sample preparation method to determine βLG in three types of baby foods purposely contaminated with dairy products before submission to thermal treatment to simulate the cross-contamination that potentially could happen during production. Once the sample is prepared, on-capillary derivatization of proteins through their amino groups is performed and CE separation with LIF detection is carried out in the same capillary (on-line) (7) (see top part of Fig. 1).

The bottom part of Fig. 1 details the procedure for determining glycoprotein isoforms in biological fluids. Protein glycosylation depends, among other factors, on the pathophysiological conditions of the individual. Glycoprotein isoforms can serve as biomarkers in different major diseases such as cancer or vascular diseases. Alpha 1-acid glycoprotein (AGP) is a 41–43 kDa protein with pI in the range 2.8–3.8 with very high heterogeneity regarding both of its glycosidic and peptidic moieties (8–10). We have previously shown that in order to analyze AGP isoforms by CE, the target glycoprotein should be previously purified from the biological matrix (11, 12). The high selectivity provided by the antibody–antigen recognition and the advantages of performing it on chromatographic format made immunoaffinity chromatography (IAC) the technique of choice for purifying AGP from serum or plasma (13). In the procedure detailed in the present chapter, AGP was purified by IAC from serum and secretome from an artery section of a patient suffering from a vascular disease. The purified glycoprotein was fluorescently derivatized via its thiol groups off-line (14) and the labeled protein was analyzed by CE-LIF.

2. Materials

2.1. Samples

1. Standards
   β-lactoglobulin A+B (βLG), alpha 1-acid glycoprotein (AGP) from human plasma and rhodamine B (RhB) were from Sigma-Aldrich (Steinheim, Germany).
2. Samples
   Three types of baby foods (MEAT, FISH, and FRUIT) were analyzed (Table 1). The main components on them were chicken with rice in the MEAT type, hake with rice in the FISH type, and orange and banana with cereal in the FRUIT type. Samples named MEAT-0, FISH-0, and FRUIT-0 corresponded to baby foods of each type guaranteed free of dairy products. These matrices were on purpose contaminated with dairy desserts in order to simulate contamination could occur during manufacturing. The contaminants were rice pudding (containing 65% of milk) for the sample MEAT-A, fresh cheese dessert with fruit (40% of fresh cheese) for the FISH-A sample, and yoghurt with pear (26% of yoghurt) for the FRUIT-A and FRUIT-B

Table 1
Composition of the samples studied

Food matrix	Dairy dessert	Dairy dessert: food matrix ratio	Dairy product: food matrix ratio[a]	Sample name[b]
Orange and banana with cereal	None	Guaranteed free of dairy product	0	FRUIT-0
Orange and banana with cereal	Yoghurt with pear	1:260	1:1,000	FRUIT-A
Orange and banana with cereal	Yoghurt with pear	1:2,600	1:10,000	FRUIT-B
Hake with rice	None	Guaranteed free of dairy product	0	FISH-0
Hake with rice	Fresh cheese dessert with fruit	1:400	1:1,000	FISH-A
Chicken with rice	None	Guaranteed free of dairy product	0	MEAT-0
Chicken with rice	Rice pudding	1:650	1:1,000	MEAT-A

[a]Ratios calculated knowing the percentage of each dairy product in the dairy dessert: the yoghurt with pear contains 26% of yoghurt, the fresh cheese with fruit contains 40% of cheese, and the rice pudding contains 65% of milk
[b]Samples of type FRUIT were heated at 105°C for 10 min, samples of type FISH were heated at 123°C for 45 min, and samples of type MEAT were heated at 121°C for 50 min
Adapted from Analytica Chimica Acta, 649, C. Pelaez-Lorenzo, J.C. Diez-Masa, I. Vasallo, M. de Frutos, A new sample preparation method compatible with capillary electrophoresis and laser-induced fluorescence for improving detection of low levels of β-lactoglobulin in infant foods, 202–210, Copyright (2009), with permission from Elsevier

samples. After the contamination, the baby foods were subject to the same thermal treatment that they would experience if contamination had occurred during the production process (50 min at 121°C for the MEAT type, 45 min at 123°C for the FISH type, and 10 min at 105°C for the FRUIT type samples). All samples were prepared for this study by Hero España, S.A. (Murcia, Spain).

Serum and secretome. Human serum (provided by Hospital Ramon y Cajal, Madrid, Spain) and secretome from an artery section (provided by Fundacion Jimenez Diaz, Madrid, Spain) were used as samples. Obtaining of the samples was approved by the ethical committees of both hospitals. Serum was obtained by standard procedure (15). Secretome was obtained from an artery section of a patient suffering from abdominal aortic aneurysm by incubation of the tissue during 24 h in serum-free RPMI medium at 37°C and collecting the supernatant (this sample will be referred as secretome henceforth).

2.2. Sample Preparation for βLG Detection in Infant Foods

1. Concentrated phosphate buffer saline (PBS×10) (see Note 1).
2. β-Mercaptoethanol, guanidine hydrochloride, and sodium tetraborate decahydrate (borax) were purchased from Sigma (St. Louis, MO, USA).
3. Sodium chloride was from Merck (Darmstadt, Germany).
4. Supernatant filtration: Whatman® 40 filter papers (Maidstone, UK) and syringe filters PVDF membrane 0.45 μm pore size from Millipore (Bedford, MA, USA).
5. Water was obtained from a Milli-Q water purification system (Millipore, Bedford, MA, USA).

2.3. Sample Preparation for Analysis of AGP Isoforms in Serum and Secretome

AGP from serum and secretome was purified by Immunoaffinity Chromatography (IAC) (12, 13, 15).

1. Polyclonal goat anti-human AGP affinity purified was purchased from Immune Systems LTD (Devon, UK).
2. Epoxy–silica (Waters Protein-Pak Affinity epoxy-activated) of 40 μm particle size and 500 Å pore diameter was from Waters (Millipore, Waters Chromatography Division, Milford, MA, USA).
3. PEEK-lined stainless steel tubing (3 cm × 4.6 mm I.D.) (Grace Davison Discovery Science, Deerfield, IL, USA) provided with 1/4–1/16 reduction end fittings and with 2 μm pore diameter frits was used as IAC column.
4. Mobile phase and binding buffer: Phosphate buffer saline (PBS) (see Note 1).
5. Desorption solution: Glycine-HCl pH 2.2 (see Note 2).
6. Neutralizing solution: 0.1 M Na_2HPO_4.
7. The HPLC system consisted of a Waters HPLC 510 pump (Milford, MA, USA) connected to two 7125 Rheodyne (Cotati, CA, USA) injection valves (named as V1 and V2) with polyetheretherketone (PEEK) injection loops of 1.4 and 0.2 mL volume, respectively. The IAC column was connected to the second injection valve and to the detector using PEEK tubing. A UV–vis LC-95 detector from Perkin Elmer (Norwalk, CO, USA) set at 280 nm was employed. Detector signal was digitalized using a 406 Analog Interface module (Beckman Instruments, Fullerton, MO, USA) and a System Gold software version 8.1 (Beckman). A PEEK tube (length 4.5 cm, 1/16″ O.D., 0.020″ I.D.) connected to the detector outlet allowed manual collection of the sample fractions. This equipment was similar to the one shown in Fig. 1 of reference (16) except that in the present study the valve V3 and the detector at 450 nm were removed.
8. Centrifugal filter devices were used with nominal M_r 10,000 cutoff membrane and 0.5 mL volume either Microcon® 10

YM-10 (Millipore, Billerica, MA, USA) or Amicon®Ultra 0.5 mL 10 K (Millipore, Carrigtwohill, Cork, Ireland) and with nominal M_r50,000 cutoff membrane and 2 mL volume Centricon®YM-50 (Amicon, Beverly, MA, USA) (see Note 3).

2.4. Fluorescent Labeling for βLG Detection in Infant Foods

1. Labeling reagent: 3-(2-Furoyl)quinoline-2-carboxaldehyde (FQ) was from Molecular Probes (Eugene, OR, USA) (see Note 4).
2. Nucleophilic agent for the derivatization with FQ: Potassium cyanide was from Sigma (see Note 5).

2.5. Fluorescent Labeling for Analysis of AGP Isoforms in Serum and Secretome

1. Reducing solutions: 0.6 mM tris(2-carboxyethyl)phosphine (TCEP) (Sigma-Aldrich, Steinheim, Germany) in 133 mM ammonium bicarbonate and 0.06 mM TCEP in 200 mM ammonium bicarbonate (see Note 6).
2. Labeling solutions: 4.9 and 0.4 mM 5-(iodoacetamido) fluorescein (5-IAF, Sigma-Aldrich) in 100 mM ammonium bicarbonate and 20% acetonitrile to derivatize samples from serum and secretome, respectively (see Note 7).
3. Dithiothreitol (DTT) (Sigma-Aldrich) solution 1 M in water.
4. Ammonium bicarbonate (Fluka, Buchs, Germany): 100 mM solution in water, daily prepared.
5. Centrifugal filter devices Amicon® Ultra 0.5 mL 10 K.

2.6. Capillary Electrophoresis Analysis for βLG Detection in Infant Foods

1. The in-house made CE apparatus with LIF detection employed previously described (17) was used with some modifications (7). Briefly, an RS/EH50R high-voltage power supply (Glassman High Voltage, Whitehouse Station, NJ, USA) was used in normal polarity configuration (anode connected at the inlet end of the capillary). Sample injection was carried out by raising the sample vial at a given height during a given time (injection by gravity). A 2060-10S Spectra Physics Ar-ion laser (9 mW) (Spectra Physics, Mountain View, CA, USA) was used as an excitation source at 488 nm in an orthogonal configuration. Fluorescence was collected using a 40× plan achromatic microscope objective, passed through a 550 nm cutoff filter and an interference filter centered at 590 nm, imaged onto an iris to block straight light, and detected with an R928 photomultiplier tube (Hamamatsu, Mamamatsu City, Japan) operated at 600 V and assembled on top of a high-precision stage for alignment (Newport, Mountain View, CA, USA). Photocurrent was processed by a 7070 photometer (Oriel, Stratford, CT, USA) and a 406 System Gold A/D converter (Beckman, Fullerton, CA, USA). A laboratory-made device was used to heat a small zone (10 cm) at the inlet of the capillary where the derivatization reaction took place,

using an F3 thermostatic bath (Haake, Karlsruhe, Germany). The heating device was made using two PVDF T-pieces model 1C300T03PV (EM-Technik, Maxdorf, Germany) interconnected by a 1 cm long Tygon tube (6 mm O.D., 5 mm I.D.) using one entrance of each of the T-pieces. This assembly formed a jacket that allowed the insertion of the first 10 cm of the separation capillary through the distal entrances of the two T-pieces. This jacket allowed the control of the temperature in this length of the capillary using water from a circulating thermostatic bath entering the device through Tygon tube connected to the two remaining entrances of the T-pieces. Water-tightness of the device was achieved using PVDF cutting rings and thumbnuts for Tygon tubing connections and silicone septa for the insertion of separation capillary.

2. Uncoated capillaries (Composite Metal Services, Worcester, UK) 60 cm length (50 cm to the detector) and 50 μm I.D. (375 μm O.D.) were used (see Note 8).
3. Separation buffer: 6 mM Borax, 6 mM SDS (pH 9.0).
4. Internal standard: Rhodamine B.

2.7. Capillary Electrophoresis for Analysis of AGP Isoforms in Serum and Secretome

1. The instrument used for CZE separations was a G1600AX CE system from Agilent Technologies (Waghäusel, Germany). Sample injection was performed by applying pressure. LIF detection was performed by means of a ZetaLIF Discovery detector (Picometrics, Toulouse, France), connected to the CE instrument, and excitation was carried out by a PC 13589 model DPSS laser (Spectra Physics) at 488 nm with a nominal output power of 20 mW. Photomultiplier tube at the detector was set at 700 V (see Note 9).
2. Fused-silica capillaries 70 cm (51 cm to the detector) and 50 μm I.D. (375 μm O.D.) were supplied by Picometrics (see Note 10).
3. Background electrolyte (BGE): 10 mM Sodium acetate, 10 mM tricine, 10 mM sodium chloride, 7 M urea, and 4.5 mM putrescine pH adjusted at 4.5 (see Note 11).

3. Methods

3.1. Sample Preparation for Detection of βLG in Baby Foods

Suspend 1 g of the baby food sample in 10 mL of concentrated buffer saline solution (PBS×10) (see Note 1). Add an aliquot of 500 μL of the extracting solution containing a final concentration of 24 mM β-mercaptoethanol, 25 mM guanidine hydrochloride, 5% (v/v) of 2.5 mM borate buffer at pH 8.3, and 30 mL of 0.15 M sodium chloride. Shake the mixture for 15 min at 80°C and centrifuge at

9,000 × *g* for 25 min at room temperature. Collect the supernatant. Finally, filter it successively through a Whatman® 40 filter paper and through a syringe filter PVDF membrane 0.45 μm pore size.

3.2. Sample Preparation for Analysis of AGP Isoforms in Serum and Secretome

The sample preparation method for AGP isoform analysis is similar to the one previously reported for plasma samples (13).

1. Immunoaffinity chromatography (IAC) anti-AGP column fabrication

   Binding of the antibody (Ab) to the packing material: Transform the epoxy groups of the supporting material into diol groups by adding 17.5 mL of 0.072 N sulfuric acid to 0.385 g Protein-Pack material and shaking for 1 h at room temperature. Discard the supernatant and wash the supporting material with water. Oxidize the formed diol groups to aldehyde by adding 5 mL of 0.23 M sodium periodate and keep for 2 h at room temperature. To stop the oxidation reaction, add 3.4 mL glycerol diluted with 15 mL of reaction solution (0.1 M sodium phosphate at pH 5.7). Discard the supernatant and wash the packing material with the reaction solution. Simultaneously, concentrate 3 mL of commercial antibody (1 mg/mL) to 0.5 mL and transfer them to the reaction solution using a centrifugal filter device (Centricon®) with a 50 kDa nominal cutoff membrane (see Note 12). Add the concentrated 0.5 mL antibody plus the same volume of sodium cyanoborohydride to the supporting material. Shake the mixture for 65 h at 4°C. Discard the supernatant and wash the Ab-packing material with 0.1 M sodium phosphate (pH 7). Reduce the unreacted aldehyde groups with 7 mg of sodium borohydride. Wash the Ab-derivatized Protein-Pack material with 0.1 M sodium phosphate at pH 7, then with a solution of pH 7 containing 0.1 M sodium phosphate and 0.5 M sodium chloride, and finally with a solution of PBS with 0.02% (w/v) sodium azide. Keep the Ab-packing material at 4°C until packing into the HPLC tubing.

   Packing the Ab support:

   Prepare a packing device consisting of a precolumn (0.63 mL volume) of the same internal diameter that of the column connected with a zero dead volume union to a larger diameter tube used as packing reservoir (13.5 mL volume). Remove the top 1/4–1/16 reduction end fitting and the frit from the column tube and connect this column tube to the precolumn with a zero dead volume union. Keep the system reservoir–precolumn–column vertical and fill the column with PBS. Mix the derivatized packing material with about 8 mL PBS and stir manually. Pour the slurry into the reservoir avoiding support settling. Fill quickly the reservoir with PBS, connect it into the HPLC pump and flush PBS through the deposit–precolumn–column system at increasing flow rate until reaching 10 mL/min. Keep this flow rate for 15 min. Afterwards, slowly decrease the flow rate until reaching 0 mL/min, stop the pump and keep it in this way for 30 min. Remove carefully the precolumn and the reservoir. Place the frit and the 1/4–1/16 reduction end fitting at the column inlet. Flush the column with PBS plus 0.02% (w/v) sodium azide and keep it at 4°C until use.

2. AGP purification from serum and secretome

   Sample pretreatment:

   Incubate 0.1 mL of sample (either serum or secretome) with 1% (v:v) Protease Inhibitor Cocktail (Sigma) for 30 min at −20°C and then dilute the incubated sample to 0.2 mL with Milli-Q water.

   Immunopurification process:

   At the beginning of the working day, right after the column is installed in the chromatographic system, condition the IAC column by pumping through PBS at 0.5 mL/min for 5 min and clean it with 3 injections of 1.4 mL each of Gly-HCl desorption solution. When the baseline is stabilized ($t=0$), inject the 0.2 mL pretreated sample using the small loop in valve V2. After 10 min ($t=10$ min), when the absorbance signal returned to baseline level (because the unbound components of the sample have been eluted), inject 1.4 mL of desorption solution using the large loop in valve V1. Collect the desorbed fraction, which contains AGP, and neutralize it with 0.1 M Na_2HPO_4 (see Note 13). At $t=30$ min, reinject this neutralized fraction in the IAC column using the large loop in valve V1. At $t=40$ min inject 1.4 mL of desorption solution. Collect the desorbed fraction, which contains purified AGP, and neutralize it with 0.1 M Na_2HPO_4. Inject 2 × 1.4 mL desorption agent and wait until the signal returns to baseline level to perform a new immunopurification process of another sample. At the beginning of each working day, clean the column by injecting 3 × 1.4 mL desorption agent. The column is stored at 4°C in PBS plus 0.02% (w/v) sodium azide overnight.

3. Salt removal and concentration are performed by using centrifuge filer devices with 10 kDa cutoff membranes (see Note 14).

3.3. On-Line Fluorescent Protein Labeling and Capillary Electrophoresis Analysis with Laser-Induced Fluorescence Detection of βLG from Baby Foods

The procedure for on-capillary derivatization is based on previously optimized method (18).

1. New capillaries were rinsed with 1 M NaOH (100 μL) followed by a rinse with Milli-Q water (100 μL). Between runs, the capillary was sequentially rinsed with Milli-Q water (100 μL), 0.1 M NaOH (100 μL), Milli-Q water (100 μL), and the separation buffer (100 μL). Rinses were made manually employing a model 1710 glass syringe from Hamilton (Bonaduz, Switzerland).
2. To properly assign the βLG peak, add rhodamine B to the sample prepared as indicated in Subheading 3.1 so that the final concentration of this internal standard (IS) in the sample is 10^{-7} M.

3. Inject a plug of a mixture of the sample spiked with rhodamine B plus a KCN solution at final concentration 10 mM into the capillary. Inject 5 mM solution of FQ in separation buffer. Both injections are performed by gravity (20 cm height) for12 s (see Note 15).
4. Mixing step: Place the vials containing the separation buffer in the inlet and outlet ends of the capillary and apply high voltage of 3 kV for 6 min. Perform the step at 65°C by using the device described in Subheading 2.6.
5. Reaction step: Switch off the power supply for 15 s. Perform the step at 65°C.
6. Separation step: Apply 15 kV at room temperature (see Note 16).
7. Collect the electropherogram showing in the *X* axis the migration time of the peaks relative to the migration time of rhodamine B.
8. Compare the relative migration time of each peak to that of a sample of standard βLG analyzed in the same way. Assign the peak of βLG in the samples to the one with the same relative migration time than standard βLG. In this way the results shown in Fig. 2 and Table 2 for identifying the βLG peak in each sample are obtained. In the figure peaks a and b migrating close to the βLG peak and arising from other components of the sample are observed; the values of migration time relative to the internal standard calculated in Table 2 permit assigning the βLG peak.

3.4. Off-Line Fluorescent Protein Labeling for AGP Isoform Analysis from Serum and Secretome

1. Warm up the samples prepared as described in Subheading 3.2 to room temperature. Gently vortex the microtubes containing the samples for homogenization.
2. Add the corresponding reducing solution to each sample: 15 μL of 0.6 mM TCEP in the case of 200 μM AGP from serum and 5 μL of 0.06 mM TCEP in the case of 7 μM AGP from secretome (see Note 6).
3. Keep the samples at 37°C for 1 h in a thermostatic bath to reduce disulfide bridges.
4. Add the corresponding labeling solution to each sample: 20 μL of 4.9 mM 5-IAF in the case of AGP from serum and 10 μL of 0.4 mM 5-IAF in the case of AGP from secretome (see Notes 7 and 17).
5. Keep the mixture at 37°C for 2.5 h in a thermostatic bath.
6. After the labeling reaction, remove the sample from the bath and add 5 μL of 1 M DTT solution.
7. Keep the mixture at room temperature for 30 min to quench the labeling reaction.
8. Apply the mixture into an Amicon® Ultra 10 K centrifugal filter device and add 100 μL of 100 mM ammonium bicarbonate solution. Centrifuge at 14,000 × *g* for 30 min at 4°C (see Note 18).

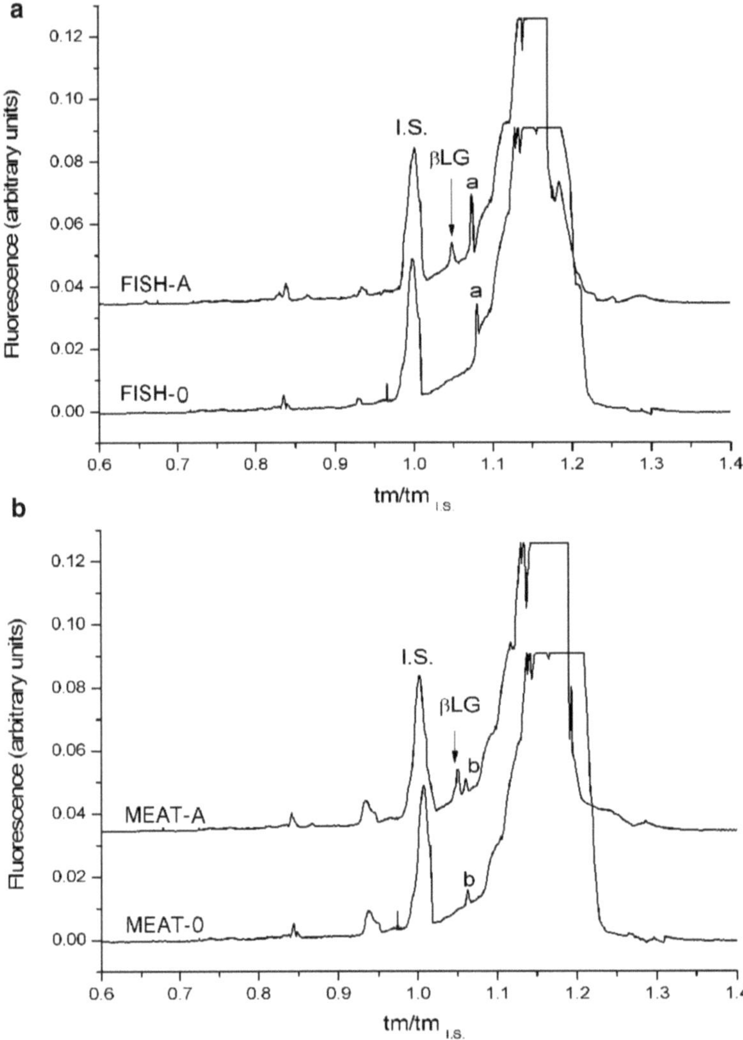

Fig. 2. CE-LIF analysis of the extracts of contaminated baby food samples (FISH-A and MEAT-A) and non-contaminated (FISH-0 and MEAT-0) with dairy products. Peaks a and b are indicated in Table 2. IS corresponding to RhB. Adapted from Analytica Chimica Acta, 649, C. Pelaez-Lorenzo, J.C. Diez-Masa, I. Vasallo, M. de Frutos, A new sample preparation method compatible with capillary electrophoresis and laser-induced fluorescence for improving detection of low levels of β-lactoglobulin in infant foods, 202–210, Copyright (2009), with permission from Elsevier.

9. Discard the filtrate and add 100 μL of 100 mM ammonium bicarbonate solution. Repeat centrifugation.
10. Discard the filtrate and wash the labeled glycoprotein by the addition of 100 μL of water to the centrifugal filter device each time, until the filtrate is colorless. Centrifugation conditions as in step 8 (see Notes 18 and 19).
11. Discard the filtrate and place the device containing the labeled glycoprotein into a new microtube at recovering position (upside down). Centrifuge at 1,000 × *g* for 3 min at 4°C.

Table 2
CE-LIF migration time (tm) and migration time relative to the I.S. ($tm/tm_{I.S.}$) for the peaks of interest in Fig. 2

Sample	$tm_{I.S.}$ (min)	$tm_{\beta LG}$ (min)	$tm_{\beta LG}/tm_{I.S.}$	$tm_{peak\ a}$ (min)	$tm_{peak\ a}/tm_{I.S.}$	$tm_{peak\ b}$ (min)	$tm_{peak\ b}/tm_{I.S.}$
FRUIT-A	8.793	9.194	1.046	n.o.	n.o.	n.o.	n.o.
FRUIT-B	8.774	9.197	1.048	n.o.	n.o.	n.o.	n.o.
FRUIT-0	9.694	n.o.	n.o.	n.o.	n.o.	n.o.	n.o.
FISH-A	8.562	8.957	1.046	9.240	1.079	n.o.	n.o.
FISH-0	8.517	n.o.	n.o.	9.195	1.080	n.o.	n.o.
MEAT-A	8.643	9.050	1.047	n.o.	n.o.	9.134	1.057
MEAT-0	8.561	n.o.	n.o.	n.o.	n.o.	9.053	1.057

$tm_{I.S.}$: migration time of RhB peak; $tm_{\beta LG}$: migration time of βLG peak; $tm_{\beta LG}/tm_{I.S.}$: migration time of βLG peak relative to migration time of RhB; $tm_{peak\ a}$: migration time of peak a; $tm_{peak\ a}/tm_{I.S.}$: migration time of peak a relative to migration time of RhB; $tm_{peak\ b}$: migration time of peak b; $tm_{peak\ b}/tm_{I.S.}$: migration time of peak b relative to migration time of RhB

Peaks a and b are shown in Fig. 2

n.o.: Not observed

Adapted from Analytica Chimica Acta, 649, C. Pelaez-Lorenzo, J.C. Diez-Masa, I. Vasallo, M. de Frutos, A new sample preparation method compatible with capillary electrophoresis and laser-induced fluorescence for improving detection of low levels of β-lactoglobulin in infant foods, 202–210, Copyright (2009), with permission from Elsevier

12. Recover the labeled AGP in water (15–20 μL), place it into a deactivated glass sample injection vial, and store at −20°C until the CZE-LIF analysis.

3.5. Capillary Electrophoresis with Laser-Induced Fluorescence Detection to Analyze the AGP Isoforms from Serum and Secretome

CZE-LIF analysis of AGP isoforms is carried out according to a previously published CZE-UV method (19), with some modifications.

1. Condition new capillaries as follows: Rinse by pressure at 1 bar with 1 M NaOH for 30 min, water for 5 min, and finally 0.1 M NaOH for 15 min.
2. Between sample runs, rinse the capillaries by pressure at 1 bar with water for 5 min, 0.1 M NaOH for 10 min, water for 5 min, and finally with the BGE for 5 min.
3. Introduce the labeled sample into the capillary by hydrodynamic injection at 35 mbar for 30 s (injected volume is approximately 32 nL).
4. Perform CZE separation at 25 kV (357 V/cm) in normal polarity (anode at injection) at 35°C. An example of the expected results is shown in Fig. 3, exhibiting the different isoforms of AGP from serum and secretome samples analyzed by CE-LIF, which are very similar to the ones observed by CE-UV for non-labeled standard AGP.

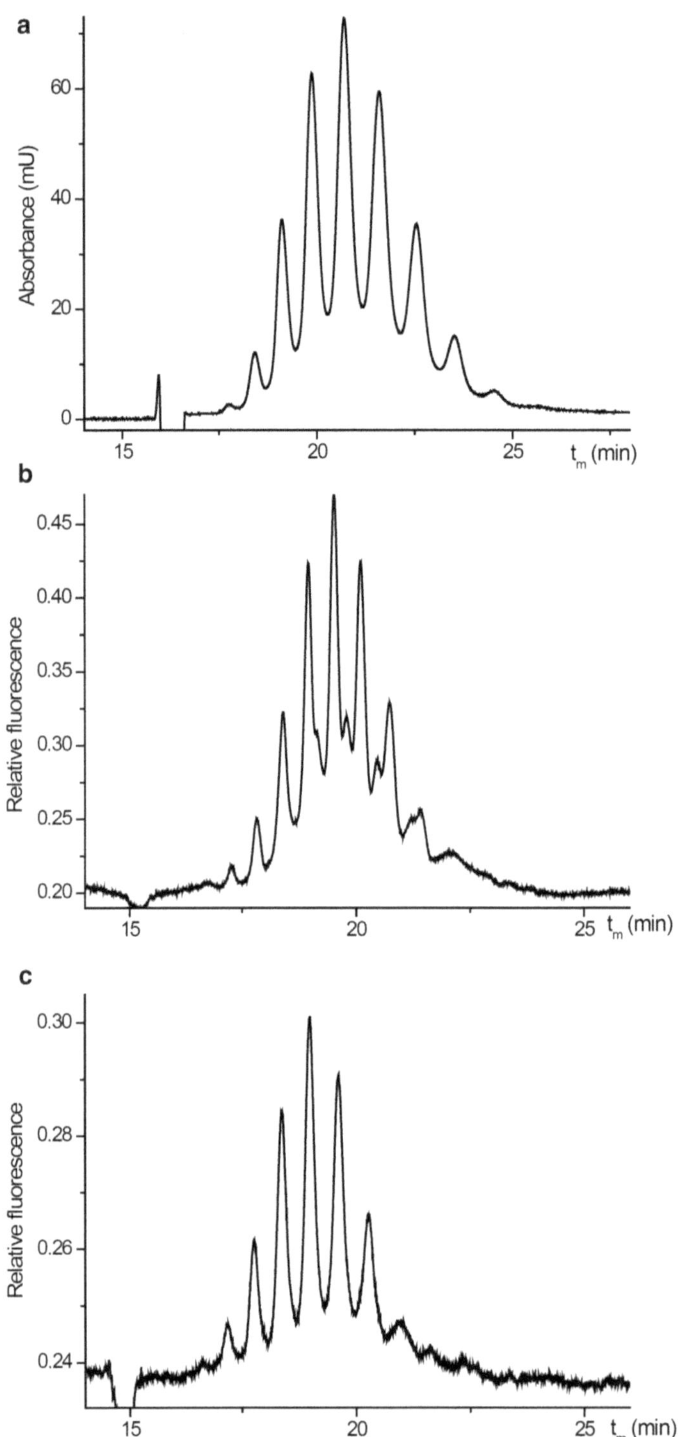

Fig. 3. CZE analysis of the isoforms of (**a**) standard AGP (0.4 mM) with UV detection, (**b**) and (**c**) 5-IAF labeled AGP purified from (**b**) serum and (**c**) secretome, both with LIF detection. BGE: 10 mM sodium acetate, 10 mM tricine, 10 mM sodium chloride, 7 M urea, and 4.5 mM putrescine, pH adjusted to 4.5. Conditions for CZE-UV: hydrodynamic injection at 35 mbar for 25 s, 59.5 cm total length, 51 cm effective length, V = 21.2 kV, 35°C, detection wavelength 214 nm. Conditions for CZE-LIF: hydrodynamic injection at 35 mbar for 30 s, 70 cm total length, 51 cm effective length, V = 25 kV, 35°C, and LIF detection. Adapted from R. Garrido-Medina, A. Puerta, Z. Rivera-Monroy, M. de Frutos, A. Guttman, J.C. Diez-Masa: Analysis of alpha-1-acid glycoprotein isoforms using CE-LIF with fluorescent thiol-derivatization. Electrophoresis. 2012, 33:1113–1119. Copyright Wiley-VCH Verlag GmbH & Co. KGaA. Reproduced with permission.

4. Notes

1. PBS consists of 0.01 M disodium hydrogen phosphate/sodium dihydrogen phosphate, 0.138 M NaCl, 2.7 mM KCl, pH 7.4. Ten times concentrated PBS (PBS×10) was prepared and from it PBS was prepared daily by diluting 1 part of (PBS×10) with 9 parts of Milli-Q water. PBS×10 was prepared as follows: A solution A consisting of 0.1 M Na_2HPO_4, 1.38 M NaCl and 27 mM KCl and a solution B consisting of 0.1 M NaH_2PO_4, 1.38 M NaCl and 27 mM KCl were prepared. Solutions A and B were mixed until reaching pH 6.85. Be aware that pH can be modified by the dilution process. For this reason (PBS×10) was made at pH 6.85 so after 10× dilution the pH was 7.4.
2. Gly-HCl pH 2.2 is made as follows: Prepare 500 mL of 0.1 M glycine containing 885 μL of pure HCl (Hydrochloric acid 35% purissimum). After that, add drop by drop 2 N HCl until pH 2.2 is reached.
3. The minimum volume in which the sample can be recovered by Microcon® 10 YM-10 and by Amicon® Ultra 0.5 mL 10 K is 5 and 15–20 μL, respectively, due to the design of the filter devices. Thus, Microcon® 10 YM-10 devices are suggested to desalt and concentrate samples with low concentration of target analyte as it is the case of secretome samples.
4. Preparation of FQ solutions: A 50 mM stock solution of FQ was prepared in methanol (HPLC grade, Scharlau Chemie, Barcelona, Spain). Since FQ degrades in solution even when stored at -20°C in darkness, small aliquots of dried FQ were prepared. To do so 10 μL aliquots of the methanolic solution were transferred to 500 μL microcentrifuge tubes. The solvent was removed under vacuum at room temperature using a model RC10-10 centrifugal evaporator (Jouan, Saint-Herblain, France). The dried FQ aliquots were stored at –20°C until use. On the day of the experiment, the aliquots were defrosted and solved in the running buffer to obtain a 5 mM FQ concentration.
5. Potassium cyanide is highly poisonous and reacts directly with acids to form lethal HCN gas. Therefore, stock solutions should be made in a basic buffer. Neutralization of waste containing KCN solution should be done by the addition of 1% NaOH solution (about 50 μL NaOH/g cyanide) followed by slow addition of bleach (about 70 μL bleach/g of cyanide).
6. Diluted solutions of TCEP in water are unstable; therefore, reducing solutions are freshly prepared for each labeling reaction. The appropriate amount of TCEP is dissolved in the respective solution of ammonium bicarbonate. Reducing agent is added to keep a ratio 2:1 in respect to the number of cysteine

residues in the glycoprotein, because if large excess is added, TCEP can react with the fluorescent dye. Solution pH is unadjusted. The amount of ammonium bicarbonate is calculated to obtain a final concentration of 100 mM bicarbonate in the mixture TCEP in bicarbonate plus AGP. For the samples in this study: 5 μL of AGP purified from serum (approximate AGP concentration 200 μM) were mixed with 15 μL of 0.6 mM TCEP in 133 mM ammonium bicarbonate. For AGP purified from secretome, 5 μL (approximate AGP concentration 7 μM) were mixed with 0.06 mM TCEP in 5 μL of 200 mM ammonium bicarbonate.

7. To prepare 4.9 mM 5-IAF labeling solution, 0.5 mg of the dye is dissolved in 200 μL medium containing 100 mM ammonium bicarbonate and 20% acetonitrile, pH unadjusted, to ensure complete solubilization of 5-IAF. The labeling solution is prepared in a microtube by vortexing. The 0.4 mM 5-IAF solution is prepared by dilution in the same medium. All the labeling solutions are prepared daily and stored in darkness until use. The 4.9 mM solution is used to derivatize the 200 μM AGP purified from serum. The 0.4 mM solution is used to label the 7 μM AGP purified from secretome.
8. Capillaries are purchased in reels and cut to the desired length. Detection window is made by burning.
9. The fluorescence signal is collected and passed through a long-pass filter (>498 nm, as specified by Picometrics) in order to filter out any excitation light.
10. Capillaries used in the Picometrics device are equipped with an ellipsoid mirror glued to the capillary window in order to collect the maximum amount of fluorescence signal from the sample.
11. BGE is weekly prepared and stored at 4°C. As putrescine is unstable at low concentrations, a 1 M stock solution is made and stored at 4°C, of which 225 μL are employed to prepare the BGE. The pH of the solution is adjusted to 4.5 by the addition of 2 M acetic acid, and finally, the volume is completed to 50 mL with water.
12. Buffer exchange and concentration of the antibody by centrifuge filter devices (Centricon®) with a 50 kDa nominal cutoff membrane (YM-50): Before using the YM-50 devices, they are passivated with a solution of 5% (w/v) Brij® 35 for 12 h to avoid any nonspecific adsorption of the antibody to the device (20). After that, rinse the filter device with Milli-Q water (3×2 mL) and wash the membrane by 2 mL Milli-Q water at 4,000×*g*, 4°C, for 15 min. Set the filter device in recovery mode (upside down) and centrifuge it at 573×*g* at 4°C for 2 min to remove the remaining water from the membrane. Set the filter device in concentrating position and add the antibody solution. As the maximum volume of the device is 2 mL and

the antibody volume is 3 mL, this concentration step is carried out by adding first 2 mL and concentrate and then adding 1 mL. Concentration is performed by centrifuging at 4,000 × *g* at 4°C for 50 min each time. After that, add reaction buffer (2 × 0.2 mL) and spin at 4,000 × *g* at 4°C for 15 min. Set the filter device in recovery mode and spin at 800 × *g* at 4°C, for 2 min, to recover the concentrated antibody. Add 0.1 mL of reaction buffer onto the membrane of the filter device and recover the remaining antibody from it by turning the device upside down (800 × *g* at 4°C for 2 min). Repeat the last step. Mix the 3 fractions containing the recovered concentrated antibody and add reaction buffer to get 0.5 mL final volume.

13. Neutralization of the fraction collected from the IAC column: The desorbed fraction is collected for 1.5 min at 0.5 mL/min flow rate and neutralized by adding 0.4–0.5 mL of 0.1 M Na_2HPO_4. Neutralization is checked by pH paper.
14. Salt removal and concentration: Place the immunopurified and neutralized AGP sample (volume about 1.2 mL) in a centrifugal filter device with a 10 kDa cut off membrane (either Amicon® Ultra 0.5 mL 10 K or Microcon® 10, though Microcon® 10 should be used for AGP from secretome, see Note 3). Because the maximum volume for the device is 0.5 mL, the process is performed by spinning 0.5 mL volumes, then adding another fraction of the sample to the same vial and spin again, and finally, adding the rest of the sample and spin it. Every time a fraction is placed in the device, centrifuge it at 13,900 × *g* for 25–30 min at 4°C. After concentrating the whole sample, rinse the AGP retained in the membrane of the filter device by centrifuging 3 × 0.5 mL of Milli-Q water at 13,900 × *g* for 30 min at 4°C. When there is almost no liquid on the membrane, set the device in the recovery mode (upside down) and centrifuge at 1,000 × *g* for 3 min at 4°C to recover the sample in the minimum volume that is usually about 20–30 μL for the Amicon® Ultra devices and 5 μL for the Microcon® devices.
15. Wash the inlet end of the capillary by immersion into a vial containing Milli-Q water, after each injection of either sample plus KCN or FQ.
16. Room temperature was set at 24°C.
17. Labeling dye 5-IAF is added at a ratio 30:1 in respect to moles of cysteine residues in AGP, assuming a complete reducing process. It is not necessary to extract the excess of TCEP from the reaction mixture, despite it could react with the dye, because the reducing agent is not added at a large excess (see Note 6). This procedure considerably simplifies the labeling process.
18. Centrifugal filter devices allow the extraction of low molecular weight compounds from the labeling mixture as TCEP and 5-IAF, which could interfere with the subsequent CZE

analysis. The addition of ammonium bicarbonate solution makes the elimination of the dye easier (higher solubilization at alkaline pH). Centrifugation time can be slightly shorter or longer, but in general, it is preferable to reduce the volume of the sample as much as possible, typically to 15–20 μL with the devices employed.

19. Normally, three cycles of washing with water are enough.

Acknowledgments

Isabel Vasallo from Hero Institute for Infant Nutrition (Hero España, S.A., Murcia, Spain) is acknowledged for preparing the infant food samples. Financial support from the Spanish Ministry of Science and Innovation (project CTQ2009-09399), CSIC (project 2010JP0013), and the Hungarian Government (project OTKA K-81839) is acknowledged. Angel Puerta acknowledges CSIC for a contract in the JAE doc program cofinanced by the European Union under the European Social Fund, and Raul Garrido-Medina acknowledges the Spanish Ministry of Education the Ph.D. fellowship (FPU) awarded.

References

1. Landers JP (ed) (2008) Handbook of Capillary and Microchip Electrophoresis and Associated Microtechniques, 3rd edn. CRC Press, Taylor & Francis Group, Boca Raton, FL
2. Veledo MT, Lara-Quintanar P, de Frutos M et al (2005) Fluorescence Detection in Capillary Electrophoresis. In: Marina ML, Rios A, Valcarcel M (eds) Analysis and Detection by Capillary Electrophoresis. Elsevier Science, Amsterdam, pp 305–374
3. Exl B-M (2001) A review of recent developments in the use of moderately hydrolyzed whey formulae in infant nutrition. Nutr Res 21:355–379
4. (2011) Regulation (EU) No 1169/2011 of the European Parliament and the Council of 25 October 2011 on the provision of food information to consumers, amending Regulations (EC) No 1924/2006 and (EC) No 1925/2006 of the European Parliament and of the Council, and repealing Commission Directive 87/250/EEC, Council Directive 90/496/EEC, Commission Directive 1999/10/EC, Directive 2000/13/EC of the European Parliament and of the Council, Commission Directives 2002/67/EC and 2008/5/EC and Commission Regulation (EC) No 608/2004, Off J Eur Union L304, 18–63
5. de Frutos M, Pelaez-Lorenzo C, Puerta A "Composition and procedure to extract beta-lactoglobulin from processed and non-processed samples and immunoassay method for beta-lactoglobulin" Patent ES200502938.
6. Pelaez-Lorenzo C, Diez-Masa JC, Vasallo I et al (2010) Development of an optimized ELISA and a sample preparation method for the detection of β-lactoglobulin traces in baby foods. J Agric Food Chem 58:1664–1671
7. Pelaez-Lorenzo C, Diez-Masa JC, Vasallo I et al (2009) A new sample preparation method compatible with capillary electrophoresis and laser-induced fluorescence for improving detection of low levels of β-lactoglobulin in infant foods. Anal Chim Acta 649:202–210
8. Fournier T, Medjoubi-N N, Porquet D (2000) Alpha-1-acid glycoprotein. Biochim Biophys Acta 1482:157–171
9. Ceciliani F, Pocacqua V (2007) The acute phase protein alpha 1-acid glycoprotein: A model for altered glycosylation during diseases. Curr Protein Pept Sci 8:91–108
10. Duche JC, Urien S, Simon N et al (2000) Expression of the genetic variants of human alpha-1-acid glycoprotein in cancer. Clin Biochem 33:197–202

11. Ongay S, Lacunza I, Diez-Masa JC et al (2010) Development of a fast and simple immunochromatographic method to purify alpha 1-acid glycoprotein from serum for analysis of its isoforms by capillary electrophoresis. Anal Chim Acta 663:206–212
12. Ongay S, Neususs C, Vaas S et al (2010) Evaluation of the effect of the immunopurification-based procedures on the CZE-UV and CZE-ESI-TOF-MS determination of isoforms of intact AGP from human serum. Electrophoresis 31:1796–1804
13. Puerta A, Martin-Alvarez PJ, Ongay S et al (2013) Immunoaffinity, capillary electrophoresis, and statistics for studying intact alpha 1-acid glycoprotein isoforms as atherothrombosis biomarker. Methods Mol Biol 919:215–30
14. Garrido-Medina R, Puerta A, Rivera-Monroy Z et al (2012) Analysis of alpha-1-acid glycoprotein isoforms using CE-LIF with fluorescent thiol-derivatization. Electrophoresis. 33:1113–1119
15. Puerta A, Diez-Masa JC, Martin-Alvarez PJ et al (2011) Study of the capillary electrophoresis profile of intact alpha-1-acid glycoprotein isoforms as biomarker of atherothrombosis. Analyst 136:816–822
16. Puerta A, Diez-Masa JC, de Frutos M (2006) Immunochromatographic determination of beta-lactoglobulin and its antigenic peptides in hypoallergenic formulas. Int Dairy J 16:406–414
17. Veledo MT, de Frutos M, Diez-Masa JC (2005) Amino acids determination using capillary electrophoresis with on-capillary derivatization and laser-induced fluorescence detection. J Chromatogr A 1079:335–343
18. Veledo MT, de Frutos M, Diez-Masa JC (2005) Development of a method for quantitative analysis of the major whey proteins by capillary electrophoresis with on-capillary derivatization and laser-induced fluorescence detection. J Sep Sci 28:935–940
19. Lacunza I, Sanz J, Diez-Masa JC et al (2006) CZE of human alpha-1-acid glycoprotein for qualitative and quantitative comparison of samples from different pathological conditions. Electrophoresis 27:4205–4214
20. Puerta A, Diez-Masa JC, de Frutos M (2004) Use of immunodotting to select the desorption agent for immunochromatography. J Immunol Methods 289:225–237

Chapter 16

Automated Capillary Electrophoresis in the Screening for Hemoglobinopathies

Frédéric Cotton, Fleur Wolff, and Béatrice Gulbis

Abstract

Hemoglobinopathies are genetic disorders of globin chains characterized by the decreased expression of α- or β-globin chains (thalassemias) or by the synthesis of an abnormal protein (hemoglobin variants in, e.g., sickle cell disease). The screening of most hemoglobinopathies relies, together with hematological results and clinical elements, on the separation and quantification of normal and abnormal hemoglobin fractions. Gel electrophoresis, isoelectric focusing, and HPLC have been the methods of choice for many years. For about 20 years, capillary electrophoresis has appeared as a strong alternative method. Since the early 2000s, automated instruments are commercially available for the analysis of Hb fractions in adult patients but also for neonatal screening.

Key words: Hemoglobin, Variant, Hemoglobinopathy, Thalassemia, Sickle cell disease, Capillary electrophoresis, Automation, Neonatal screening

1. Introduction

Hemoglobin (Hb) is a tetramer composed of 2 α-type and 2 β-type globin chains. During the fetal life, red blood cells mainly contain Hb F ($\alpha_2\gamma_2$). At birth, it still represents about 80% of total Hb, the rest being Hb A ($\alpha_2\beta_2$). After 6 months, Hb A reaches 95% and Hb F is less than 1.5%. At that time, a third fraction, Hb A_2 ($\alpha_2\delta_2$), is seen at a level of around 2.0–3.5%. Hemoglobinopathies are genetic disorders due to mutations of α- or β-globin genes and leading to a decreased expression of these genes (quantitative defects) or to the synthesis of an abnormal protein called Hb variant (qualitative defects).

Quantitative defects are mainly represented by α- and β-thalassemias, where the synthesis of normal α- or β-globin chain is decreased (1). Patients with β-thalassemia major (two mutated β-globin genes) have very severe anemia with many complications

Nicola Volpi and Francesca Maccari (eds.), *Capillary Electrophoresis of Biomolecules: Methods and Protocols*, Methods in Molecular Biology, vol. 984, DOI 10.1007/978-1-62703-296-4_16, © Springer Science+Business Media, LLC 2013

including growth retardation, bone marrow expansion, and iron overload. β-Thalassemia minor (one mutated β-globin gene) is a mild condition with microcytosis and hypochromia only. It is important to make the differential diagnosis between this defect and other causes of microcytic anemia (i.e., iron deficiency, α-thalassemia …), especially for genetic counselling. An elevated level of Hb A_2, and sometimes of Hb F, is the best diagnostic test for β-thalassemia minor. Different forms of α-thalassemia exist, depending on the number of α-globin genes defect. Clinical expression is only seen with the suppression of three genes, reflected by the presence of Hb H, a tetramer of β-globin chains (β_4). Hb Bart's (γ_4) is seen in newborns with one to three α-globin genes defects but also in healthy newborns, at low levels (<0.2%).

Until now, 1,200 Hb variants have been described. About 10% of them display an abnormal function (polymerization, instability, oxidation, increased affinity…) but only a few dozens are clinically significant. Hb S is the most frequent variant, causing sickle cell disease at the homozygous state (2). Due to the replacement of the glutamic acid in position 6 by a valine, Hb S tends to polymerize when deoxygenated and red blood cells then take an abnormal sickle shape. The lack of deformability of sickle cells is the cause of severe hemolytic anemia and life-threatening vaso-occlusive crises. Other Hb variants such as Hb C, Hb D-Punjab, or Hb O-Arab may copolymerize with Hb S. Therefore, compound heterozygotes, i.e., Hb SC, Hb SD, and Hb SO, suffer from sickle cell syndromes. Hydroxycarbamide induces the expression of Hb F which limits Hb S polymerization and clinical events. The Hb F level is thus monitored in sickle cell patients treated with this drug.

Hb analysis relies on the separation of different Hb fractions based on their charge difference. Cation-exchange chromatography has long been the method of choice for Hb A_2 and Hb F quantification, but since the late 1990s, capillary zone electrophoresis (CZE) is an alternative method. For Hb variants, gel electrophoresis, isoelectric focusing (IEF), and HPLC have been used. More recently, CZE has appeared as a powerful tool (3). The presumptive identification and accurate quantification of Hb variants are mandatory to make the difference between traits and clinically significant forms. However, secondary and sometimes tertiary methods, including genetic analysis, are needed to confirm the variant identity (4–6).

Newborn screening for hemoglobinopathies is recommended in order to start preventive treatments. In sickle cell syndromes, it was proved that early antipneumococcal vaccination and prophylactic penicillin therapy dramatically decreases child mortality. Another advantage of newborn screening is the detection of healthy carriers for genetic counselling in the family. Hb analysis can be done on liquid cord blood or heel blood collected and dried on paper filter (Guthrie card) (7).

2. Hemoglobinopathies and Capillary Electrophoresis

CZE at alkaline pH is an efficient method for the separation of Hb fractions. In 1991, Chen developed a CZE method using fused silica capillaries filled with a pH 8.6 proprietary buffer that could separate Hb A, Hb S, and Hb C within a few minutes (8). In 1997, Jenkins published a similar method using a pH 9.98 borate buffer that was able to separate and quantify common variants (Hb S, Hb D, Hb E, and Hb A but not Hb F) and Hb A_2 with a precision of 8.4% (9). Later, Shihabi described a method using a pH 8.5 Tris buffer showing a precision of 5.7% for Hb A_2 quantification (10). At the same time, a seducing approach was developed by the Analis company for two commercial kits. Their patented dynamic coating of the capillary wall with charged polymers in alkaline or acidic conditions allowed to control and increase the electroosmotic flow in the capillary and also prevented protein adhesion to the capillary wall. The advantages were short run times, high reproducibility, and long capillary lifetime. We could show that these methods allowed the separation of most common Hb variants and the accurate and precise quantification of Hb A_2, Hb F, and other Hb fractions (11–13). Capillary IEF and micellar electrokinetic chromatography can also be used for Hb analysis but are limited to specialized investigations (14, 15).

3. Automated CZE

3.1. Instrument

In the early 2000s, automated instruments such as the Capillarys 2 system (Sebia Benelux, Vilvoorde, Belgium) appeared on the market. This instrument is a fully automated CZE instrument using eight parallel fused silica capillaries (100 μm ID × 17 cm length) and handling eight-tube racks with barcode reading. The Capillarys hemoglobin assay is dedicated to adult blood samples. Hemolysates are automatically prepared from red cells pellets. After hydrodynamic injection, Hb fractions are separated under a voltage of 7.8 kV in a pH 9.4 buffer and detected at 415 nm. Migration times are given in arbitrary units, the retention time of Hb A being fixed at 150. The throughput is 34 samples per hour. Liquid cord blood can be analyzed in "cord blood mode" where the retention time of Hb F is fixed at 180. Typical results are shown on Figs. 1 and 2.

The Capillary 2 Neonat Fast instrument is similar to the Capillarys 2 but it includes an automated puncher for dried blood on filter paper with barcode reading of the cards. With the Neonat Fast Hb assay, punched spots are automatically eluted and Hb fractions are then separated in the same manner as for the Capillarys Hemoglobin assay, with an adjustment of the Hb F peak to 180. The throughput is 48 samples per hour.

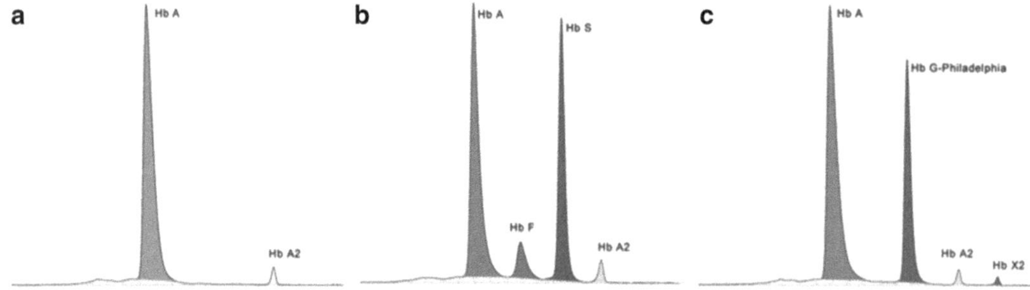

Fig. 1. Separation profiles obtained on blood samples analyzed with the Capillarys Hemoglobin assay. (**a**) Normal profile, (**b**) Hb S heterozygote, and (**c**) Hb G-Philadelphia heterozygote. The mutated α-chain of Hb G-Philadelphia is also included in Hb A_2, which appears as a double peak (Hb A_2 + Hb X_2)

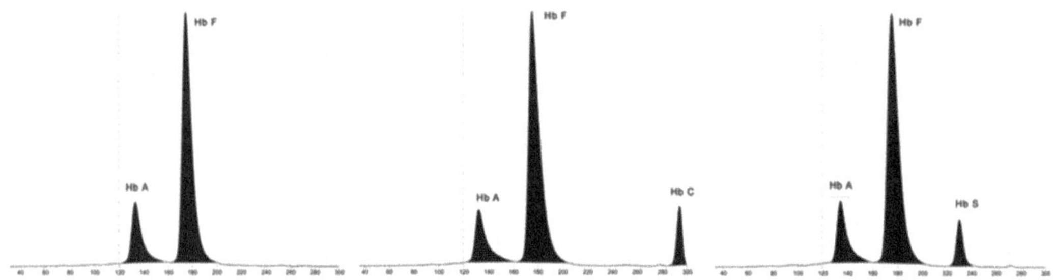

Fig. 2. Separation profiles obtained on liquid cord blood samples analyzed with the Capillarys Hemoglobin assay in "cord blood mode." (**a**) Normal profile, (**b**) Hb C heterozygote, and (**c**) Hb S heterozygote.

3.2. Methods

3.2.1. Diagnosis of Hemoglobinopathies in Adult Patients

Many studies evaluated the ability of the Capillarys Hemoglobin assay to detect Hb variants and to quantify Hb fractions.

Louahabi et al. showed that the analytical performances of the method were good with imprecision lower than 5 and 10% for Hb A_2 and Hb F quantification, respectively, and lower than 2.5% for migration times (16). In other studies, better results with lower imprecision were given (17–19). Several studies showed that Hb A_2 and Hb F results obtained with CZE were slightly lower from those obtained with other methods such as HPLC. Therefore, reference ranges had to be adapted to 2.1–3.2% for Hb A_2 and <0.8% for Hb F (16). However, CZE was superior to HPLC for quantifying Hb A_2 in the presence of various Hb variants except Hb C (17, 20, 21).

Several papers showed that the Capillarys Hemoglobin method was suitable for the presumptive identification and quantification of common Hb variants, including Hb H, although quantitative results were slightly different from those obtained with HPLC in some cases (16, 22, 23). Several researchers showed that this method was better than HPLC for the detection of Hb Constant Spring, an important unstable α-globin variant (18, 24). Fucharoen et al. studied the identification of 14 α- and β-globin chain variants frequent in Southeast Asia, such as Hb Q-Thailand, Hb Siam, Hb Paksé, Hb Tak, and Hb Pyrgos (24). We previously published the migration times of 18 variants identified by genetic analysis (17). An updated list is now presented on Table 1. Some variants (Hb Anderlecht, Hb Köln, Hb Loire, Hb Marseille, Hb Raleigh, and Hb Saclay) were not separated from Hb A. Hb Köln is unstable and could be detected thanks to an abnormally broad peak shape. The other variants not separated from Hb A have no clinical consequence. Some variants such as Hb S and Hb Stanleyville II had close migration times, but α-globin chain mutants could be detected as Hb A_2 was split into two peaks (Fig. 1c).

3.2.2. Newborn Screening for Hemoglobinopathies

Several studies have evaluated the potential of the Capillary system for the newborn screening of sickle cell disease and other hemoglobinopathies. They all focused on the Neonat Fast Hb assay. Renom et al. analyzed 6,766 newborn samples with IEF and CZE and showed that all significant variants were detected and presumably identified, including Hb Bart's. The absence of Hb A was well observed in four cases of β-thalassemia major (25). Mantikou et al. evaluated the system on 1,600 samples analyzed with HPLC and CZE. They also compared liquid cord blood and dried blood collected on filter paper. All results were perfectly concordant although some degradation of Hb fractions was seen when blood was collected on paper (26). Murray et al. made a similar evaluation on 347 dried blood samples analyzed with HPLC and CZE. Their conclusion was identical (27). We performed a validation of the Capillarys Hemoglobin assay in "cord blood mode" by analyzing 962 liquid cord blood samples with IEF and CZE (Fig. 2). No discrepancy between the results provided by both methods was observed (28). Table 2 shows the migration times of the five most common Hb variants and of Hb Bart's.

The CZE methods being quantitative, some authors studied the possibility to detect β- and α-thalassemia traits according to the Hb Bart's and Hb A level, respectively. Mantikou et al. showed a relation between Hb A level and gestational age at birth and suggested that below 15%, a β-thalassemia trait should be suspected (26, 29). Srivorakun et al. correlated the Hb Bart's level with the genotype of 226 newborns displaying 17 different thalassemic genotypes. Hb Bart's was detected in all patients with two or three α-globin genes defects. However, Hb Bart's was detected only in a few cases with one α-globin gene defect (30). Munkongdee et al. made a similar study on 587 newborns including 158 cases with one to three α-globin genes defects. Hb Bart's was detected at high levels in all cases of two or three α-globin genes defects but there was an overlap between Hb Bart's levels in normal newborns and those carrying one α-globin gene defect. The sensitivity and specificity for diagnosing α-thalassemia were 95 and 98%, respectively, using a value of 0.2% for Hb Bart's (31).

Table 1
Mean migration time of Hb variants with the Capillarys Hemoglobin assay (arbitrary units)

Hb	Migration time
H	29
J-Norfolk[a]	75
J-Baltimore	80
J-Toronto[a]	91
Hofu	93
Grady[a]	94
Ube II[a]	103
K-Woolwich	123
Camden	130
Anderlecht[a]	150
Köln	150
Loire	150
Marseille	150
Raleigh	150
Saclay	150
G-San Jose	181
G-Coushatta	197
Korle Bu	201
G-Philadelphia[a]	204
D-Ouled Rabah	205
G-Philadelphia	205
D-Punjab	206
Lepore	206
Osu Christiansborg	208
S	214
Stanleyville II[a]	214
O-Indonesia	215
Ottawa[a]	217
Arya[a]	218

(continued)

Table 1 (continued)

Hb	Migration time
E	229
C-Harlem	238
O-Arab	240
C	251
Constant Spring[a]	256

[a]α-Globin chain mutant

Table 2
Mean migration time of frequent Hb variants with the Capillarys Hemoglobin assay in "cord blood mode" (arbitrary units)

Hb	Migration time
Bart's	47
D-Punjab	217
S	227
C	281
E	245
O-Arab	262

4. Discussion and Conclusion

Capillary electrophoresis is a powerful tool for the analysis of normal and abnormal Hb fractions and thus for the screening and diagnosis of severe diseases like thalassemias and sickle cell disease.

The fully automated Capillarys Hemoglobin assay allows a very good and reproducible separation of normal and abnormal Hb fractions in short times. The quantification of Hb A_2 and Hb F is precise, even in the presence of most Hb variants. Results are significantly lower than with HPLC and adapted reference ranges should be used. The method detects and distinguishes most common and clinically significant Hb variants.

The Neonat Fast Hb and Capillarys Hemoglobin in "cord blood mode" assays are fully automated and can handle dried blood on paper filter and liquid cord blood. All clinically significant Hb variants are easily detected and the accurate quantification of Hb A and Hb Bart's is useful for the detection of β- and α-thalassemia.

Thanks to all these advantages, the Capillarys instrument is probably one of the most powerful first-line solutions for hemoglobinopathies screening in newborn and adult patients.

References

1. Higgs DR, Engel JD, Stamatoyannopoulos G (2012) Thalassaemia. Lancet 379:373–383
2. Rees DC, Williams TN, Gladwin MT (2010) Sickle-cell disease. Lancet 376:2018–2031
3. Wang J, Zhou S, Huang W, Liu Y, Cheng C, Lu X, Cheng J (2006) CE-based analysis of hemoglobin and its applications in clinical analysis. Electrophoresis 27:3108–3124
4. Clarke GM, Higgins TN (2000) Laboratory investigation of hemoglobinopathies and thalassemias: review and update. Clin Chem 46: 1284–1290
5. Wajcman H, Prehu C, Bardakdjian-Michau J, Prome D, Riou J, Godart C et al (2001) Abnormal hemoglobins: laboratory methods. Hemoglobin 25:169–181
6. Bain BJ (2011) Haemoglobinopathy diagnosis: algorithms, lessons and pitfalls. Blood Rev 25: 205–213
7. Gulbis B, Cotton F, Ferster A, Ketelslegers O, Dresse MF, Rongé-Collard E, Minon JM, Lé PQ, Vertongen F (2009) Neonatal haemoglobinopathy screening in Belgium. J Clin Pathol 62:49–52
8. Chen FT, Liu CM, Hsieh YZ, Sternberg JC (1991) Capillary electrophoresis - a new clinical tool. Clin Chem 37:14–19
9. Jenkins MA, Hendy J, Smith IL (1997) Evaluation of hemoglobin A2 quantitation assay and hemoglobin variant screening by capillary electrophoresis. J Capillary Electrophor 4:137–143
10. Shihabi ZK, Hinsdale ME, Daugherty HK Jr (2000) Hemoglobin A2 quantification by capillary zone electrophoresis. Electrophoresis 21:749–752
11. Cotton F, Lin C, Fontaine B, Gulbis B, Janssens J, Vertongen F (1999) Evaluation of a capillary electrophoresis method for routine determination of hemoglobins A2 and F. Clin Chem 45:237–243
12. Lin C, Cotton F, Fontaine B, Gulbis B, Janssens J, Vertongen F (1999) Capillary zone electrophoresis: an additional technique for the identification of hemoglobin variants. Hemoglobin 23:97–109
13. Gulbis B, Fontaine B, Vertongen F, Cotton F (2003) The place of capillary electrophoresis techniques in screening for haemoglobinopathies. Ann Clin Biochem 40:659–662
14. Mario N, Baudin B, Aussel C, Giboudeau J (1997) Capillary isoelectric focusing and high-performance cation-exchange chromatography compared for qualitative and quantitative analysis of hemoglobin variants. Clin Chem 43:2137–2142
15. Lin C, Gulbis B, Delobbe E, Cotton F, Vertongen F (1998) Separation of human globin chains by micellar electrokinetic capillary chromatography. J Chromatogr B Biomed Sci Appl 719:47–54
16. Louahabi A, Philippe M, Lali S, Wallemacq P, Maisin D (2006) Evaluation of a new Sebia kit for analysis of hemoglobin fractions and variants on the Capillarys system. Clin Chem Lab Med 44:340–345
17. Cotton F, Malaviolle X, Vertongen F, Gulbis B (2009) Evaluation of an automated capillary electrophoresis system in the screening for hemoglobinopathies. Clin Lab 55:217–221
18. Winichagoon P, Svasti S, Munkongdee T, Chaiya W, Boonmongkol P, Chantrakul N, Fucharoen S (2008) Rapid diagnosis of thalassemias and other hemoglobinopathies by capillary electrophoresis system. Transl Res 152:178–184
19. Yang Z, Chaffin CH, Easley PL, Thigpen B, Reddy VV (2009) Prevalence of elevated hemoglobin A2 measured by the CAPILLARYS system. Am J Clin Pathol 131:42–48
20. Higgins TN, Khajuria A, Mack M (2009) Quantification of HbA2 in patients with and without beta-thalassemia and in the presence of HbS, HbC, HbE, and HbD Punjab hemoglobin variants: comparison of two systems. Am J Clin Pathol 131:357–362
21. Paleari R, Gulbis B, Cotton F, Mosca A (2012) Interlaboratory comparison of current high-performance methods for HbA2. Int J Lab Hematol 34(4):362–368
22. Higgins T, Mack M, Khajuria A (2009) Comparison of two methods for the quantification and identification of hemoglobin variants. Clin Biochem 42:701–705

23. Keren DF, Hedstrom D, Gulbranson R, Ou CN, Bak R (2008) Comparison of Sebia Capillarys capillary electrophoresis with the Primus high-pressure liquid chromatography in the evaluation of hemoglobinopathies. Am J Clin Pathol 130:824–831
24. Fucharoen G, Srivorakun H, Singsanan S, Fucharoen S (2011) Presumptive diagnosis of common haemoglobinopathies in Southeast Asia using a capillary electrophoresis system. Int J Lab Hematol 33:424–433
25. Renom G, Mereau C, Maboudou P, Périni JM (2009) Potential of the Sebia Capillarys neonat fast automated system for neonatal screening of sickle cell disease. Clin Chem Lab Med 47: 1423–1432
26. Mantikou E, Harteveld CL, Giordano PC (2010) Newborn screening for hemoglobinopathies using capillary electrophoresis technology: Testing the Capillarys Neonat Fast Hb device. Clin Biochem 43:1345–1350
27. Murray C, Hall SK, Griffiths P (2011) An evaluation of the Sebia capillarys Neonat Haemoglobin FAST™ system for routine newborn screening for sickle cell disease. Int J Lab Hematol 33:533–539
28. Wolff F, Cotton F, Gulbis B (2012) Screening for haemoglobinopathies on cord blood: a report of laboratory and clinical experience. J Med Screen 19:116–122
29. Mantikou E, Arkesteijn SG, van Beckhoven JM, Kerkhoffs JL, Harteveld CL, Giordano PC (2009) A brief review on newborn screening methods for hemoglobinopathies and preliminary results selecting beta thalassemia carriers at birth by quantitative estimation of the HbA fraction. Clin Biochem 42:1780–1785
30. Srivorakun H, Fucharoen G, Changtrakul Y, Komwilaisak P, Fucharoen S (2011) Thalassemia and hemoglobinopathies in Southeast Asian newborns: diagnostic assessment using capillary electrophoresis system. Clin Biochem 44:406–411
31. Munkongdee T, Pichanun D, Butthep P, Klamchuen S, Chalermpolprapa V, Winichagoon P, Svasti S, Fucharoen S (2011) Quantitative analysis of Hb Bart's in cord blood by capillary electrophoresis system. Ann Hematol 90: 741–746

Chapter 17

Protein Fingerprinting of *Staphylococcus aureus* by Capillary Electrophoresis with On-Capillary Derivatization and Laser-Induced Fluorescence Detection

Cristina Pelaez-Lorenzo, Maria Teresa Veledo, Ramon Gonzalez, Mercedes de Frutos, and Jose Carlos Diez-Masa

Abstract

This chapter describes a complete procedure for obtaining protein fingerprints of microorganisms using capillary electrophoresis (CE) with laser-induced fluorescence detection (LIF). *Staphylococcus aureus*, a human pathogen responsible of frequent and resistant infections, is used as model microorganism to show the feasibility of this procedure. Bacteria are grown in different culture media or submitted to temperature or nitrosative stress conditions. After the growth of the bacteria, the protein extracts are obtained by cell lysis using sonication. The water-soluble fraction of these lysates is derivatized on-capillary with the fluorogenic reagent 3-(2-furoyl)quinoline-2-carboxaldehyde. The fluorescent products are analyzed by CE and detected by LIF. Practical advices for the interpretation of the electropherograms are given. To do so, the variations of the protein fingerprints of the bacteria with the culture conditions, such as growth medium, or the stressing conditions, such as heat shock or nitrosative stress, are used as example.

Key words: Capillary electrophoresis, Laser-induced fluorescence detection, On-column derivatization, Protein fingerprinting, *Staphylococcus aureus*, Growth conditions, Stressing conditions

1. Introduction

Capillary electrophoresis (CE) has demonstrated to be a powerful tool, either to separate intact microorganisms (1, 2) or to obtain protein fingerprints of different microorganisms (3, 4). When combined with laser-induced fluorescence detection (LIF), minute amounts of proteins can be analyzed, even at the level of single cells (5, 6). In addition, the development of on-capillary, covalent (5, 6) or non-covalent (7, 8) derivatization methods for CE-LIF

Nicola Volpi and Francesca Maccari (eds.), *Capillary Electrophoresis of Biomolecules: Methods and Protocols*, Methods in Molecular Biology, vol. 984, DOI 10.1007/978-1-62703-296-4_17,

allows the reduction of the reaction time and the consumption of reagents, the use of a small sample volume, and automation of the analytical process. Several investigations can be carried out using protein fingerprints of bacteria such as adaptation of microorganisms to modification in their environment, i.e., in their culture media or in their accommodation to the stressing conditions. Furthermore, the coupling of CE techniques with mass spectrometry (9) has shown to have a great potential in these studies due to the identification capabilities of the MS techniques in protein analysis.

Bacteria are able to rapidly adapt to changing environments. In fact, bacteria deploy only those factors required for growth and survival within a particular environment and they show a number of genetic responses to environmental conditions (10). In particular, under stress conditions bacterial cells synthesize proteins which are different from those produced under normal conditions. Most of them are highly conserved proteins which appear as response to different stimuli (11). This is the case of the proteins known as heat shock proteins (hsp) (12) or those known as cold shock proteins (csp) (13). For instance, Laport et al. (12) have reported the synthesis of three new proteins with a molecular weight higher than 90 kDa when the growth temperature of a culture of *Staphylococcus aureus* was increased from 37 to 42°C. Also, Richardson et al. (14) reported the stress responses of *S. aureus* to the nitric oxide produced by mammalian hosts as a defense against the bacteria. The protein maps obtained by two-dimensional gel electrophoresis (2D-PAGE) (15) or by transcriptomic analysis (16) of the nitrosative stress response of *S. aureus* have allowed the demonstration of changes in protein patterns under stressing conditions. However, these techniques are laborious and time consuming, and the protein maps obtained are usually complex. Similarly bacteria can modify the proteins expressed to adapt themselves to modifications in culture media (17).

Besides, protein fingerprinting obtained by CE-LIF has allowed establishing differences between species that can contribute to their identification and classification (18, 19). Therefore, the identification of bacterial species based on these protein profiles could complement the information obtained by classical microbiological assays.

This chapter reports a detailed procedure for obtaining the protein fingerprinting of *S. aureus* using CE-LIF. The procedure has the following major steps: (1) Bacteria growth at different culture conditions. (2) Protein extraction after cell lysis. (3) On-capillary derivatization of the extracts of proteins with the fluorogenic reagent 3-(2-furoyl)quinoline-2-carboxaldehyde together with the analysis of the fluorescent products by CE-LIF. A section dealing with interpretation of results has been introduced in which the interpretation of the electropherograms of the protein fingerprints obtained by

CE-LIF is carried out. In this section, the electropherograms obtained when the culture media are modified and those obtained as a response of the bacteria to stressing conditions (temperature and nitrosative stress) are used as examples. Although in this chapter the procedure is demonstrated to be feasible for *S. aureus*, it can also be used for other microorganisms.

2. Materials

2.1. Bacterial Strains and Culture Media

1. The strain employed was *Staphylococcus aureus* CECT 4465 from the Spanish Type Culture Collection.
2. *S. aureus* was maintained in Nutrient Agar plates (NA) (Pronadinasa, Madrid, Spain).
3. Bacteria were grown in nutrient broth (NB) or tryptic soy broth (TSB) at different times and/or different stressing conditions (temperatures and NO concentration) depending on the investigation carried out (see Subheading 3).
4. The *S. aureus* growth curves were determined by measuring the OD at 600 nm using an S-22 Boeco spectrophotometer (Boeckel, Hamburg, Germany).
5. In nitrosative stress studies of *S. aureus*, the nitric oxide donor diethylammonium (Z)-1-(*N*,*N*-diethylamino)diazen-1-ium-1,2-diolate (DEA-NONOate) from Alexis Biochemicals (San Diego, CA, USA) was used.
6. To evaluate NO formation in the cultures, the Griess assay was used. For this assay sulfanilamide and *N*-(1-naphtyl)ethylene-diamine dihydrochloride used for the preparation of Griess reagents were obtained from Sigma Chemical (St. Louis, MO, USA) (see Note 1).

2.2. Lysis of Bacteria and Protein Extraction

1. For washing the microorganism prior to lysis, PBS was used (see Note 2).
2. Lysis of bacteria was carried out by ultrasonication using a Misonix (Farmingdale, NY, USA) cell disruptor in 15 mL falcon tubes.
3. Total protein concentration was estimated in the supernatant of the bacterial lysates by the Bio-Rad (Hercules, CA, USA) protein assay based on the Bradford dye-binding procedure (20).

2.3. On-Column Derivatization of Proteic Extracts

1. Labeling reagent was 3-(2-furoyl)quinoline-2-carboxaldehyde (FQ) from Molecular Probes (Eugene, OR, USA) (see Note 3).
2. The nucleophilic agent for the derivatization with FQ was potassium cyanide from Sigma (see Note 4).

2.4. CE-LIF Instrumentation

1. The in house made CE apparatus with LIF detection described in a previous publication (21) was used with some modifications (22). Briefly, an RS/EH50R high voltage power supply (Glassman High Voltage, Whitehouse Station, NJ, USA) was used in normal polarity configuration (anode connected at the inlet end of capillary). Sample injection was carried out by raising the sample vial at a given high during a given time (injection by siphoning). A 2060-10S Spectra Physics Ar-ion laser (9 mW) (Spectra Physics, Mountain View, CA, USA) was used as an excitation source at 488 nm in the LIF detector with an orthogonal configuration. Fluorescence was collected using a 40× plan-achromatic microscope objective, passed through a 550 nm cut-off filter and an interference filter centered at 590 nm, imaged onto an iris to block straight light, and detected with an R928 photomultiplier tube (Hamamatsu, Hamamatsu City, Japan) operated at 600 V and mounted on top of a high-precision stage for alignment (Newport, Mountain View, CA, USA). Photocurrent was processed by a 7070 photometer (Oriel, Stratford, CT, USA) and a 406 System Gold A/D converter (Beckman, Fullerton, CA, USA). A laboratory-made device was used to heat a small zone (10 cm) at the inlet of the capillary (see Note 5) where the derivatization reaction took place, using an F3 water thermostatic bath (Haake, Karlsruhe, Germany).
2. Uncoated capillaries (Composite Metal Services, Worcester, UK) 60 cm length (50 cm to the detector) and 50 μm I.D. (375 μm O.D.) were used (see Note 6).
3. CE running buffer was 50 mM phosphate, 15 mM sodium pentyl sulfate (SPS) (pH 7.0).
4. For sample solvent preparation a volume of approximately 20 μL of 1 M NaOH was added to 10 mL of the running buffer to make a solution of pH 11.0.
5. A 200 mM KCN stock solution was made in 2.5 mM borax (pH 9.3) (see Note 4).

3. Methods

3.1. Culture Conditions

1. For getting the protein fingerprinting of *S. aureus* grown in different culture media (see Note 7):
   - Inoculate one or two colonies per tube of *S. aureus* from the agar plate in 16 mm internal diameter tubes containing 5 mL of either TSB or NB media. Incubate for 24 h with orbital shaking at 200 rpm at 20°C.
   - At the indicated time take a 200 μL aliquot of the culture and perform bacteria lysis and protein extraction.
   - Each culture must be done at least in duplicate.

2. For getting the protein fingerprinting of *S. aureus* grown at different temperatures:
   - Inoculate one or two colonies per tube of *S. aureus* from the agar plate in 16 mm internal diameter tubes containing 5 mL of TSB.
   - Incubate with orbital shaking at 200 rpm for 60 h at 20°C and for 15 h at 42°C.
   - At the indicate time take a 200 μL aliquot of the culture and perform bacteria lysis and protein extraction.
   - Each culture must be done at least in duplicate.
3. For getting the protein fingerprinting in nitrosative stress conditions:
   - Prepare a preinoculum by inoculating one of two colonies of *S. aureus* from the agar plate in a 16 mm internal diameter tube containing 5 mL of NB media and incubate overnight with orbital shaking at 200 rpm and 20°C.
   - Inoculate 100 μL of this preinoculum in 10 mL of NB culture broth pH 7.3 in 50 mL Erlenmeyer flask and incubate with orbital shaking at 200 rpm and 37°C.
   - Measure the optical density of the *S. aureus* culture at intervals of 30 min using a spectrophotometer to follow the growth of the *S. aureus* culture (see Note 8).
   - When OD_{600nm} ~ 1, that is mid-log phase of *S. aureus* growing (after approximately 4 h of growing), add to the culture a mixture of 22 μL of 500 mM DEA-NONOate solution (see Note 9). Measure the pH in every culture before adding the NO donor to be certain of working around the physiological pH (see Note 10).
   - Take a 200 μL aliquot of the culture 1 h after the addition of the NO donor (approximately after 5 h of growing). Perform bacteria lysis and protein extraction of this aliquot.
   - Each experiment must be done at least in duplicate.

3.2. Lysis of Bacteria and Protein Extraction

1. Eliminate the culture medium by centrifugation at 5,000 × *g* for 5 min at room temperature.
2. Wash the pellets three times in PBS solution and resuspend in Milli-Q water.
3. Disrupt the bacteria on an ice bath by ultrasonication (4×, 60 s).
4. Centrifuge the crude sonicate at 6,000 × *g* for 20 min at room temperature and collect the supernatant.
5. Determine the total protein concentration in an aliquot of each extract by the Bradford dye-binding procedure.
6. Store the rest of the supernatant at −20°C until performing the analysis by CE-LIF.

3.3. Sample Preparation for On-Capillary Derivatization

1. Thaw the aliquot of protein extract to be studied.
2. Mix it with the corresponding volume of a 200 mM KCN solution, and then dilute it with a 50 mM phosphate, 15 mM SPS (pH 11.0) buffer solution to obtain a final concentration of 5 mM KCN and approximately 5×10^{-7} M of total protein in the final mixture of all the extracts studied.

3.4. On-Column Derivatization of the Extracts and CE-LIF Analysis

The procedure for on-capillary derivatization is based on a previously optimized method (23).

1. New capillaries were rinsed with 1 M NaOH (100 μL) followed by a rinse with Milli-Q water (100 μL). Between runs, the capillary was sequentially rinsed with Milli-Q water (100 μL), 0.1 M NaOH (100 μL), Milli-Q water (100 μL), and the separation buffer (100 μL). Rinses were made manually employing a model 1710 glass syringe from Hamilton (Bonaduz, Switzerland).
2. Inject a plug of a mixture of the sample added with KCN solution into the capillary. Inject 5 mM solution of FQ in separation buffer. Both injections are performed by siphoning (20 cm height) for12 s (see Note 11).
3. Mixing step: Place the vials containing the separation buffer in the inlet and outlet ends of the capillary and apply high voltage of 3 kV for 6 min. Perform this step at 65°C by using the device described in Note 5.
4. Reaction step: Switch off the power supply for 15 s (note that reaction step is carried out also at 65°C).
5. Separation step: Apply 15 kV at room temperature (see Note 12).
6. Collect the electropherograms. Analyze each sample at least in duplicate.

4. Interpretation of Results

4.1. Fingerprint Repeatability

The repeatability of the protein patterns can be analyzed for injections of the same culture and for different cultures grown under identical conditions. Figure 1 shows that the protein profiles were very similar, the number and the position of the peaks in the protein patterns were always reproducible, however, slight differences in the height of some of the peaks were observed. Therefore, in our experience the peak height cannot be considered as a differentiation factor between conditions. Probably, these differences observed among cultures could be due to variations in bacterial growth and the consequent protein content (compare Fig. 1B with Fig. 1C, D, or E), even though the total amount of protein in the different samples was the same (6.4×10^{-7} M according to the Bradford assay). No peaks were observed in the electropherograms

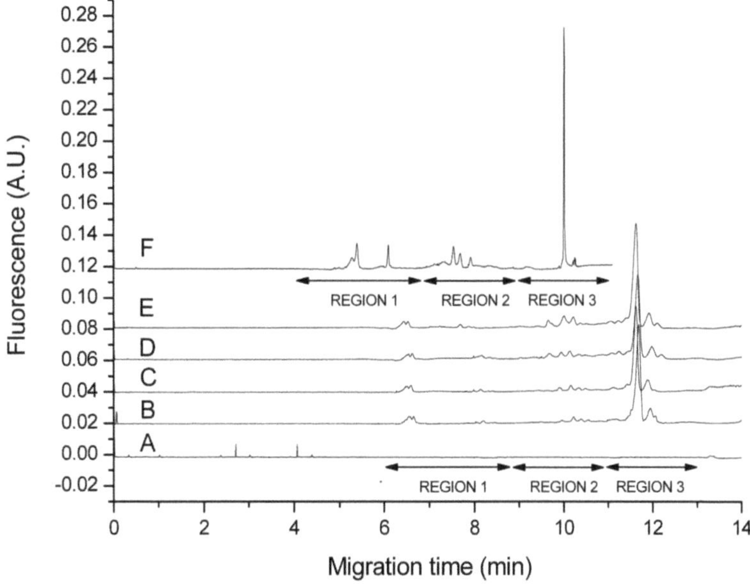

Fig. 1. Capillary electrophoresis patterns of a blank assay (***A***), patterns of three consecutive injections of the same culture of *S. aureus* (*B–D*), and two cultures of *S. aureus* grown in the same conditions (*D* and *E*). Growth conditions: temperature 37°C, medium NB, and time 18 h. CE-LIF conditions: running buffer 50 mM phosphate, 15 mM SPS (pH 7.0), uncoated capillary ($L=60$ cm, $l=50$ cm, 50 μm I.D.), injection by siphoning ($h=20$ cm, 12 s) of a mixture of the protein extract with a 5 mM KCN solution followed by an injection of 5 mM FQ in running buffer, mixing 6 min at 3 kV, reaction 15 s at 0 kV, mixing and reaction temperature 65°C, separation performed at 15 kV and room temperature. Trace *F* represents the electrophoresis pattern of a culture of *S. aureus* obtained in the same condition than traces *B–E* (for both, culture and the CE-LIF separation) carried out in a new capillary. "Adapted from Analytica Chimica Acta, 659, Veledo M.T., Pelaez-Lorenzo C., Gonzalez R., de Frutos M., Diez-Masa J.C., Protein fingerprinting of *Staphylococcus* species by capillary electrophoresis with on-capillary derivatization and laser-induced fluorescence detection, 81–86, Copyright (2010), with permission from Elsevier".

of a blank, i.e., culture media where no microorganisms were grown (see Fig. 1A).

It is worth mentioning that for pattern repeatability it is of special importance to work, whenever possible, with the same fused silica capillary for CE-LIF experiments. Protein fingerprints obtained with other different capillaries from the same brand but different batch are shown in Fig. 1F (compare electropherogram in trace F with any of the traces B–E), where it can be observed that migration time of the peaks and the peak heights are different from capillary to capillary. It is well known that small modifications of the EOF of the capillary cause slight differences in the migration time of analytes, thus influencing repeatability. Moreover, we have observed that a capillary that has been used for a large number of runs (>20) is degraded and different migration times are obtained with respect to new capillaries. However the three different regions (indicated as Region 1, 2, and 3 in the figures) of the electropherogram in

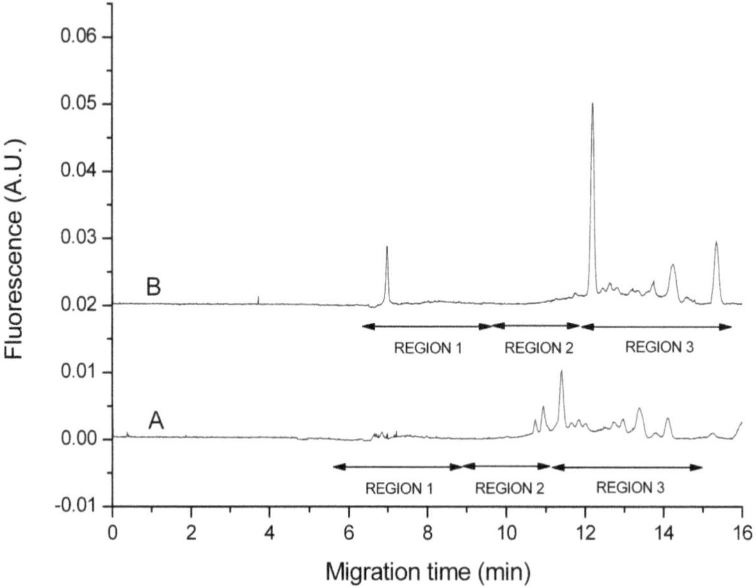

Fig. 2. Capillary electrophoresis patterns of proteins for cultures of *S. aureus* grown in (*A*) NB medium and (*B*) TSB medium. Other growth conditions: temperature 20°C, time 24 h. Rest of the conditions as in Fig. 1. "Reprinted from Analytica Chimica Acta, 659, Veledo M.T., Pelaez-Lorenzo C., Gonzalez R., de Frutos M., Diez-Masa J.C., Protein fingerprinting of *Staphylococcus* species by capillary electrophoresis with on-capillary derivatization and laser-induced fluorescence detection, 81–86, Copyright (2010), with permission from Elsevier".

which the peaks seem to cluster can be easily recognized from capillary to capillary.

4.2. Applications

Using the procedure described in this chapter the effect of different growth conditions of the microorganisms can be studied. As an example, the effect on proteins patterns of *S. aureus* grown in two different culture media, NB and TSB, were compared.

Figure 2 shows that the profiles corresponding to both cultures were different. Some peaks could correspond to over-expressed proteins while others are related to infra-expressed proteins. For instance, the profile corresponding to the culture grown in NB presents three peaks in region 3 that are not observed in the profile of the culture grown in TSB; however, a shift in migration time of the peaks probably due to differences in EOF could not be ruled out. In consequence, the method developed allows the differentiation of cultures of *S. aureus* grown in different media on the basis of these protein patterns.

CE-LIF technique can also be used to study protein fingerprints of microorganisms in different stressing conditions. We have studied (19) the effect of temperature and NO on the fingerprint of *S. aureus*. As an example of the interpretation of these electrophoretic patterns, we discuss briefly some details on both studies.

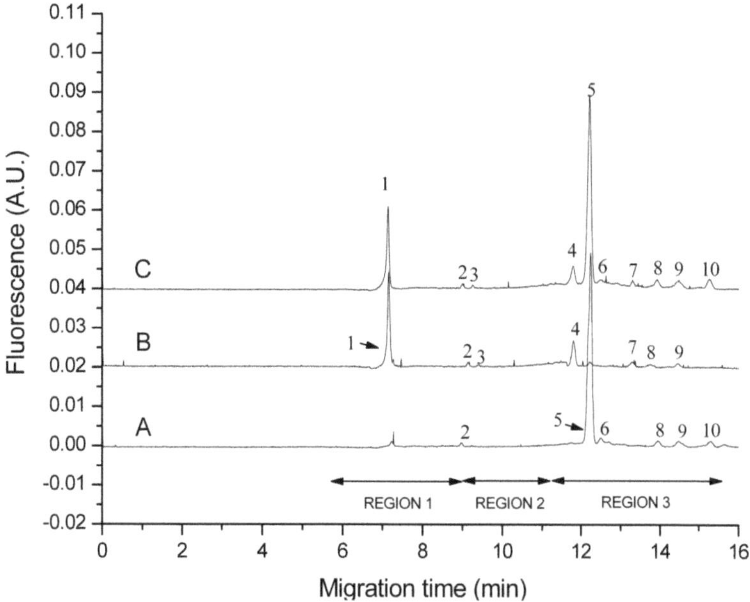

Fig. 3. Capillary electrophoresis patterns of proteins for (*A*) a culture of *S. aureus* grown at 20°C for 60 h, (*B*) a culture of *S. aureus* grown at 42°C for 15 h, and (*C*) a mixture 1:1 of the protein extracts of both cultures shown in (*A*) and in (*B*). Medium: TSB. Other conditions as in Fig. 1. "Reprinted from Analytica Chimica Acta, 659, Veledo M.T., Pelaez-Lorenzo C., Gonzalez R., de Frutos M., Diez-Masa J.C., Protein fingerprinting of *Staphylococcus* species by capillary electrophoresis with on-capillary derivatization and laser-induced fluorescence detection, 81–86, Copyright (2010), with permission from Elsevier".

To study the temperature effect two cultures of *S. aureus* were grown at 20 and 42°C (in TSB medium for 60 and 15 h, respectively). These are suboptimal temperatures for *S. aureus* which has an optimum growth temperature of 37°C (12). In that experiments two different growing times (60 and 15 h) were used to obtain approximately the same number of bacterial cells at the end of the same growing phase of the microorganism than that obtained at the optimal growth temperature. In Fig. 3A and B the protein profiles obtained for cultures of this microorganism grown at these two temperatures are shown.

Although, several peaks could differentiate both electrophoretic profiles, only the peaks showing major differences will be considered here. For instance, by comparing Fig. 3A and B with Fig. 3C it can be readily seen that peak 1 is only shown in the profiles of the cultures grown at 42°C and peak 5 is the major peak in the cultures at 20°C, while it is scarcely present when the microorganism is grown at 42°C. Minor peaks, such as peaks 3, 4, and 10, could in addition serve to differentiate both cultures, but these minor components are more sensitive to variations among culture, as indicated before. Temperatures of around 42°C are considered extreme for *S. aureus* and it could be hypothesized that these peaks could be related to proteins induced in stress conditions, such as hsp (12) or csp (13).

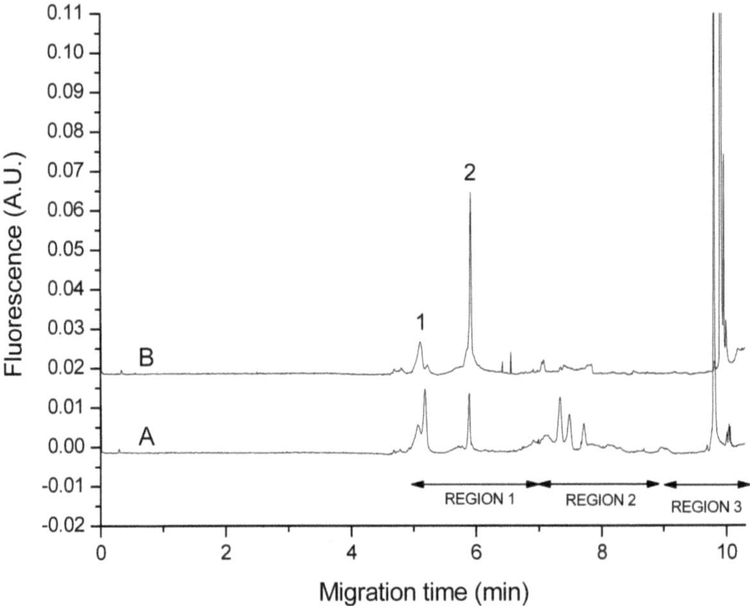

Fig. 4. Capillary electrophoresis patterns of proteins for (*A*) a culture of *S. aureus* grown in NB and (*B*) *S. aureus* grown in NB supplemented with 22 μL of 500 mM DEA-NONOate. Culture conditions 5 h at 37°C. See Subheading 3.1 for the time at which the NO donor was added. Other conditions as in Fig. 1. "Reprinted from Analytica Chimica Acta, 659, Veledo M.T., Pelaez-Lorenzo C., Gonzalez R., de Frutos M., Diez-Masa J.C., Protein fingerprinting of *Staphylococcus* species by capillary electrophoresis with on-capillary derivatization and laser-induced fluorescence detection, 81–86, Copyright (2010), with the permission from Elsevier".

Finally, the effect of the exposure of *S. aureus* to NO on the electrophoretic fingerprint of proteins has been studied. Nitric oxide (NO) is an important antimicrobial agent and it has been shown to be indispensable for the clearance of diverse pathogens. Several studies have suggested that NO is bacteriostatic and bactericidal for *S. aureus* (24–26) although adaptative response of these bacteria to nitrosative stress is not clear. The optimum growth temperature of the bacteria (37°C), which had been grown for 18 h, was selected and the effect of the NO donor DEA-NONOate on the fingerprint of *S. aureus* was investigated. As shown in Fig. 4 for cultures grown in the presence of the donor, the height of the peaks marked as 1 and 2 in the electropherogram has increased; however, peaks in region 2 of the electropherogram have decreased or even some peaks are not observed.

Therefore, the procedure described allows one to differentiate the profiles *S. aureus* exposed and not exposed to the effect of NO. Transcriptomic analysis of the nitrosative stress response of *S. aureus* has identified both upregulated and downregulated genes (66 and 18 genes, respectively) in response to NO donors (16).

5. Notes

1. Griess assay (27) measures the nitrite ion (NO_2^-) concentration, a rapid, stable, and non-volatile breakdown product of NO in solution. The test uses two reagents. Reagent 1 consists of a solution of 1% (w/v) of sulfanilamide in 5% (v/v) of phosphoric acid. Reagent 2 is prepared by adding 0.1% (w/v) of *N*-(1-naphthyl)ethylenediamine dichlorhydrate in Milli-Q water. Griess reagents in presence of nitrite ions produce a diazo compound (with purple color) whose absorbance can be measured at 540 nm.

2. PBS consists of 0.01 M disodium hydrogenphosphate/sodium dihydrogenphosphate, 0.138 M NaCl, and 2.7 mM KCl (pH 7.4).

3. Preparation of FQ solutions: A 50 mM stock solution of FQ was prepared in methanol (HPLC grade, Scharlau Chemie, Barcelona, Spain). Since FQ degrades in solution even when stored at −20°C in darkness, small aliquots of dried FQ were prepared. To do so, 10 μL aliquots of the methanolic solution were transferred to 500 μL microcentrifuge tubes. The solvent was removed under vacuum at room temperature using a model RC10-10 centrifugal evaporator (Jouan, Saint-Herblain, France). The dried FQ aliquots were stored at −20°C protected from light until use. On the day of the experiment, the aliquots were defrosted and solved in the running buffer to obtain a 5 mM FQ solution.

4. Potassium cyanide is highly poisonous. It reacts with acids to form lethal HCN gas. Therefore, stock solutions should be made in a basic buffer. Neutralization of waste containing KCN solution should be done by the addition of 1% NaOH solution (about 50 μL NaOH/g cyanide) followed by slow addition of bleach (about 70 μL of bleach/g of cyanide).

5. The heating device was made using two PVDF T-pieces model 1C300T03PV (EM-Technik, Maxdorf, Germany) interconnected by a 1-cm-long Tygon tube (6 mm O.D., 5 mm I.D.) using one entrance of each of the T-pieces. This assembly formed a jacket that allowed the insertion of the first 10 cm of the separation capillary through the distal entrances of the interconnected T-pieces. The capillary was retained in the T-pieces by inserting it through silicone septa. This jacket allowed the control of the temperature in this length of the capillary using water from an external circulating thermostatic bath entering the device through Tygon tube connected to the two remaining entrances of the T-pieces. Water-tightness of the device was achieved using PVDF cutting rings and thumb-nuts

provided with the T-pieces for Tygon tubing connections and silicone septa used for the insertion of capillary.

6. Capillaries are purchased in reels and cut to the desired length. Detection window is made by burning off the polyimide layer.
7. Although in this chapter the whole procedure is described only for the species *S. aureus*, we have also used it for studying other species, such as *Staphylococcus epidermidis* (19).
8. To assess the effect of the addition of DEA-NONOate on the physiology of the *S. aureus* the growth curves of this microorganism (variation of the OD_{600nm} value of the cultures with time) were determined in nitrosative stress conditions and in the control cultures. To do so, three cultures were grown simultaneously at 37°C in NB culture media. Culture 1 (supplemented with DEA-NONOate) was prepared as follows: when the OD_{600nm} value of the *S. aureus* culture was around 1 (at about 4 h after the start of the culture), it was added with 22 μL of a 500 mM solution of the NO donor in 0.01 M of NaOH in water. Culture 2 (culture supplemented with 0.01 M NaOH) was prepared as Culture 1, but it was added with only 22 μL of an 0.01 M aqueous solution of NaOH (the solvent used for the NO donor) and no NO donor was added. Culture 3 (untreated culture) was prepared as Cultures 1 and 2, but nothing was added to the culture. In the three types of cultures 100 μL aliquots were sampled at intervals of 30 min since the start of the culture. The samples were diluted (1:4) in Milli-Q water and the OD_{600nm} was determined. An example of the growth curves obtained in these experiments is given in Fig. 5. It can be observed that the release of NO in the culture causes a decrease in the growth rate of the microorganism. This alteration can potentially produce a change in protein expression and therefore in the protein fingerprint obtained by CE-LIF.
9. In Griess assay, measurements were carried out by duplicate in 96-well plates. In each well 50 μL of sample and 50 μL of the Griess reagent 1 (see Note 1) were added to the well and sample was incubated for 5 min at room temperature in the darkness. Then, 50 μL of the Griess reagent 2 (see Note 1) was added and the solution was incubated again for other 5 min in the same conditions. Finally, the absorbance at 540 nm was measured in a plate reader. Solutions of sodium nitrite in NB solution (concentration range 3–100 μM) placed in other wells of the same plate were used to obtain the calibration curve. The concentration of nitrite ion in the sample was determined by interpolation of the average values of the reading of two samples in the calibration curve. As controls, aliquots containing only the (clean) nutrient broth NB, culture containing bacteria, and NB solution added with NO donor were used. As an example, variation of the NO_2^- present in the cultures in

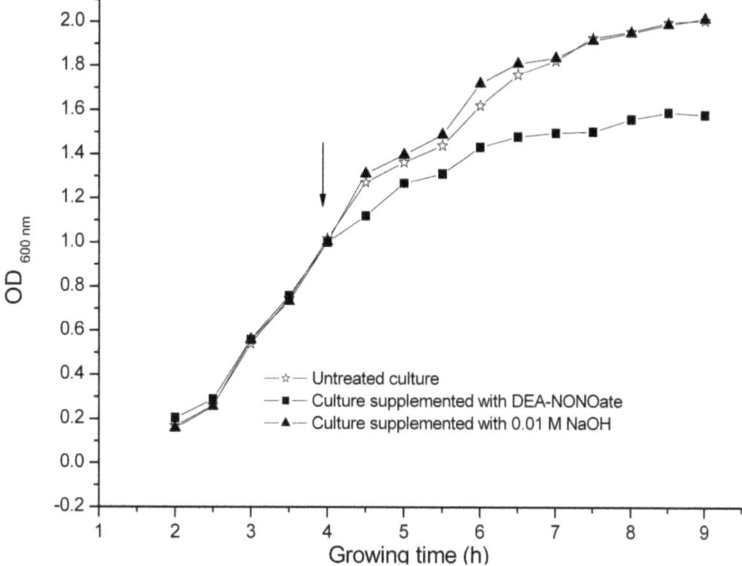

Fig. 5. *S. aureus* growth curves for the untreated culture (Culture 3 in Note 8), the cultures supplemented with 0.01 M NaOH in water (the solvent of the NO donor) (Culture 2 in Note 8), and culture supplemented with 22 μL of 500 mM DEA-NONOate (Culture 1 in Note 8). The *arrow* indicates the time at which the NO donor was added.

the time elapsed after the addition of the NO donor is given in Fig. 6. The addition of 1.1 mM DEA-NONOate, a rapid releasing NO donor, produced a NO_2^- concentration of 1.1 mM in 30 min, which decreased slowly until a concentration of 0.9 mM after 4 h of its addition to the culture As expected, the controls (the clean culture medium and the bacteria cultures in the absence of NO donors) in two individual assays did not produce absorbance at 540 nm, which indicates the absence of their interference in Griess assay. The third control, the culture medium in the absence of bacteria supplemented with the NO donor, produced a concentration of NO_2^- higher than the maximum value reached in the presence of bacteria when DEA-NONOate was added, probably due to the absence of cells that would react with NO (results not shown in Fig. 6).

10. This control was performed due to the fact that, although insensitive to most buffer constituents and contaminants, DEA-NONOate decomposition is extremely sensitive to small changes in pH (28).
11. After each injection of sample plus KCN or FQ, the inlet end of the capillary was washed by immersing it three times in a vial containing Milli-Q water to avoid cross-contamination of the sample and the FQ solution.
12. Room temperature was set at 24°C.

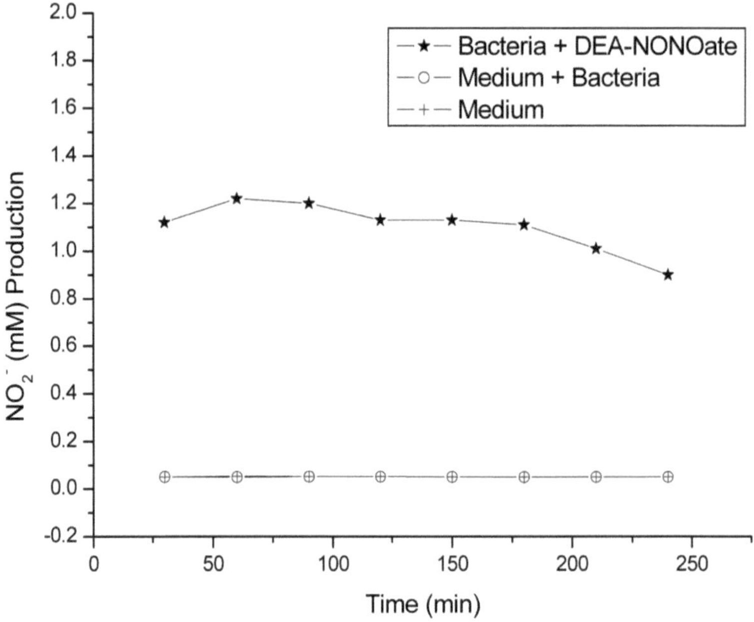

Fig. 6. Measurements by Griess assay of NO production (expressed as NO_2^- concentration in the culture medium) after the addition of the DEA-NONOate to a *S. aureus* culture. The NO donor was added to the cultures at the mid-log phase of the growing microorganism (see *arrow* in Fig. 5). Measurements were made every 30 min for a 6 h period.

Acknowledgements

The Spanish Ministry of Science and Innovation (Project CTQ2009-09399), CSIC (Project 2010JP0013), and Domingo Martinez Foundation are acknowledged for financial support. Dr. A. Martinez (Department of Immunology, Hospital de la Princesa, Madrid, Spain) and Dr. A. Puerta (Institute of Organic Chemistry, CSIC, Spain) are acknowledged for helpful discussion, and Dr. J.M. Barcenilla (Institute of Food Science, Technology, and Nutrition, CSIC, Spain) for technical assistance.

References

1. Armstrong DW, Schulte G, Schneiderheinze JM et al (1999) Separating microbes in the manner of molecules. 1. Capillary electrokinetic approaches. Anal Chem 71:5465–5469
2. Petr J, Maier V (2012) Analysis of microorganisms by capillary electrophoresis. Trends Anal Chem 31:9–22
3. Shen Y, Xiang F, Veenstra TD et al (1999) High-resolution capillary isoelectric focusing of complex protein mixtures from lysates of microorganisms. Anal Chem 71:5348–5353
4. Hu S, Lee R, Zhang Z et al (2001) Protein analysis of an individual *Caenorhabditis elegans* single-cell embryo by capillary electrophoresis. J Chromatogr B 752:307–310
5. Zhang Z, Krylov S, Arriaga EA et al (2000) One-dimensional protein analysis of an HT29 human colon adenocarcinoma cell. Anal Chem 72:318–322

6. Hu S, Jiang J, Cook LM et al (2002) Capillary sodium dodecyl sulfate-DALT electrophoresis with laser-induced fluorescence detection for size-based analysis of proteins in human colon cancer cells. Electrophoresis 23:3136–3142
7. Benito I, Marina ML, Saz JM et al (1999) Detection of bovine whey proteins by on-column derivatization capillary electrophoresis with laser-induced fluorescence monitoring. J Chromatogr A 841:105–114
8. Csapo Z, Gerstner A, Sasvari-Szekely M et al (2000) Automated ultra-thin-layer SDS gel electrophoresis of proteins using non-covalent fluorescent labeling. Anal Chem 72:2519–2525
9. Smith RD, Pasa-Tolic L, Lipton MS et al (2001) Rapid quantitative measurements of proteomes by Fourier transform ion cyclotron resonance mass spectrometry. Electrophoresis 22:1652–1668
10. VanBolegen RA, Abshire KZ, Pertsemlidis A et al (1999) Gene-protein database of *Escherichia coli* K-12, edition 6. In: Neihardt FC, Curtiss R III, Ingraham JL et al (eds) *Escherichia coli* and *Salmonella*: cellular and molecular biology, 2nd edn. ASM Press, Washington, pp 2067–2116
11. Ohta T, Honda K, Kuroda M et al (1993) Molecular characterization of the gene operon of heat-shock proteins hsp 60 and hsp10 in methicillin-resistant *Staphylococcus-aureus*. Biochem Biophys Res Commun 193:730–737
12. Laport MS, de Castro ACD, Villardo A et al (2001) Expression of the major heat shock proteins DnaK and GroEL in *Streptococcus pyogenes*: a comparison to *Enterococcus faecalis* and *Staphylococcus aureus*. Curr Microbiol 42:264–268
13. Katzif S, Danavall D, Bowers S et al (2003) The major cold shock gene, *cspA*, is involved in the susceptibility of *Staphylococcus aureus* to an antimicrobial peptide of human cathepsin G. Infect Immun 71:4304–4312
14. Richardson AR, Dunman PM, Fang FC (2006) The nitrosative stress response of *Staphylococcus aureus* is required for resistance to innate immunity. Mol Microbiol 61:927–939
15. Hecker M, Engelmann S, Cordwell SJ (2003) Proteomics of *Staphylococcus aureus*—Current state and future challenges. J Chromatogr B 787:179–195
16. Fuchs S, Pane-Farre J, Kohler C et al (2007) Anaerobic gene expression in *Staphylococcus aureus*. J Bacteriol 189:4275–4289
17. Madigan MT, Martinko JM, Parker J (2003) Brock biology of microorganisms. Prentice Hall, Upper Saddle River, NJ
18. Zhang Z, Carpenter E, Puyan X et al (2001) Manipulation of protein fingerprints during on-column fluorescent labeling: Protein fingerprinting of six *Staphylococcus* species by capillary electrophoresis. Electrophoresis 22:1127–1132
19. Veledo MT, Pelaez-Lorenzo C, Gonzalez R et al (2010) Protein fingerprinting of *Staphylococcus* species by capillary electrophoresis with on-capillary derivatization and laser-induced fluorescence detection. Anal Chim Acta 658:81–86
20. Bradford MM (1976) Rapid and sensitive method for quantitation of microgram quantities of protein utilizing principle of protein-dye binding. Anal Biochem 72:248–254
21. Veledo MT, de Frutos M, Diez-Masa JC (2005) Amino acids determination using capillary electrophoresis with on-capillary derivatization and laser-induced fluorescence detection. J Chromatogr A 1079:335–343
22. Pelaez-Lorenzo C, Diez-Masa JC, Vasallo I et al (2009) A new sample preparation method compatible with capillary electrophoresis and laser-induced fluorescence for improving detection of low levels of β-lactoglobulin in infant foods. Anal Chim Acta 649:202–210
23. Veledo MT, de Frutos M, Diez-Masa JC (2005) Development of a method for quantitative analysis of the major whey proteins by capillary electrophoresis with on-capillary derivatization and laser-induced fluorescence detection. J Sep Sci 28:935–940
24. DeGroote MA, Fang FC (1995) NO Inhibitions: antimicrobial properties of nitric oxide. Clin Infect Dis 21(Suppl 2), S162
25. Malawista SE, Montgomery RR, Van Blaricom G (1992) Evidence for reactive nitrogen intermediates in killing of *staphylococci* by human neutrophil cytoplasts. J Clin Invest 90:631–636
26. Kaplan SS, Lancaster JR, Basford RE et al (1996) Effect of nitric oxide on *staphylococcal* killing and interactive effect with superoxide. Infect Immun 64:69–76
27. Griess P (1879) Bemerkungen zu der Abhandlung der HH. Weselsky und Benedikt, "Uber einige Azoverbindungen". Ber Deutsch Chem Ges 12:426–428
28. Davies KM, Wink DA, Saavedra JE et al (2001) Chemistry of the diazeniumdiolates. 2. Kinetics and mechanism of dissociation to nitric oxide in aqueous solution. J Am Chem Soc 123:5473–5481

Chapter 18

Capillary Electrophoresis of Seed Storage Proteins: The Separation and Identification of Microheterogeneous Rice Glutelin Subunits

Tomoyuki Katsube-Tanaka

Abstract

Glutelin, the major seed storage protein of rice (*Oryza sativa* L.), consists of multiple polymeric and monomeric subunits. Each subunit is composed of an α and a β polypeptide that are covalently linked by a disulfide bond. To analyze the microheterogeneous glutelin subunits using capillary electrophoresis (CE), the author identified the appropriate sample preparation procedures as well as optimal CE conditions. The glutelin was dissociated into its component α and β polypeptides by denaturation and reduction with low urea and 2-mercaptoethanol for a long incubation time at room temperature. The molecular species of the completely dissociated α and β polypeptides were identified and quantitatively analyzed by CE and SDS-PAGE using glutelin mutants. The measured CE migration times of the polypeptides correlated well with the calculated charge-to-size parameter values. Therefore, the rapid, simple, and precise separation and quantification of microheterogeneous proteins by CE required not only optimal CE conditions but also adequate protein pretreatment based on the molecular nature of the protein tested.

Key words: Capillary electrophoresis, Storage protein, Microheterogeneous subunit, Rice seed, Glutelin

Abbreviations

CE	Capillary electrophoresis
ME	2-Mercaptoethanol
SDS-PAGE	Sodium dodecyl sulfate-polyacrylamide gel electrophoresis

1. Introduction

Capillary electrophoresis (CE) is a promising, highly automatable methodology with a high peak efficiency and capacity that requires only a small amount of sample (1). It is widely used for the analyses of small-molecule compounds and polymeric macromolecules such

Nicola Volpi and Francesca Maccari (eds.), *Capillary Electrophoresis of Biomolecules: Methods and Protocols*, Methods in Molecular Biology, vol. 984, DOI 10.1007/978-1-62703-296-4_18, © Springer Science+Business Media, LLC 2013

as nucleic acids and proteins. Reports of the analyses of cereal proteins by CE are accumulating steadily; for example, recent studies have investigated wheat ω-gliadin (2), wheat LMW-/HMW-glutenin (3), a number of soybean and azuki bean proteins (4), and soybean protein hydrolysates (5). However, the CE analysis of cereal proteins is challenging due to their complexity, dissimilar nature, and large dynamic concentration range (6). Moreover, the adsorption of proteins to capillary walls is a general challenge in the CE analysis of proteins. Thus, insufficient reproducibility hinders the detailed CE analysis of cereal proteins.

The major components of cereal proteins are storage proteins, which usually consist of multiple microheterogeneous subunits encoded by a multigene family. Because each subunit has its own unique characteristics and differs from the other subunits in terms of its nutritional, physicochemical, and physiological qualities, the differentiation and quantification of the multiple microheterogeneous subunits are prerequisites for cereal protein improvement.

In this chapter, the author describes the separation and identification of the subunits of glutelin, one of the major storage proteins in rice (*Oryza sativa* L.) seeds. The successful CE separation of its component α and β polypeptides can only be achieved by preincubation under appropriate denaturing and reducing conditions. The pre-extraction of the other storage proteins also seems to promote reproducible analysis. Here, the author discusses the usefulness of the CE analytical system based on the molecular nature of the target protein for basic and applied research, such as the kinetic characterization of higher-order structures and a quantitative analysis for the protein improvement.

2. Materials

All chemicals and equipments are standard analytical grade for CE in free-zone electrophoresis (FZE) mode.

2.1. CE Instrument, Capillary, and Reagents

1. CAPI-2000 T instrument (Otsuka Electronics, Osaka, Japan).
2. Uncoated fused silica capillaries, 75 μm i.d. × 50 cm (37.5 cm effective length) (Otsuka Electronics, Osaka, Japan).
3. Capillary cutter (Otsuka Electronics, Osaka, Japan).
4. Glass vial for CE (Otsuka Electronics, Osaka, Japan).
5. Microcentrifuge tube, 500 μl.
6. Separation buffer: 50 mM iminodiacetic acid, 20 % (v/v) acetonitrile, 0.05 % (w/v) hydroxypropyl methylcellulose, and 26 mM lauryl sulfobetaine (SB3-12) (7) (see Note 1).
7. 500 mM acetic acid.
8. Deionized water.

2.2. Plant Materials and Reagents

1. Rice seeds from wild-type (cultivar Koshihikari) and glutelin mutant (cultivars Type 1, Type 2, Type 3, a-123 less, and LGC-1) plants.
2. Glutelin extraction buffer: 1 % (v/v) lactic acid, 1 mM ethylenediaminetetraacetic acid (EDTA).
3. Albumin and globulin pre-extraction buffer: 35 mM potassium phosphate (pH 7.6), 0.4 M NaCl.
4. Prolamin pre-extraction buffer: 60 % (v/v) n-propanol.

3. Methods

3.1. Sample Preparation

1. Extract and purify protein samples to the highest purity possible (see Note 2).
2. Unfold the protein to eliminate higher-order structures and reduce the disulfide bonds (if any). Each component polypeptide should be analyzed in a solution containing the minimum level of denaturing and/or reducing agents using incubation under specific conditions (see Note 3). If the denaturation and/or reduction are incomplete at the sample preparation step, the polypeptides may not be sufficiently separated from each other in the CE analysis. An example of the effect of sample preparation on the CE profile is shown in Fig. 1 (see Note 4).

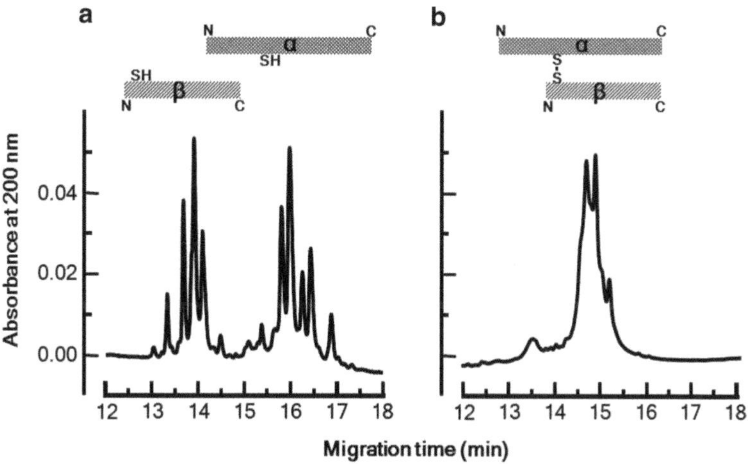

Fig. 1. CE profiles of standard japonica rice glutelin (cultivar Koshihikari) and schematic representations of glutelin polypeptides. Glutelin polypeptides, isolated from a standard japonica rice cultivar, Koshihikari, were incubated at room temperature (20 °C) in a lactic acid solution supplemented with 0.5 M urea and 10 % ME for (**a**) 10 h and (**b**) 1 h. The CE profiles of each glutelin preparation are shown in the lower panel. The predicted structures of each glutelin preparation are represented schematically in the upper panel (Reproduced from ref. 14 with permission from John Wiley and Sons).

3. Centrifuge the sample solution in a 500 μl microcentrifuge tube to remove any insoluble debris prior to CE injection.
4. Place the microcentrifuge tube containing the sample onto a glass vial for CE.

3.2. CE Setup

1. Cut uncoated fused silica capillaries with a capillary cutter and remove the polyimide coating by burning in a small flame to create a detector window area on the capillary.
2. Set the prepared capillary in the cartridge of the CE instrument and load the cartridge into the CE instrument according to the CE instruction manual.
3. Condition the new capillary by filling with the separation buffer under 15 kV electrical pressure for at least 6 h (see Note 5).
4. Ensure that the electrical current is appropriate and stable (see Note 6).
5. After conditioning, store the capillary filled with the separation buffer (by dipping the capillary ends into the separation buffer) until just before use.

3.3. CE Analysis

1. Prepare all buffer, sample, and empty vials at the appropriate positions according to the CE instruction manual. Program all settings in the CE-operating computer.
2. Inject samples using the hydrodynamic, siphoning, or electrokinetic method (see Note 7).
3. Run the program. Examples of CE separations and corresponding SDS-PAGE patterns are shown in Fig. 2 (see Note 8).
4. Detect the migrated polypeptides at a wavelength of 200 nm with a photodiode array (see Note 9).
5. Calculate the electric charge and the size of the target polypeptide and compare these values with the actual migration time values. An example of an accurate correlation between the calculated electric charge and molecular weight and the measured migration time is shown in Fig. 3 (see Note 10).
6. Perform a sequential rinse procedure with the separation buffer, 500 mM acetic acid, deionized water, and the separation buffer for 10 min each (see Note 11).

4. Notes

1. Separation buffer should be freshly prepared because changes in electrolyte concentration lead to changes in the CE profile. When the electric current is increased due to buffer contamination, the migration of the proteins appears to be slow. The author

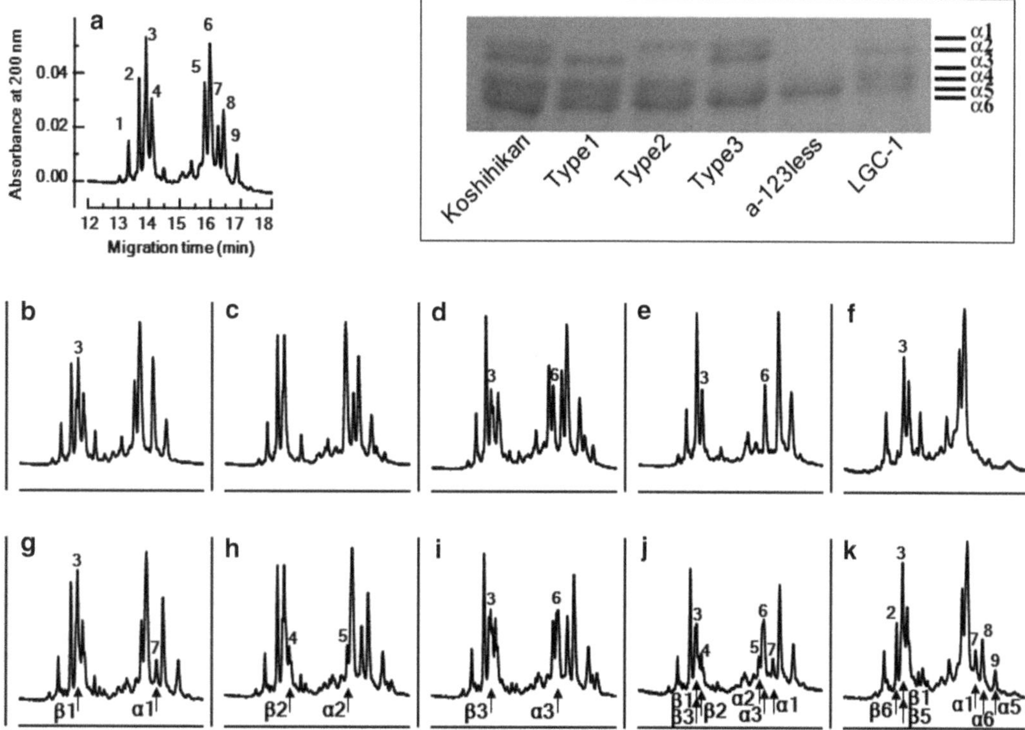

Fig. 2. Comparison of glutelin CE profiles and SDS-PAGE patterns. Glutelins extracted from the Type 1, Type 2, Type 3, a-123 less, and LGC-1 glutelin mutants were individually analyzed by capillary electrophoresis (CE) (panels **b**, **c**, **d**, **e**, and **f**, respectively) or were mixed and coanalyzed with glutelin from a standard japonica cultivar, Koshihikari (panels **g**, **h**, **i**, **j**, and **k**, respectively). The major peaks detected in the case of glutelin from Koshihikari were numbered according to migration time from the shortest to the longest duration (panel **a**). The deletion of bands in the mutants were confirmed by SDS-PAGE of the α polypeptide (*upper-right inlet*) and matched with CE peaks (panels **g**–**k**) (Reproduced from ref. 14 with permission from John Wiley and Sons).

prepares the separation buffer by mixing the following stock solutions: 0.2 M iminodiacetic acid, 1 % (w/v) hydroxypropyl methylcellulose, 0.2 M lauryl sulfobetaine (SB3-12), and acetonitrile. All stock solutions except the acetonitrile are stored in a refrigerator.

2. The analytic target used by the author is glutelin, the major storage protein of rice seeds. Rice glutelin, which is soluble in diluted acid or basic solutions, is extracted from seed endosperms of wild-type (cultivar Koshihikari) and glutelin mutant cultivars (Type 1, Type 2, Type 3, a-123 less, and LGC-1) with an extraction buffer (1 % (v/v) lactic acid including 1 mM EDTA). The other rice seed storage proteins, such as water-soluble albumins, neutral saline-soluble globulins, and alcohol-soluble prolamins, are sequentially pre-extracted and removed with a 35 mM potassium phosphate buffer (pH 7.6) containing 0.4 M NaCl followed by 60 % (v/v) n-propanol (8, 9).

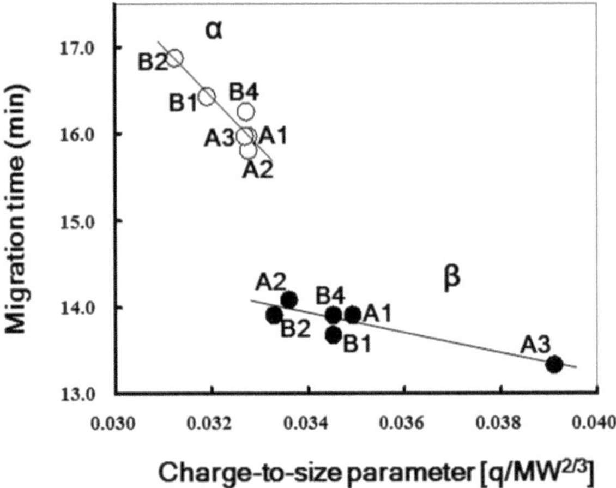

Fig. 3. Correlation between the measured CE migration time and the calculated charge-to-size parameter ($q/MW^{2/3}$) of glutelin α and β polypeptides. Charge values at pH 2.3 and the molecular weight for each of α and β polypeptides of the six major glutelin molecular species (GluA1, GluA2, GluA3, GluB1, GluB2, and GluB4) were calculated using rice genomic sequences. The charge-to-size parameter $q/MW^{2/3}$ was then estimated and compared with the measured value of migration time in capillary electrophoresis. Open circles represent α polypeptides, and closed circles represent β polypeptides (Reproduced from ref. 14 with permission from John Wiley and Sons).

The advantages of pre-extraction for CE include less peak tailing and excellent repeatability, as documented by Bean and Lookhart (2001) (10).

3. A high concentration of urea is suggested to be problematic because it promotes crystallization, exhibits low-UV range absorbance, and may introduce protein modifications (11). The author uses a low urea concentration and relatively long incubation times and high temperatures.
4. Rice glutelin is composed of α and β polypeptides that are covalently linked by a disulfide bond. Glutelin dissolved in the extraction buffer is incubated with 0.5 M urea and 10 % (v/v) 2-mercaptoethanol (ME) at 20 °C for at least 10 h to completely dissociate into the two components, α and β polypeptides. If the dissociation of the glutelin polypeptides is incomplete, one group of poorly separated peaks consisting of α and β polypeptides will be observed (Fig. 1). It is notable that this rapid CE analysis system enables us to analyze kinetic structural changes in glutelin.
5. Conditioning is generally performed with 0.1 M HCl and/or 0.1 M NaOH, from which the author refrains.
6. If the electric current is unstable, try the rinse procedure described above and continue the conditioning.
7. The author uses the siphoning method.

Table 1
Relationships between the properties of glutelin α and β polypeptides measured by CE and SDS-PAGE

CE				
Peak #	Time (min)	Area (%)	SDS-PAGE	Subunit
1	13.33	4.62	β4	GluA3
2	13.67	10.54	β6	GluB1
3	13.90	20.35	β3 β1 β5	GluA1 GluB4 GluB2
4	14.08	11.39	β2	GluA2
5	15.80	12.28	α2	GluA2
6	15.97	20.26	α3 α4	GluA1 GluA3
7	16.25	7.57	α1	GluB4
8	16.42	10.01	α6	GluB1
9	16.87	2.98	α5	GluB2

Reproduced from ref. 14 with permission from John Wiley and Sons

8. The author uses a program at 25 °C with a voltage of 15 kV for 20 min. Analysis using this optimal condition results in the detection of two groups of well-separated peaks within 13–17 min (Fig. 1). In the wild-type cultivar Koshihikari, which has six major glutelin subunits (*GluA1, GluA2, GluA3, GluB1, GluB2,* and *GluB4*), four and five major peaks representing the β and α polypeptides, respectively, are detected (Fig. 1 and Table 1). Meanwhile, five glutelin mutants—Type 1, lacking α1 (*GluB4*); Type 2, lacking α2 (*GluA2*); Type 3, lacking α3 (*GluA1*); a-123 less, lacking α1-α3; and LGC-1, in which the accumulation of α1 (*GluB4*), α5 (*GluB2*), and α6 (*GluB1*) are drastically decreased because of RNA interference—show different CE profiles. The differential CE profiles in the mutants are comparable to the observed SDS-PAGE patterns, permitting the identification of the CE peaks (Table 1). Because the CE system is more convenient and accurate than SDS-PAGE, it may be useful for the high-throughput quantitative analysis of rice glutelin.
9. The detected peak area is calibrated using the migration time.
10. The electrophoretic mobility of peptides and proteins is correlated with the charge-to-size parameter, $q/MW^{2/3}$, where q

is the charge and MW is the molecular weight (12). The amino acid sequences of the α and β polypeptides of the six major glutelin subunits are deduced from the genomic DNA sequences in the clones numbered AP003274 (*GluA1*), AC021891 (*GluA2*), AC133398 (*GluA3*), AP005511 (*GluB1* and *GluB2*), and AP005428 (*GluB4*). The electric charge of the glutelin polypeptides at the pH value of the CE separation buffer (pH 2.3) is calculated from the amino acid sequences as proposed by Manabe (1990) (13). The migration time shows a good inverse correlation with the charge-to-size parameter within the same type of polypeptides (Fig. 3), confirming the relationship summarized in Table 1. However, the slope of the regression lines is different between α and β polypeptides, suggesting the formation of different higher-order structures.

11. If several samples are analyzed consecutively, the sequential rinse procedure may be replaced with a single rinse with the separation buffer between samples.

Acknowledgments

This work was supported in part by grants to T.K.-T. from the Ministry of Education, Culture, Sports, Science, and Technology, Japan.

References

1. Elhamili A, Wetterhall M, Sjödin M, Sebastiano R, Bergquist J (2010) Analysis of peptides using N-methylpolyvinylpyridium as silica surface modifier for CE-ESI-MS. Electrophoresis 31:1151–1156
2. Wang A, Gao L, Li X, Zhang Y, He Z, Xia X, Zhang Y, Yan Y (2008) Characterization of two 1D-encoded ω-gliadin subunits closely related to dough strength and pan bread-making quality in common wheat (*Triticum aestivum* L.). J Cereal Sci 47:528–535
3. Di Luccia A, Lamacchia C, Mamone G, Picariello G, Trani A, Masi P, Addeo F (2009) Application of capillary electrophoresis to determine the technological properties of wheat flours by a glutenin index. J Food Sci 74:C307–311
4. García-Ruiz C, García MC, Cifuentes A, Marina ML (2007) Characterization and differentiation of diverse transgenic and nontransgenic soybean varieties from CE protein profiles. Electrophoresis 28:2314–2323
5. Simó C, Domínguez-Vega E, Marina ML, García MC, Dinelli G, Cifuentes A (2010) CE-TOF MS analysis of complex protein hydrolyzates from genetically modified soybeans—a tool for foodomics. Electrophoresis 31:1175–1183
6. Herrero M, García-Cañas V, Simo C, Cifuentes A (2010) Recent advances in the application of capillary electromigration methods for food analysis and Foodomics. Electrophoresis 31:205–228
7. Bean SR, Lookhart GL (2000) Ultrafast Capillary Electrophoretic Analysis of Cereal Storage Proteins and Its Applications to Protein Characterization and Cultivar Differentiation. J Agric Food Chem 48:344–353
8. Katsube-Tanaka T, Duldulao JBA, Kimura Y, Iida S, Yamaguchi T, Nakano J, Utsumi S (2004) The two subfamilies of rice glutelin differ in both primary and higher-order structures. Biochim Biophys Acta 1699:95–102
9. Takemoto Y, Coughlan SJ, Okita TW, Satoh H, Ogawa M, Kumamaru T (2002) The rice mutant esp2 greatly accumulates the glutelin precursor and deletes the protein disulfide isomerase. Plant Physiol 128:1212–1222

10. Bean SR, Lookhart GL (2001) Recent developments in high-performance capillary electrophoresis of cereal proteins. Electrophoresis 22:1503–1509
11. Wehr T, Rodriguez-Diaz R, Zhu M (1999) Capillary Electrophoresis of Proteins. In: Cazes J (ed) Chromatographic Science Series, vol 80. Marcel Dekker, New York, pp 1–286
12. Rickard EC, Strohl MM, Nielsen RG (1991) Correlation of electrophoretic mobilities from capillary electrophoresis with physicochemical properties of proteins and peptides. Anal Biochem 197:197–207
13. Manabe T (1990) Electrophoresis. In: Seikagakkai N (ed) New Biochemical Experiment Series, Vol. 1 Protein I. Tokyo Kagaku Dojin, Tokyo, pp. 329–387. In Japanese
14. Katsube-Tanaka T, Iida S, Yamaguchi T, Nakano J (2010) Capillary electrophoresis for analysis of microheterogeneous glutelin subunits in rice (*Oryza sativa* L.). Electrophoresis 31:3566–3572

Chapter 19

Evaluating Amyloid Beta (Aβ) 1–40 Degradation by Capillary Electrophoresis

Benjamin J. Alper and Walter K. Schmidt

Abstract

Here, we describe a capillary electrophoresis method for evaluating proteolysis of the amyloid beta peptide (Aβ) 1–40. This method is suitable for kinetic studies, demands little specialized equipment, and consumes only small quantities of a commercially available substrate whose physiological accumulation is thought to underlie the development of Alzheimer's disease.

Key words: Capillary electrophoresis, Amyloid beta (Aβ) 1–40, Proteolysis, Insulin-degrading enzyme

1. Introduction

The neuronal accumulation of amyloid beta (Aβ) peptides has been linked to Alzheimer's disease (1). According to the amyloid hypothesis, an imbalance between the production and elimination of Aβ peptides leads to their accumulation within the brain and resultant neurotoxicity (2). Degradation of Aβ by proteolytic enzymes such as the insulin-degrading enzyme (IDE) represents one pathway by which these peptides may be eliminated from the body, potentially slowing or preventing the onset of Alzheimer's disease (3–7).

Capillary electrophoresis (CE) has previously been used for the electrophoretic separation and detection of Aβ peptides, to study their aggregation, and to detect Aβ peptides within biological samples (8–10). Here, we describe in extensive procedural detail our reported method for applying CE to evaluate the proteolytic degradation of Aβ 1–40 in vitro (11). This method has the advantages

Nicola Volpi and Francesca Maccari (eds.), *Capillary Electrophoresis of Biomolecules: Methods and Protocols*, Methods in Molecular Biology, vol. 984, DOI 10.1007/978-1-62703-296-4_19, © Springer Science+Business Media, LLC 2013

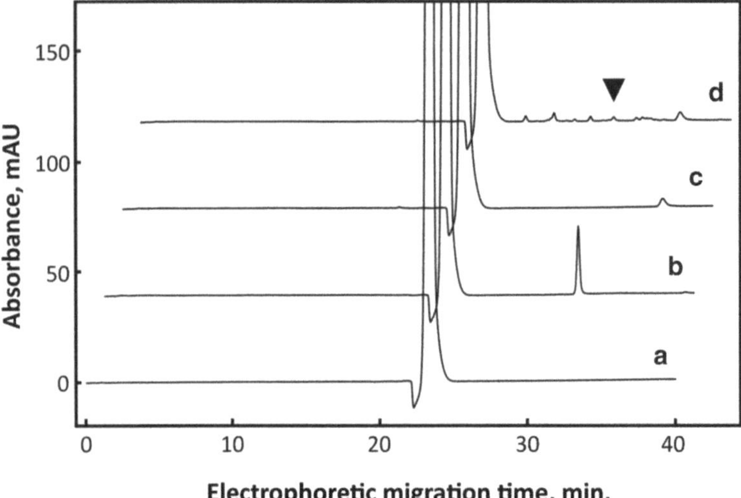

Fig. 1. IDE-mediated Aβ 1–40 proteolysis can be evaluated using CE. The components of the Aβ 1–40 degradation reaction were evaluated alone and in combination by capillary electrophoresis (CE). Shown are the chromatographic profiles of proteolysis buffer alone (**a**) or containing 12 μM Aβ 1–40 (**b**), 11 nM IDE (**c**), and a mixture of 12 μM Aβ 1–40 and 11 nM IDE (**d**). The relative position of intact Aβ, had it not been degraded, is marked (*arrowhead*). All samples contained 2.5% DMSO, which is responsible for the peak observed immediately after migration of a trough caused by bulk flow of the injection plug, and were incubated at 37 °C for 60 min prior to analysis. Chromatograms depict sample absorbance at 200 nm. Electrophoretic migration time and sample absorbance axes have been offset between chromatograms to facilitate data presentation. This figure and corresponding legend have been previously published (see Alper, B.J. and Schmidt, W.K. (2009), A capillary electrophoresis method for evaluation of Aβ proteolysis in vitro, *J Neurosci Methods* 178, 40–45).

of consuming minimal quantities of commercially available substrate relative to some other methods, and requires little specialized equipment other than a standard CE instrument outfitted with a fused silica capillary. Representative chromatograms and kinetic data obtained using the reported method are presented in Figs. 1 and 2, and have been reproduced from (11), with permission of the publisher.

2. Materials

Use autoclaved Millipore 18 Ω deionized sterile water or equivalent unless stated otherwise. Buffers should be made and stored at room temperature using clean, sterile glassware.

2.1. Buffers and Solvents

1. Capillary electrophoresis background electrolyte: 100 mM potassium phosphate, pH 7.6. Weigh out 3.6 g dibasic potassium phosphate (K_2HPO_4) and 30.2 g monobasic potassium phosphate (KH_2PO_4) and dissolve in 1.8 L water in a 2 L container. Adjust pH using minimal amount of NaOH or HCl as

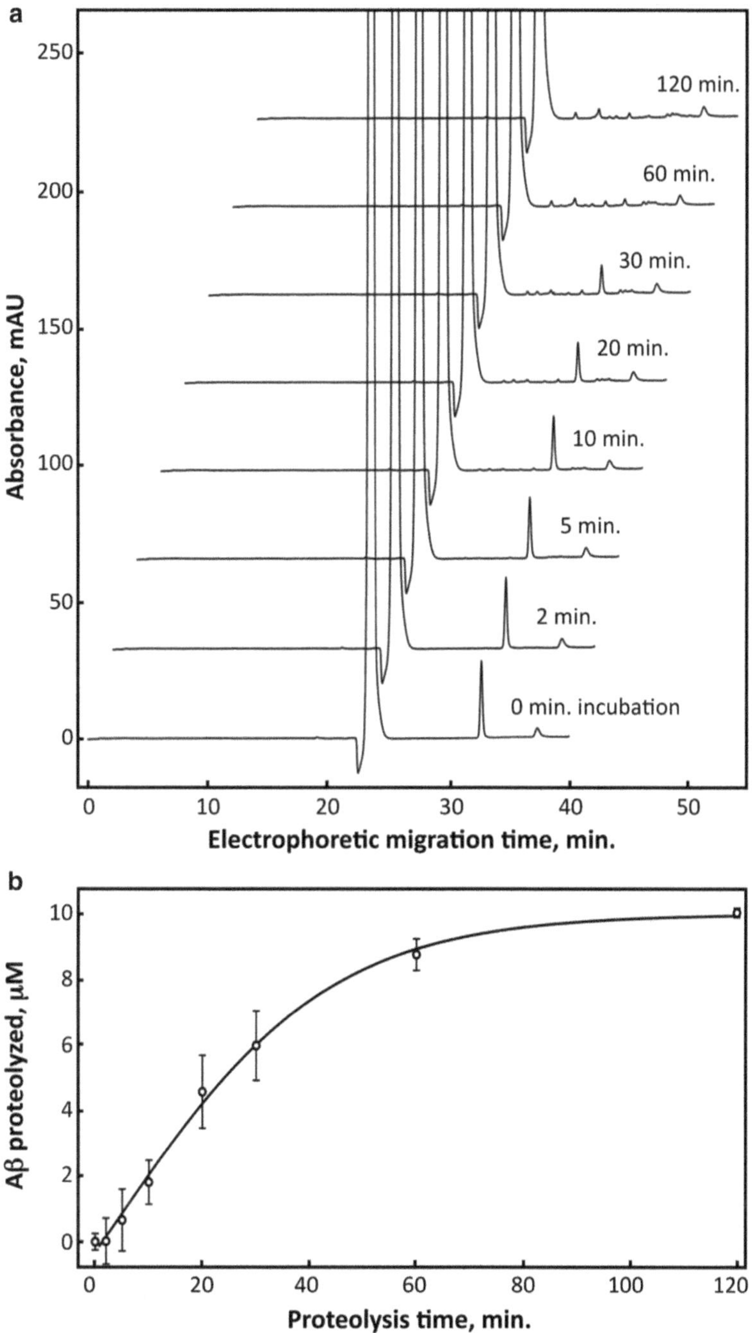

Fig. 2. CE permits kinetic analysis of IDE-mediated Aβ 1–40 proteolysis. Reaction mixtures initially containing 12 μM Aβ 1–40 and 11 nM IDE were incubated at 37 °C over a 120 min time course. Aliquots were removed at specific times indicated and enzyme activity was quenched. Chromatogram traces (**a**) were obtained for each sample as described in Fig. 1. A standard curve that correlates Aβ concentration with peak area was used to derive the amount of Aβ degraded at each time point, which is depicted graphically (**b**). The graphical representation of data derives from the averaged values of three experimental replicates, each having single points for each time point condition. Error bars represent standard deviation from the mean. This figure and corresponding legend have been previously published (see Alper, B.J. and Schmidt, W.K. (2009), A capillary electrophoresis method for evaluation of Aβ proteolysis in vitro, *J Neurosci Methods* 178, 40–45).

needed, and make up to 2 L final volume with water (see Note 1). Filter out particulates using a 0.22 μm pore size membrane and vacuum filtration.

2. Capillary electrophoresis injection and Aβ 1–40 proteolysis (CEP) buffer: 10 mM potassium phosphate, pH ~7.6. Prepare 100 mL by diluting 10 mL capillary electrophoresis background electrolyte with 90 mL water.
3. Enzyme storage buffer: 50 mM HEPES (4-(2-hydroxyethyl)-1-piperazineethanesulfonic acid), 140 mM NaCl, pH 7.4. Prepare by dissolving 2.4 g HEPES and 1.6 g NaCl in 140 mL water. Adjust pH using minimal amount of NaOH or HCl as needed. Add 40 mL glycerol with gentle stirring and make up to 200 mL final volume with water (see Note 2).
4. Hexafluoroisopropanol (HFIP: 1,1,1,3,3,3-Hexafluoro-2-propanol) (Sigma Chemical Company, St. Louis, MO, USA). Obtain at least 50 μL.
5. Methanol (CH_3OH). HPLC grade, ~100 mL.
6. Hydrochloric acid (HCl), 1 N. Prepare by diluting HPLC-compatible stock solution with filtered, sterile 18 Ω deionized water.
7. Sodium hydroxide (NaOH), 0.1 N. Prepare by dissolving 0.4 g NaOH in 80 mL water and make up to 100 mL final volume with water. Filter by vacuum filtration using a 0.22 μm pore size membrane.

2.2. Enzymes and Substrates

1. Insulin-degrading enzyme (IDE), 1 μg/μL in enzyme storage buffer. Prepare IDE, or an alternative source of enzymatic activity that can degrade Aβ 1–40, for cold storage and subsequent proteolytic assay by diluting to 1 μg/μL in enzyme storage buffer. Mix thoroughly, and freeze immediately as aliquots at −80 °C (see Notes 2 and 3).
2. Amyloid beta 1–40 peptide (Aβ 1–40), 4 μg/μL in dimethylsulfoxide (DMSO). In a fume hood, dissolve Aβ 1–40 by resuspending completely in approximately 50 μL HFIP, then purge under nitrogen gas until organic solvent is fully evaporated (see Notes 4 and 5). After purging, resuspend dried monomeric peptide in dry DMSO to a final concentration of 4 μg/μL (i.e., 500 μg Aβ 1–40 in 125 μL DMSO; see Note 6).

2.3. Capillary Electrophoresis Instrumentation and Accessories

1. Agilent G1600AX CE instrument or functional equivalent (Agilent Technologies Inc., Santa Clara, CA, USA). The CE instrument should be capable of performing automated pressurized injection and maintaining a thermostated capillary cassette.
2. Uncoated fused silica extended path length capillary, 56 cm effective length, 50 μm inner diameter (recommended: Agilent part number G1600-61211).

3. 1 mL polypropylene buffer vials and 100 μL sample vials with resealable polyurethane caps (recommended: Agilent part numbers 5182-0567, 9301-0978, and 5181-1512, respectively).

2.4. Other Equipment Needed

1. PCR thermocycler with heated bonnet. Should accommodate at least 24 PCR tubes.
2. Empty pipette tip rack insert (96 × 200 μL tip format) or similarly arrayed plastic PCR tube holder.
3. Forceps.
4. Liquid nitrogen.
5. Dewar flask.
6. Insulated, wide-mouthed working container for liquid nitrogen. The ideal container should be wide enough to allow an empty pipette tip rack insert (96 × 200 μL tip format) to freely float on the liquid nitrogen (recommended: foam Dewar vessel model # FD-800, Spearlab, Inc., San Francisco, CA, USA).
7. Bath sonicator.
8. Microcentrifuge, maximum force 16,000 × *g*.

3. Methods

The Aβ 1–40 degradation assay has three distinct parts. First, the proteolytic reaction is performed in a temperature-controlled environment and stopped by flash freezing in liquid nitrogen. Second, the proteolytic activity is eliminated from the assay by heat inactivation, and samples are prepared for CE analysis. Third, the CE instrument is conditioned, standardized, and used to evaluate the extent of Aβ 1–40 degradation that has occurred during the course of the proteolysis experiment. All procedures are carried out at ambient temperature unless stated otherwise.

3.1. Enzymatic Proteolysis of Aβ 1–40

1. Set up for the proteolytic assay. Preheat and hold temperature of thermocycler at 37 °C. Thaw IDE (1 μg/μL in enzyme storage buffer) or alternative source of proteolytic activity on ice, and thaw Aβ 1–40 (4 μg/μL in DMSO) at room temperature. Obtain liquid nitrogen, and half-fill a shallow, wide-mouthed, insulated container. Float an empty pipette tip rack insert (96 × 200 μL tips) in the liquid nitrogen (see Note 7). This will allow you to flash freeze samples while keeping them organized. Have forceps handy.
2. Prepare 2× IDE (enzyme) and 2× Aβ 1–40 (substrate) premixes. The enzyme and substrate premixes are prepared, respectively, by diluting IDE (1 μg/μL in enzyme storage

buffer) and Aβ 1–40 (4 μg/μL in DMSO) with an appropriate amount of proteolysis buffer (10 mM KPi, pH 7.6). Enough of each premix should be made to allow for a final reaction volume of 40 μL (i.e., 20 μL of each component) in duplicate. To reduce experiment-to-experiment variation and facilitate quick assembly and initiation of reactions, we recommend aliquoting the components of the reaction into an arrayed set of uncapped PCR tubes placed in a pre-warmed thermocycler. The arrayed organization allows use of a multichannel pipettor for rapid assembly of the proteolytic reactions. In our preferred approach, the substrate mix (20 μL per tube) is pre-dispensed into the appropriate number of reaction tubes (e.g., 1A, 1B, 2A, 2B; two tubes per time point), and the enzyme mix (50 μL per tube) is temporarily reserved in a separate set of tubes (1C, 2C, etc.), yielding three tubes per numbered set (i.e., 1A-C, 2A-C, etc.). This ordering minimizes substrate consumption and variation in the quantity of substrate input to the reaction, but can be reversed if enzyme is the more costly component.

3. Initiate the proteolytic reaction. Using a multichannel pipettor, simultaneously transfer 20 μL of warmed 2× enzyme premix (i.e., tube C contents) to the first set of tubes containing 2× substrate premix (i.e., set A). Mix quickly and thoroughly, and cap PCR tubes containing the proteolytic reaction mixture. Immediately repeat for the second set of substrate-containing tubes (i.e., set B). Discard tubes containing residual 2× enzyme premix (i.e., set C).
4. Terminate the proteolytic reaction. At each predetermined time point (i.e., 0, 5 min), remove the appropriate tubes using forceps, and flash freeze the samples in liquid nitrogen. Samples may be organized using an empty pipette tip rack insert that is kept floating on liquid nitrogen.
5. Adjust the thermocycler to hold temperature at 80 °C once all time points have been collected and flash frozen.
6. Heat kill the proteolytic activity. Transfer flash-frozen samples directly from liquid nitrogen to thermocycler heated to 80 °C. Do not thaw samples in the interim between freezing and heat denaturing. Heat samples at 80 °C for 3 min, then return samples to liquid nitrogen.
7. Store samples at −80 °C until ready to proceed with sample preparation for analysis by capillary electrophoresis (see Note 8).

3.2. Sample Preparation for CE

1. Thaw samples. Remove samples from −80 °C, place in an empty pipette tip rack insert (96 × 200 μL format), and allow samples to thaw gradually at room temperature (see Note 9).

2. Sonicate thawed samples to ensure solubility of Aβ 1–40. In a bath sonicator, float the pipette tip rack insert containing your samples in water such that contact is maintained between the PCR tubes and the floor of the bath sonicator. Sonicate for 10 min.
3. Centrifuge samples at 16,000 × *g* for 1 min in a standard microcentrifuge to remove any precipitate or insoluble material from the sonicated samples. Use a rotor sized for PCR tubes or incorporate adapters into a standard rotor so that PCR tubes can be accommodated.
4. Transfer the supernatant of the sonicated and clarified sample into a graduated CE sample vial, and place this sample in the loading tray of the capillary electrophoresis instrument. Analyze immediately (see Note 9).

3.3. Capillary Conditioning, Standardization, and Sample Analysis by CE

Prior to thawing and analyzing frozen samples from the proteolysis experiment, it is imperative to condition the capillary of the capillary electrophoresis instrument to ensure that electrophoretic separation and detection of Aβ 1–40 can be performed in a reproducible matter. Similarly, it is critical to establish a standard curve that can be used to equate empirically determined Aβ 1–40 peak areas with peptide concentrations of the solutions being analyzed, as described below.

1. Set up the capillary electrophoresis instrument. Outfit an Agilent G1600AX or similar instrument with a 56 cm effective length, 50 μm inner diameter uncoated fused silica extended light path capillary. Place 500 mL capillary electrophoresis background electrolyte (100 mM potassium phosphate, pH 7.6) in sterile glassware in the buffer reservoir after rinsing pressurized intake lines with ethanol. Thermostat the capillary cassette to 25 °C.
2. Designate and fill 1 mL polypropylene buffer vials as follows, then apply polyurethane caps:

   Vial 1: Electrophoresis inlet reservoir. Fill with 1 mL capillary electrophoresis background electrolyte from buffer reservoir and replenish after each run or conditioning step.

   Vial 2: Electrophoresis outlet reservoir. Fill with 1 mL capillary electrophoresis background electrolyte from buffer reservoir and replenish after each run or conditioning step.

   Vial 3: HPLC grade methanol. Add 1 mL.

   Vial 4: Hydrochloric acid (HCl), 1 N. Add 1 mL.

   Vial 5: Autoclaved 18 Ω deionized water. Add 1 mL.

   Vial 6: Sodium hydroxide (NaOH), 0.1 N. Add 1 mL.

   Vial 7: Liquid waste. Leave empty.

3. Condition the capillary. Apply the following rinses sequentially at ~1 bar:

   Rinse 1: HPLC grade methanol, 8 min.

   Rinse 2: 1 N HCl, 8 min.

   Rinse 3: Autoclaved 18 Ω deionized water, 8 min.

   Rinse 4: 0.1 N NaOH, 8 min.

   Rinse 5: Autoclaved 18 Ω deionized water, 2 min.

   Rinse 6: Capillary electrophoresis background electrolyte from electrophoresis inlet reservoir, 8 min.

4. Prepare 6 twofold serial dilutions of Aβ 1–40 4 μg/μL in DMSO within CEP buffer, beginning with a maximal concentration of 0.2 μg/μL Aβ 1–40. Each sample from the dilution series should be prepared to a total volume of 40 μL following serial dilution.

5. Load the most concentrated sample from your Aβ 1–40 dilutions onto the capillary by pressurized injection, perform electrophoretic separation, and repeat until peptide elution time and peak area are consistent and reproducible (see Note 10). Sample loading should be performed as 20 s × 20 mbar pressurized injection, followed by 40 min × 50 μA positive polarity electrophoretic separation, while analyte detection should be monitored at 200 nm. A major DMSO solvent peak should be detected around 20–25 min run time, and a smaller Aβ 1–40 peak should be detected between 30 and 35 min. Between runs, replenish the capillary electrophoresis background electrolyte in the 1 mL electrophoresis inlet reservoir (Vial 1) and 1 mL electrophoresis outlet reservoir (Vial 2), and sequentially rinse the capillary with 0.1 N NaOH and background electrolyte for 8 min each at ~1 bar, emptying to liquid waste container (Vial 7).

6. Analyze the chromatographic data obtained from Step 5, and use this to construct a standard curve equating Aβ 1–40 concentration and peak area. Also generate a standard curve equating DMSO peak areas with volume input for each sample. Normalizing Aβ 1–40 peak area proportional to the DMSO solvent peak area provides an internal standard and loading control during subsequent proteolysis experiments.

7. Thaw proteolytic samples and prepare according to Subheading 3.2. Perform sample analysis in the manner described within Step 5, and use standard curve to determine the amount of Aβ 1–40 remaining in sample. The amount of degradation is obtained by subtracting the amount of Aβ 1–40 remaining in the sample from the original input amount, which is determined from the 0 min proteolysis time point.

4. Notes

1. Even relatively small changes in background electrolyte pH can have noticeable effects on CE run times and elution profiles. It is thus important to prepare enough buffer to ensure that CE standardization and subsequent analysis can be conducted without changing background electrolyte or other buffer preparations. A 2 L total volume of background electrolyte was more than sufficient for the analysis presented in (11). Preparation of phosphate buffers is essentially as described by G. Gomori (12).
2. Enzyme preparation and storage buffers are described here on the basis of precedent (11). Alternative sources of enzyme activity and storage buffers have not been exhaustively evaluated. Such alternatives may not be detrimental to experimental analysis, especially if small volumetric amounts of the enzymatic activity are sufficient to permit ample Aβ 1–40 proteolysis. The effects of different enzyme preparations and storage buffers on Aβ 1–40 proteolysis and CE resolution efficiency should be determined empirically before conducting exhaustive studies.
3. Purified recombinant *Rattus norvegicus* (rat) IDE provides a suitable source of proteolytic activity towards Aβ 1–40. It can be cloned, expressed, and purified by immobilized nickel affinity chromatography in accordance with established methods (11). Other sources of enzymatic activity are compatible (13) or theoretically compatible with this method (see Note 2).
4. We recommend using HFIP-treated Aβ 1–40 from rPeptide (catalog number A-1153-1, rPeptide, Inc., Athens, GA, USA). Peptide from this source has been pretreated with HFIP, ensuring its initial monomeric state upon resuspension in DMSO. Any amount of aggregated Aβ 1–40 in the starting sample can serve to nucleate additional aggregation. Aggregates might complicate CE analysis and could potentially develop while samples are awaiting sample injection.
5. Alternatively, non-treated Aβ 1–40 can be treated with HFIP to ensure the initial monomeric state of the peptide. After dissolving Aβ 1–40 in HFIP, remove the organic solvent completely by purging under an inert atmosphere such as nitrogen gas. rPeptide, Inc., provides non-treated Aβ 1–40 in a glass container that is capped with a rubber stopper. This container is compatible with both HFIP resuspension and nitrogen purging. During gas purging, two syringe needles can be used to puncture the rubber stopper, providing both an inlet and outlet for nitrogen gas. HFIP is corrosive and presents a respiratory

hazard. It should be stored appropriately and handled with care under proper ventilation.

6. DMSO, a weakly acidic aprotic solvent, exhibits a solvent elution peak that does not overlap with Aβ 1–40 or its degradation products during CE analysis. Further, the solvent is not volatile, and we have observed that electrophoretic solvent peak area is relatively constant throughout a 12 h injection period at ambient temperature.
7. A pipette tip rack insert allows the samples to remain submerged in liquid nitrogen due to its wider hole size. When a PCR tube rack is used instead, samples have a tendency to float out from the smaller hole and become disorganized. These observations may be impacted by the design of the racks available to us, so any plastic organizer planned for use should be tested beforehand.
8. We have stored samples at −80 °C for up to 6 weeks without noticeable detrimental effects. Longer storage periods may also be acceptable but should be validated.
9. Do not thaw more samples than can be evaluated by CE within a 12 h period at any one time. During assay standardization, little or no detectable reduction in Aβ 1–40 peak area was observed following analysis of a sample that was repeatedly injected over a 12 h loading period. A longer period may be permissible but should be validated.
10. Reproducibility in electrophoretic migration of Aβ 1–40 should be observed after 5–6 runs maximum.

Acknowledgment

We are grateful to Drs. Nandu Menon, Alan Przybyla, and Zachary Wood (all of the University of Georgia, Department of Biochemistry and Molecular Biology) for advice and assistance, and for access to experimental equipment and materials. Aβ 1–40 consumed during method development and analysis was a gift of rPeptide, Inc. (Athens, GA, USA).

References

1. Glenner GG, Wong CW (1984) Alzheimer's disease: initial report of the purification and characterization of a novel cerebrovascular amyloid protein. Biochem Biophys Res Comm 120:885–890
2. Hardy J, Selkoe DJ (2002) The amyloid hypothesis of Alzheimer's disease: progress and problems on the road to therapeutics. Science 297:353–356
3. McDermott JR, Gibson AM (1997) Degradation of Alzheimer's beta-amyloid protein in human and rat brain peptidases: involvement of insulin-degrading enzyme. Neurochem Res 22:49–56
4. Vekrellis K, Ye Z, Qiu WQ, Walsh D, Hartley D, Chesneau V et al (2000) Neurons regulate extracellular levels of amyloid-beta protein via proteolysis by insulin-degrading enzyme. J Neurosci 20:1657–1665

5. Mukherjee A, Song E, Kihiko-Ehmann M, Goodman JP, St Pyrek J, Estus S et al (2000) Insulysin hydrolyzes amyloid-beta peptides to products that are neither neurotoxic nor deposit on amyloid plaques. J Neurosci 20:8745–8749
6. Miller BC, Eckman EA, Sambamurti K, Dobbs N, Chow KM, Eckman CB et al (2003) Amyloid beta peptide levels in brain are inversely correlated with insulysin activity levels in vivo. Proc Natl Acad Sci USA 100:6221–6226
7. Leissring MA, Farris W, Chang AY, Walsh DM, Wu X, Sun X et al (2003) Enhanced proteolysis of beta-amyloid in APP transgenic mice prevents plaque formation, secondary pathology, and premature death. Neuron 40:1087–1093
8. Varesio E, Rudaz S, Krause K-H, Veuthey J-L (2002) Nanoscale liquid chromatography and capillary electrophoresis coupled to electrospray mass spectrometry for the detection of amyloid beta peptide related to Alzheimer's disease. J Chr A 974:135–142
9. Sabella S, Quaglia M, Lanni C, Racchi M, Govoni S, Caccialanza G et al (2004) Capillary electrophoresis studies on the aggregation process of beta-amyloid 1–42 and 1–40 peptides. Electrophoresis 25:3186–3194
10. Verpillot R, Otto M, Klafki H, Taverna M (2008) Simultaneous analysis by capillary electrophoresis of five amyloid peptides as potential biomarkers of Alzheimer's disease. J Chr A 1214:157–164
11. Alper BJ, Schmidt WK (2009) A capillary electrophoresis method for evaluation of Aβ proteolysis in vitro. J Neurosci Methods 178:40–45
12. Gomori G (1955) Preparation of buffers for use in enzyme studies. Methods Enzymol 1:138–146
13. Alper BJ, Rowse JW, Schmidt WK (2009) Yeast Ste23p shares functional similarities with mammalian insulin-degrading enzymes. Yeast 11:595–610

Chapter 20

Capillary Electrophoresis for Protein Profiling of the Dimorphic, Pathogenic Fungus, *Penicillium marneffei*

Julie M. Chandler, Heather R. Trenary, Gary R. Walker, and Chester R. Cooper

Abstract

Penicillium marneffei is an endemic, dimorphic fungus that exhibits very significant morbidity among immune compromised persons living or having traveled in Southeast Asia. The dimorphic nature of *P. marneffei*, which is believed to be a major contributing factor to infection by this fungus, is thermally regulated. At 25 °C, the fungus grows as a mold, but converts to a yeast phase when incubated at 37 °C. Hence, protein profiling of these developing forms will help ascertain the underpinning molecular mechanisms associated with this phase transition, and perhaps provide clues to virulence in this pathogenic fungus. This chapter outlines the basic procedures previously used to demonstrate distinct differences in protein expression between the mold and yeast phases of *P. marneffei*.

Key words: Isoelectric focusing, Two-dimensional gel electrophoresis, Protein profiling

1. Introduction

Penicillium species are considered cosmopolitan fungi that grow as multicellular, multinucleate molds (1, 2). Under the appropriate conditions, the hyphae comprising the mold phase give rise to specialized structures bearing uninucleate spores, better known as conidia. The latter are the means by which this fungus reproduces asexually since sexual reproduction is infrequent or not known for many *Penicillium* species. Moreover, members of this genus are considered benign (3). Since the late nineteenth century, fewer than 100 cases of infection (penicilliosis) by *Penicillium* species have been recorded in the medical literature. This is a remarkably low morbidity rate given the worldwide distribution of the fungus and the increasing prevalence of immune compromised individuals. In fact, only four cases of penicilliosis have been recorded among individuals having AIDS (3–5). However, there is one

Nicola Volpi and Francesca Maccari (eds.), *Capillary Electrophoresis of Biomolecules: Methods and Protocols*, Methods in Molecular Biology, vol. 984, DOI 10.1007/978-1-62703-296-4_20, © Springer Science+Business Media, LLC 2013

species of *Penicillium* that does not wholly adhere to the above description of the genus. That species is *P. marneffei*.

Penicillium marneffei emerged in the late 1980s as an infectious agent associated with the advent of AIDS in Southeast Asia, particularly in Thailand (3, 6, 7). To date, more than 7,000 cases of penicilliosis due to *P. marneffei* have been recorded in Thailand, especially in the country's northern region. Infections due to this pathogen, however, have been restricted to humans and bamboo rats. No other natural cases of disease in other animals are known, nor have studies been able to ascertain the reservoir of infection. Epidemiological investigations dismiss the bamboo rats as the reservoir and suggest that the rodent is an accidental sentinel for infection. Moreover, infections by *P. marneffei* are restricted to portions of Southeast Asia stretching from southern China to Malaysia and from Vietnam to the eastern reaches of India. Other than a single laboratory-acquired infection, cases of disease diagnosed outside the endemic region have always been associated with persons having traveled in or emigrated from the endemic area.

Another unique feature of *P. marneffei* relative to other *Penicillium* species is that the former is dimorphic (3, 6, 7) . When cultured at 25 °C, *P. marneffei* exhibits the morphology typically associated with the genus. However, in vivo, *P. marneffei* is observed as single-celled yeasts that divide by schizogony (binary fission). The yeast cells most often reside within phagocytes, indicating that *P. marneffei* has evolved the ability to evade host defenses. The yeast phase can be replicated by incubating conidia or hyphae at 37 °C. Moreover, the thermally regulated nature of dimorphism in *P. marneffei* is reversible. Yeast cells produced at 37 °C that are subsequently incubated at 25 °C will convert to the mold phase. By comparison, in those rare cases of penicilliosis caused by species other than *P. marneffei*, the fungi grow filamentously in tissue and do not form a yeast phase either in vitro or in vivo.

Clearly, the ability of *P. marneffei* to undergo phase transition, i.e., reversible conversions between the mold and yeast phases, is strongly associated with pathogenicity. In earlier work, we observed that morphological differences at 25 and 37 °C occur early during the germination of conidia (3). This suggested that the molecular signals responsible for dimorphism operate early in the differentiation process. We hypothesized that such signals ought to be reflected in protein expression patterns. Therefore, we developed methods to examine the protein profiles of the developing mold and yeast phases of *P. marneffei* (8). Our previously published results clearly demonstrate not only the utility of proteomic approaches to studying dimorphism in *P. marneffei*, but also that such strategies complement and extend those observations using genetic-based methodologies.

The protocol detailed below was successfully employed in our laboratory to develop protein profiles from *P. marneffei*. Although our investigations have been based upon equipment obtained from a single vendor (Bio-Rad Laboratories), there are very comparable systems available from other sources. The protocol that we describe herein can be readily modified for use with these other systems.

Finally, we discuss the use of immobilized pH gradient gel (IPG) strips to separate proteins having isoelectric points from pH 5 to pH 8. This range seems to cover most of the available proteome in *P. marneffei*. However, we have effectively used IPG strips covering other pH ranges to assess proteins expressing different isoelectric points. Subsequently, the proteins were separated by mass by second-dimension polyacrylamide gel electrophoresis, then proteins of interest were recovered from the gel and submitted to a external service center for sequencing by mass spectrometry. The latter procedure is not described below, but has been previously detailed (8). Like many smaller institutions, due to resource and infrastructure limitations, it is more effective and cost efficient to have our samples processed by a service center.

2. Materials

2.1. Organism and Media

1. *Penicillium marneffei* strain F4 (CBS 119456).
2. Potato dextrose agar (PDA) and Sabouraud dextrose broth (SAB) (see Note 1).

2.2. Equipment and Materials: (See Note 2)

1. Mini-BeadBeater® (Biospec Products, Bartlesville, OK).
2. Glass wool.
3. Hemocytometer.
4. Screened caps (Cat. No. 170-3711; Bio-Rad Laboratories, Hercules, CA).
5. Acid-washed glass beads (0.5 mm dia; Biospec Products).
6. Immobilized pH gradient gel (IPG) strips—ReadyStrip™ IPG strips, pH 5–8 (Recommended source: Bio-Rad Laboratories) (see Note 3).

2.3. Buffers and Chemicals: (See Note 4)

1. Tris–EDTA buffer (TE)—10 Mm tris(hydroxymethyl)aminomethane (Tris); 1 mM ethylenediaminetetraacetic acid (EDTA), pH 8.0.
2. Lysis buffer—20 mM Tris–HCl, pH 7.6; 10 mM NaCl; 0.5 mM deoxycholate; 40 μL/mL of protease inhibitor cocktail.

3. Modified sample buffer (MSB)—2 M thiourea; 7 M urea; 4 % (w/v) 3[(3-cholamidopropyl)dimethylammonio]-propanesulfonic acid (CHAPS); 1 % w/v dithiothreitol (DTT); 2 % carrier ampholytes pH 3–10.
4. Trichloroacetic acid (TCA).
5. 2DE buffer—8.4 M urea; 2.4 M thiourea; 5 % CHAPS; 25 mM spermine base; 50 mM DTT.
6. Bradford dye—0.01 % Coomassie Brilliant Blue G-250; 8.5 % (v/v) phosphoric acid.
7. Rehydration buffer (RB)—8 M urea; 1 % CHAPS; 15 mM DTT; 0.2 % BioLytes (recommended source: Bio-Rad Laboratories); 0.001 % bromophenol blue.
8. Equilibration buffer I—6 M urea; 2 % sodium dodecyl sulfate (SDS); 0.375 M Tris–HCl pH 8.8; 20 % glycerol; 130 mM DTT.
9. Equilibration buffer II—6 M urea; 2 % SDS; 0.375 M Tris–HCl pH 8.8; 20 % glycerol; 135 mM iodoacetamide.
10. Tris–Glycine–SDS buffer (TGS)—3.02 g/L Tris; 18.8 g/L glycine; 1.0 g/L SDS).
11. Overlay agarose—0.5 % (w/v) low melt agarose in TGS buffer with 0.001 % bromophenol blue.
12. Second-dimension polyacrylamide gels—Prepare the desired concentration in accord with Table 1.
13. Electrophoresis buffer—0.375 M Tris (pH 8.8); 0.1 % SDS.
14. Coomassie stain—0.25 % (w/v) Coomassie brilliant blue R-250; 45 % methanol; 10 % acetic acid.
15. High destain /SYPRO fixing solution—40 % methanol, 10 % acetic acid.
16. Low destain—10 % methanol; 6 % acetic acid.
17. SYPRO Ruby protein gel stain (recommended source: Bio-Rad Laboratories) (Table 1).

3. Methods

3.1. Cell Growth and Collection

1. *Penicillium marneffei*, F4 strain, is inoculated onto 30 ml of PDA layered onto the bottom of 150 cm^2 cell culture flasks with vented caps, then incubated for 10 days at 25 °C.
2. On the tenth day, gently scrape off the surface growth from 2 plates using 10 mL of sterile water and a cell scraper.
3. Filter the culture suspension through glass wool between screened caps (sterilized by autoclaving) at 150–200 × g for 15–30s (see Note 5). This apparatus filters out hyphae and allows only conidia to be present in the filtrate.

Table 1
Components of polyacrylamide gels differing in concentration

Solution components	Total gel volume	
	5 ml	100 ml
6 % gel		
H_2O	2.85 ml	58.0 ml
40 % acrylamide	0.75 ml	15.0 ml
1.5 M Tris (pH 8.8)	1.3 ml	25.0 ml
10 % SDS	0.05 ml	1.0 ml
10 % ammonium persulfate	0.05 ml	1.0 ml
TEMED	0.004 ml	0.08 ml
8 % gel		
H_2O	2.6 ml	53.0 ml
40 % acrylamide	1.0 ml	20.0 ml
1.5 M Tris (pH 8.8)	1.3 ml	25.0 ml
10 % SDS	0.05 ml	1.0 ml
10 % ammonium persulfate	0.05 ml	1.0 ml
TEMED	0.003 ml	0.06 ml
10 % gel		
H_2O	2.35 ml	48.0 ml
40 % acrylamide	1.25 ml	25.0 ml
1.5 M Tris (pH 8.8)	1.3 ml	25.0 ml
10 % SDS	0.05 ml	1.0 ml
10 % ammonium persulfate	0.05 ml	1.0 ml
TEMED	0.002 ml	0.04 ml
12 % gel		
H_2O	2.1 ml	43.0 ml
40 % acrylamide	1.5 ml	30.0 ml
1.5 M Tris (pH 8.8)	1.3 ml	25.0 ml
10 % SDS	0.05 ml	1.0 ml
10 % ammonium persulfate	0.05 ml	1.0 ml
TEMED	0.002 ml	0.04 ml
15 % gel		
H_2O	1.725 ml	35.5 ml
40 % acrylamide	1.875 ml	37.5 ml
1.5 M Tris (pH 8.8)	1.3 ml	25.0 ml
10 % SDS	0.05 ml	1.0 ml
10 % ammonium persulfate	0.05 ml	1.0 ml
TEMED	0.002 ml	0.04 ml

4. Count conidia using a hemocytometer (see Note 6). Calculate the volume of the conidial solution that will yield 1×10^7 conidia/mL in 50 mL of SAB contained in 500 mL Erlenmeyer flasks. Transfer this volume to these flasks pre-warmed to either 25 or 37 °C. Incubate flasks in orbital shakers (120 rpm) at 25 °C (to initiate mold development) or 37 °C (to initiate yeast development).

5. After incubation for the selected period of time, collect the cells by centrifugation at 17,600 × *g* for 15 min at 4 °C, followed by two TE washes. Again, collect the cells by centrifugation at 17,600 × *g* for 15 min at 4 °C. The washed cell pellet can be kept on ice briefly or frozen in small aliquots at −80 °C prior to protein extraction.

3.2. Homogenization and Sample Preparation

1. Place approximately 400 mg wet pellet of cells in an ice-cold 2.0 ml *O*-ring screw-capped microfuge tube, add an equal amount of acid-washed glass beads as well as 800 μL lysis buffer. Bead beat cells in a Mini-BeadBeater operating at 5,000 rpm for 4 min in 30s increments, cooling in an ice bath in between for 30s.
2. Centrifuge the resulting slurry at 6,000 × *g* for 10 min at 4 °C. Transfer supernatant to a pre-weighed 1.5 mL microcentrifuge tube and precipitate proteins by adding 20 % TCA. Cool tubes on ice for 20 min and centrifuge at 2,000 × *g* for 20 min at 4 °C. The resulting protein pellet will be loose.
3. Carefully discard the supernatant, and then wash pellet three times with 500 μL cold acetone. Collect the pellet each time by centrifugation at 850 × *g* for 1 min at 4 °C.
4. Air-dry the pellet in a biological safety hood. Calculate the dry weight of the protein by determining the difference in weight between the empty tube and the one now containing the pellet. Dried protein pellets can be stored at −80 °C.
5. Resuspend the dry protein pellet in MSB using the following scale: for a 5–15 mg pellet, use 500 μL MSB; for a 16–30 mg pellet, use 750 μL MSB (see Note 7). Store the suspensions at −80 °C.

3.3. Protein Quantification

1. A modified Bradford assay (10) is used to determine the concentration of protein in each sample. Specifically, each reaction tube contains 80 μl water, 10–20 μl 0.1 M HCl, 10 μl 2DE buffer, and 4.0 ml of Bradford dye. A standard curve should be established using serial dilutions of known concentrations of a protein standard, e.g., bovine serum albumin. Sample concentrations are determined from this standard curve by using 5–10 μl of protein extract added to a reaction tube. The absorbencies of all reaction tubes are recorded at 595 nm using a standard spectrophotometer.

3.4. Isoelectric Focusing (IEF)

1. Use passive rehydration to prepare IPG strips to separate proteins by isoelectric point. In particular, use 125 μL rehydration buffer for 7 cm strips and 300 μL for 17 cm strips. Add the buffer to a rehydration tray and lay the IPG strip over it, cover with mineral oil, and place on an orbital shaker for 12–24 h.
2. For protein loading prior to IEF, load either 100–150 μg protein onto 7 cm IPG strips or 200–255 μg protein onto 17 cm IPG strips. Use electrode wicks (dampened with deionized,

distilled water), then place the protein-loaded IPG strips into the focusing tray according to instructions provided by manufacturer (Bio-Rad Laboratories). Cover the IPG strips with mineral oil.

3. Perform IEF as follows: Focus IPG strips (both 7 cm and 17 cm) at 20 °C starting at 250 V for 15 min using a linear voltage ramp. For the 7 cm strips, it takes over 2 h to reach 4,000 V with the completion of focusing at 40,000 V-h. For the 17 cm strips, it takes over 3 h to reach 10,000 V with the completion of focusing at 60,000 V-h.
4. Equilibrate strips in equilibration buffer I for 10 min with agitation using an orbital shaker operating at room temperature. Repeat this procedure once, but replace the buffer with equilibration buffer II.

3.5. SDS-Polyacrylamide Gel Electrophoresis

1. Rinse the IPG strips once with TGS, then place and seal the strip onto polyacrylamide gels (typically 10 % or 12 % concentration) with melted overlay agarose. Allow the agarose to solidify.
2. Load gels into either a Bio-Rad Mini-Protean (7 cm) or Bio-Rad Protean II XL (17 cm) electrophoretic apparatus. Apply a constant current of 14–16 mA to the 7 cm gels and of 10–24 mA to the 17 cm gels. Continue electrophoresis until the dye front reaches the bottom of the gel.
3. Remove the gels from the gel plates and stain according to the two options below:
   (a) Coomassie method: Add enough Coomassie stain to sufficiently cover the gel. Stain the gel with gentle agitation at room temperature for at least 1 h. Remove the Coomassie stain, prior to adding the high destain solution for 1 h. Destain the gel using gentle agitation at room temperature. Subsequently, remove the solution from the gel and add low destain solution. Continue destaining at room temperature using gentle agitation until dark spots are visible against a light blue background.
   (b) Fluorescent staining: Place gels in fixing solution for 1 h using gentle agitation at room temperature. Remove this solution and add enough SYPRO Ruby protein gel stain to cover the gel. Stain for 4–24 h using gentle agitation at room temperature. Remove SYPRO Ruby stain and rinse the gel in deionized, distilled water before imaging.

3.6. Image Analysis and Protein Sequencing

1. Image the Coomassie or SYPRO Ruby stained gels using the appropriate apparatus (see Note 8). Using PDQuest (Bio-Rad) imaging software, or an equivalent product, analyze the gels using triplicate runs of yeast and mold samples for comparison. Replicate gels were used to produce match sets and then match

sets from mold and yeast were compared to identify unique and shared proteins.

2. Excise protein spots of interest from the gel by using a sterile Pasteur pipette to cut and retrieve the gel piece from a gel illuminated by UV light or visible light (depending on stain employed).
3. Place the gel piece in a sterile microcentrifuge tube and process as recommended by the service center that will perform the mass spectrometric sequencing (see Note 9).

3.7. Data Processing

1. Most mass spectrometric service centers will provide the sequencing data in a usable form. All data processing should be performed following recommended guidelines either by the service center or the originating laboratory (11).

4. Results

Figure 1 depicts a typical protein profile of the mold and yeast phases from *P. marneffei*. Both gel images show labels indicating particular protein spots that had been isolated and subsequently sequenced by mass spectrometry. The details of these sequences can be found elsewhere (8). However, for illustration purposes, two particular proteins are highlighted. Spot B28 (indicated by the box in both gel images) was determined to be malate dehydrogenase after analysis of its derived mass spectrometric sequence. As shown by the densitometry reading of the area surrounding Spot B28, this protein is more highly expressed in the mold phase. In contrast, Spot A1 (indicated by the arrow in the right gel image) appears to be expressed only by the yeast phase of *P. marneffei*. The densitometry reading confirms this observation. From sequence analysis, Spot A1 was determined to be a Ran GTPase.

These results demonstrate the utility of proteomic approaches to studying dimorphism in *P. marneffei*. Such strategies are being used to complement and extend those observations we are currently generating through genetic-based methodologies. Collectively, both strategies are helping us address our investigations of the molecular signals regulating phase transition, and perhaps virulence, in this pathogenic fungus.

5. Notes

1. We use the formulation of SAB that contains 2 % glucose.
2. Although suitable equipment to perform the procedures described herein can be obtained from a number of vendors,

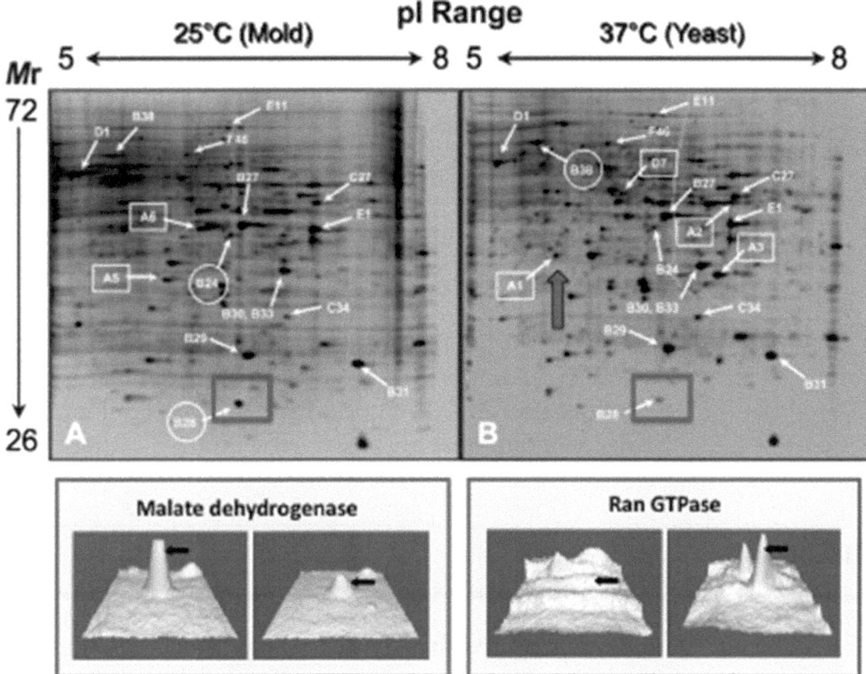

Fig. 1. Typical protein profile of the mold and yeast phases of *P. marneffei*. The left gel image represents the mold phase profile from developing hyphae collected after 24 h of incubation at 25 °C. The right gel image represents the profile from developing yeasts collected after 24 h of incubation at 37 °C. The densitometry readings of specific protein spots from these gels are depicted in the two lower images. The lower left image indicates the relative protein concentration of Spot B28 (indicated by a box in both gel images) at 25 °C (*left inset*) and 37 °C (right inset). The lower right image indicates the relative protein concentration of Spot A1 (indicated by an *arrow* in the yeast phase gel image) at 25 °C (left inset) and 37 °C (*right inset*). Respectively, mass spectrometric sequencing results determined these proteins to be malate dehydrogenase and Ran GTPase (Image modified from Chandler et al. (8)).

most of the laboratory equipment used in our studies was obtained from Bio-Rad Laboratories (Hercules, CA). In particular procedures, this is so noted in the protocol.

3. Although there are a number of sources for purchasing immobilized pH gradient gel strips, we employed ReadyStrip™ IPG strips. For the protocol described above, we employed strips having pH 5–8 range. This range seems to cover most of the available proteome in *P. marneffei*. However, we have successfully used strips covering other pH ranges to assess proteins expressing different isoelectric points.
4. In some cases, a specific source for a reagent is recommended. Also, all solutions were prepared using distilled, deionized water.
5. Details regarding the construction and use of this apparatus can be found in Gifford and Cooper (9).
6. Methods for using a hemocytometer and performing the proper calculations to determine cell concentration can be

readily found on a number of qualified web pages (e.g., see http://www.whitelabs.com/beer/cell_count.html).

7. To completely resuspend the protein pellet in MSB, continuous and vigorous shaking may be required for 24–48 h. We have performed this procedure at room temperature without any significant affect upon the quality of the protein extract.
8. We employ an appropriate Bio-Rad Laboratories imaging system, but other suitable systems exist that will perform in a similar manner.
9. We typically use the services provided by the Mass Spectrometry and Proteomics Facility at Ohio State University (www.ccic.ohio-state.edu/MS/index.htm) having developed a productive working relationship.

Acknowledgment

Funding provided by the National Science Foundation (DBI 0330883) supported the development of this methodology.

References

1. Pitt JI (1979) The genus Penicillium and its teleomorphic states Eupenicillium and Talaromyces. Academic, London, p 634, vi
2. Moss MO (1987) Morphology and physiology of *Penicillium* and *Acremonium*. In: Peberdy JF (ed) *Penicillium* and *Acremonium*. Plenum Press, New York, pp 37–72
3. Cooper CR Jr, Haycocks NG (2000) *Penicillium marneffei*: an insurgent species among the penicillia. J Eukaryot Microbiol 47:24–28
4. Alvarez S (1990) Systemic infection caused by *Penicillium decumbens* in a patient with acquired immunodeficiency syndrome. J Infect Dis 162:283
5. Hoffman M, Bash E, Berger SA, Burke M, Yust I (1992) Fatal necrotizing esophagitis due to *Penicillium chrysogenum* in a patient with acquired immunodeficiency syndrome. Eur J Clin Microbiol Infect Dis 11:1158–1160
6. Cooper CR, Vanittanakom N (2008) Insights into the pathogenicity of *Penicillium marneffei*. Future Microbiol 3:43–55
7. Vanittanakom N, Cooper CR Jr, Fisher MC, Sirisanthana T (2006) *Penicillium marneffei* infection and recent advances in the epidemiology and molecular biology aspects. Clin Microbiol Rev 19:95–110
8. Chandler JM, Treece ER, Trenary HR, Brenneman JL, Flickner TJ, Frommelt JL, Oo ZM, Patterson MM, Rundle RT, Valle OV, Kim TD, Walker GR, Cooper CR Jr (2008) Protein profiling of the dimorphic, pathogenic fungus. Penicillium marneffei. Proteome Sci 6:17
9. Gifford TD, Cooper CR Jr (2009) Karyotype determination and gene mapping in two clinical isolates of *Penicillium marneffei*. Med Mycol 47:286–295
10. Bradford MM (1976) A rapid and sensitive method for the quantitation of microgram quantities of protein utilizing the principle of protein-dye binding. Anal Biochem 72:248–254
11. Carr S, Aebersold R, Baldwin M, Burlingame A, Clauser K, Nesvizhskii A (2004) The need for guidelines in publication of peptide and protein identification data: Working Group on Publication Guidelines for Peptide and Protein Identification Data. Mol Cell Proteomics 3: 531–533

Chapter 21

Capillary Electrophoretic Enzyme Assays

Gerhard K.E. Scriba, Hans Abromeit, Martina Hense, and Yi Fan

Abstract

In the past years, capillary electrophoresis has become a frequently used technique for enzyme assays due to the high separation efficiency and versatility as well as small sample size and low consumption of chemicals. The capillary electrophoresis assays can be divided into two general categories: pre-capillary (or offline) assays and in-capillary (or online) assays. In pre-capillary assays, the incubation is performed offline and substrate(s) and product(s) are subsequently analyzed by capillary electrophoresis. In in-capillary assays enzyme reaction and separation of the analytes are performed inside the same capillary. In such assays the enzyme is either immobilized or in solution. The latter techniques is also referred to as electrophoretically mediated microanalysis (EMMA) indicating that the individual steps of the incubation as well as analysis are performed via electrophoretic phenomena. This chapter describes both techniques using the deacetylation of acetyl-lysine residues in model peptides by sirtuin enzymes as well as the hydrolysis of acetylthiocholine by acetylcholinesterase as examples.

Key words: Enzyme assay, Electrophoretically mediated microanalysis, Sirtuin, Capillary electrophoresis

1. Introduction

Capillary electrophoresis (CE) has been applied to the analysis of biological important molecules due to advantages including high separation efficiency, small sample size, automation, and rapid method development. The first CE enzyme assay was published about 20 years ago when Banke and colleagues monitored alkaline protease activity by CE (1). In recent years, all aspects of enzyme-related analysis have been investigated by CE including the determination of enzymatic activity, enzyme kinetics, evaluation of enzyme substrates or inhibitors, and investigation of metabolic pathways. CE enzyme assays can be categorized into (1) pre-capillary or offline enzyme assays and (2) in-capillary or online enzyme assays.

Nicola Volpi and Francesca Maccari (eds.), *Capillary Electrophoresis of Biomolecules: Methods and Protocols*, Methods in Molecular Biology, vol. 984, DOI 10.1007/978-1-62703-296-4_21, © Springer Science+Business Media, LLC 2013

Both approaches will be briefly outlined below and subsequently illustrated using the deacetylation of a N_ε-acetyl-lysine non-peptide substrate or a tripeptide derivative by recombinant human sirtuin 1 enzyme (SIRT1) as well as the hydrolysis of acetylthiocholine by acetylcholinesterase as examples. So-called post-capillary assays have also been described but will not be considered here. For in-depth summaries of CE enzyme assays including EMMA and chip electrophoresis, the reader is referred to recent reviews (2–8).

1.1. Pre-capillary Enzyme Assays

In pre-capillary enzyme assays, also termed offline assays, the incubation of substrate(s), cofactor(s), and enzyme is carried out in a separate vial. Only the analysis of substrate(s) and/or product(s) is carried out by using CE. Typically, all components are mixed and the reaction is initiated either by the addition of the enzyme or the substrate. After incubation for the intended period of time, the reaction is stopped, and the sample is injected into the CE system for analysis of the substrate and/or the reaction product.

The advantage of offline assays is the ease of method development because both steps, enzyme reaction and CE analyte separation, can performed under the respective optimum conditions. Often, the background electrolyte of the CE analytical separation is not compatible with the enzyme reaction, e.g., enzyme reactions are typically performed at physiological pH while CE separations may require a low pH or a high pH background electrolyte which can lead to denaturation of the enzyme. However, some problems may arise upon direct injection of the incubation mixture into the CE system. In case of a high ionic strength of the incubation solution or a high protein content broad peaks and variability of the electroosmotic flow (EOF) and, subsequently, a low sensitivity and reproducibility of the method may result. Deproteinization prior to sample injection in order to avoid protein adsorption onto the capillary wall or even blocking of the capillary can be achieved by addition of an acid or acetonitrile in the quenching step of the reaction followed by centrifugation of the precipitated proteins. However, the use of strong acids to quench the reaction results in a strongly acidic sample solution which may result in peak distortion in the CE analysis. Protein adsorption to the capillary wall may be reduced using permanently polymer-coated capillaries (9) or employing a dynamic coating of the capillary wall (10, 11). Pre-capillary assays also allow repeated sampling from the same vial so that in combination with sequential CE analyses kinetic assays can be performed.

Besides the most often applied capillary zone electrophoresis (CZE) mode, the micellar electrokinetic chromatography (MEKC) mode has also been applied as an alternative method for analyte separation. MEKC has the advantage of a higher tolerance towards samples with high ionic strength or protein concentrations. Moreover, MEKC may also provide higher resolution of analytes in some cases. For example, Inagaki and coworkers investigated the activity of dihydrodiol dehydrogenase 1 using prostaglandin D_2 (PGD_2) as substrate (12). Upon addition of 20 mM SDS to a 100 mM borate buffer, pH 9.5, the separation of the substrate PGD_2, the hydrogenation product 9α,11β-prostaglandin F_2, and two structural isomers present in tissues could be accomplished within 10 min. No sample workup prior to the injection was required. Moreover, the cofactor NADPH as well as the reaction product $NADP^+$ could also be separated allowing the simultaneous determination of all compounds involved and, thus, a direct, comprehensive, and reliable analysis of the enzyme reaction.

Due to the short optical path length because of the small diameter of the capillary UV detection may not always be sensitive enough for the determination of low concentrations of substrate(s) or product(s). This may be accomplished using more sensitive detection modes such as laser-induced fluorescence (LIF) detection or mass spectrometric (MS) detection. In case of LIF detection, labeling of the substrates with a suitable chromophore is required. Labeled substrates can be used or the compounds have to be derivatized following the enzyme reaction using suitable reagents (13, 14). Alternatively, stacking techniques can be applied as online preconcentration steps. A simple way to achieve stacking is the injection of low ionic strength samples which can be obtained using acetonitrile for quenching of the reaction. Several techniques are available and have been summarized in recent reviews (15–17).

Pre-capillary assays have been applied to the determination of enzyme activity and kinetics, substrate and inhibitor screening as well as in vitro drug metabolism studies. Due to the versatility this approach has been used in the majority of CE enzyme assays. Chiral separations of the reaction products are possible. Some examples of pre-capillary assays are compiled in Table 1; further examples have been compiled in the reviews (2–8).

1.2. In-Capillary Enzyme Assays

In the in-capillary assays both the enzyme reaction and the subsequent analytical separation of product and substrate are performed in the same capillary. The first assay was reported by Bao and Reigner (42) and was later referred to as electrophoretically mediated microanalysis (EMMA) because different mobilities of the reaction partners, i.e., enzyme, substrate(s), and product(s), are exploited. In this technique enzyme, substrate and cofactors are introduced into the capillary as consecutive plugs. Mixing of the zones is achieved either by diffusion or by application of a voltage. Following the enzymatic conversion, the substrate and the resulting product are separated by CE. This approach is especially interesting considering the low sample consumption of enzyme, cofactors, and substrate. In addition, all steps can be automated and integrated in one instrument. Furthermore, no sample workup is required. Two general modes have been used to load and mix enzyme and substrate, the so-called continuous mode (long contact mode)

Table 1
Examples of pre-capillary CE enzyme assays

Enzyme	Substrate	CE mode	Sample preparation	Detection	Application	Reference
5′-Nucleotidase	AMP	CZE	Centrifugation	UV 259 nm	Enzyme activity	(18)
Inosine triphosphate pyrophosphohydrolase	Inosine triphosphate	CZE	Centrifugation	UV 250 nm	Enzyme activity in erythrocytes	(19)
Nucleotide pyrophosphatases/ phosphodiesterases	ATP	CZE	Direct injection	UV 210 nm	Enzyme kinetics and screening of inhibitors	(20)
Arylsulfatase B	Glycosaminoglycans	CZE	Fluorescence labeling and centrifugation	LIF 488 nm/ 520 nm	Enzyme activity	(21)
Human dipeptidyl peptidase III	Endomorphin-1	CZE	Heating	UV 191 nm	Enzyme kinetics	(22)
Angiotensin-converting enzyme	Hip-His-Leu	CZE	Direct injection	UV 228 nm	Inhibitory activity of marine protein hydrolysates	(23)
Neutral protease	Bovine casein	CZE	Centrifugation	UV 214 nm	Enzyme activity	(24)
Proteases	Fluorescence-labeled peptides	CZE	Addition of DNA standards	LIF 488 nm/ 532 nm	Monitoring proteolysis of substrates by proteases in biological mixtures	(25)
Trypsin	Myoglobin, ACTH (1–14), angiotensin 1	CZE	Direct injection	DAD/MS	Kinetics of proteolytic reaction	(26)
Serum cholinesterase	Benzoylcholine	CZE	Dilution	UV 233 nm	Enzyme activity	(27)
Sirtuin1, sirtuin 2	Fmoc-labeled peptides	CZE	Deproteination, centrifugation	UV 220 nm	Substrate evaluation, enzyme kinetics, inhibitor screening	(28)
Sirtuin 1	Dns-Lys(Ac)NH_2	CZE	Centrifugation	UV 220 nm	Enzyme kinetics, inhibition	(29)
Sirtuin 1	DMAC-Lys(Ac)NH_2	MEKC	Centrifugation	UV 220 nm	Enzyme kinetics, inhibition	(29)

Adenosine kinase	Adenosine and derivatives	MEKC	Direct injection	UV 260 nm	Substrate characterization, inhibitor screening	(30)
Sphingosine kinase	Fluorescein-labeled sphingosine	CZE	Dilution	LIF 488 nm/ 518 nm	Kinetics study, kinase activity in cells	(31)
D-amino acid oxidase	D,L-Trp, D,L-Phe	CZE	Direct injection	UV 214 nm	Chiral assay, enzyme kinetics	(32)
Cytochrome P450 2C9	Diclofenac	MEKC	Direct injection	UV 200 nm	Enzyme kinetics, inhibitor screening	(33)
Cytochrome P450 2D6, Cytochrome P450 3A4	Propafenone	CZE	Liquid–liquid extraction	UV 195 nm	Enzyme kinetics, enantioselective metabolism	(34)
Protoporphyrinogen oxidase	Protoporphyrin IX	CZE	Direct injection	UV 400 nm	Enzyme kinetics and inhibition	(35)
Monoamine oxidase	Kynuramine	CZE	Centrifugation	UV 214 nm	Enzyme activity	(36)
Adenylyl cyclase	ATP	CZE	Direct injection	UV 254 nm	Enzyme kinetics, stimulation, and inhibition	(37)
L-amino acid oxidase	Dns-amino acids	CZE	Centrifugation	UV 254 nm	Enzyme kinetics	(38)
Tyrosine kinase	Fluorescein-labeled peptide	CZE	Direct injection	LIF 488 nm/ 520 nm	Enzyme kinetics and inhibition	(39)
Protein kinase CK2	EDANS-labeled peptide	CZE	Direct injection	UV 214 nm	Enzyme kinetics and inhibition	(40)
Sphingosine kinase 2	Sphingosine fluorescein	CZE	Direct injection	LIF 488 nm/ 520 nm	Enzyme kinetics, inhibition and activity in cell lysate	(41)

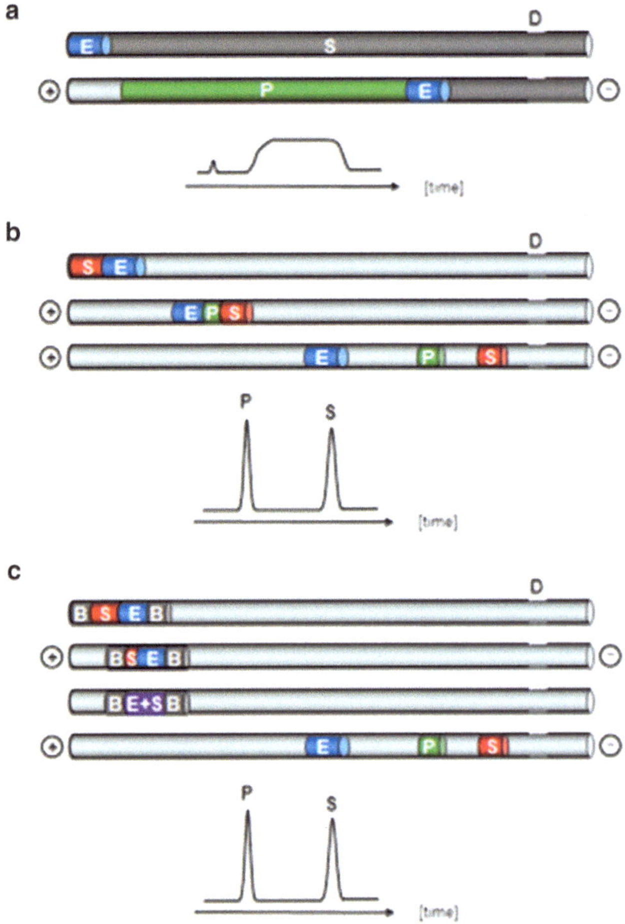

Fig. 1. Schematic representation of EMMA modes. (**a**) continuous mode, (**b**) plug–plug mode and (**c**) partial filling mode.

and the plug–plug EMMA (short contact mode) (3, 6). In the continuous mode the entire capillary is filled with either the substrate or the enzyme. Subsequently the second reactant (enzyme or substrate) is injected as a plug (zonal sample introduction) as shown in Fig. 1a. Upon application of a voltage the reactant moves through the capillary resulting in continuous formation of the product which is recorded as a flat broad sample zone. Alternatively, the second reactant is introduced continuously from the inlet vial (moving boundary sample introduction). In the classical plug–plug EMMA (short contact mode) the reactants are introduced as consecutive plugs (Fig. 1b). Upon application of an electric field, the reactants merge due to their electrophoretic mobilities. Enzymatic conversion takes place and substrate and product are subsequently separated and detected as narrow peaks. In this mode the first injected reactant has to possess the lower electrophoretic mobility in order to assure mixing of the zones. One limitation of the classical plug–plug approach is arising when the separation buffer inactivates the enzyme due to incompatible (extreme) pH or composition. In this case background electrolyte and enzyme as well as substrate can be separated by injecting an additional plug of incubation buffer (Fig. 1c). This mode has been termed "partial filling mode" (43). Mixing of enzyme and substrate is achieved by applying a low voltage so that the reactants mix by electromigration. Subsequently, the voltage is switched off to let the reaction proceed for a specified period of time, also called "zero potential amplification time." Finally, an electrical field suitable to separate substrate and product is applied. In a variant, the so-called at-inlet mode, the consecutively injected substrate and enzyme plugs mix by simple diffusion letting the reaction proceed for a given period of time before applying an electrical field for analyte separation. This is useful in the case of enzymes that are sensitive to an electric field. Furthermore, mixing can be achieved by the transverse diffusion of laminar flow profiles (44, 45). The solutions are injected by pressure as a series resulting in plugs with parabolic profiles with longitudinal interfaces between the plugs. The plugs mix by transverse diffusion while longitudinal diffusion is negligible.

Methods relying on electrophoretic mixing of the reactants require knowledge of their electrophoretic mobilities. Parameters that have to be optimized during method development include substrate and enzyme concentration, plug length, injection sequence, mixing voltage, online incubation time in the absence of voltage, separation buffer, and separation voltage. When the electrophoretic motilities of the substrate and the enzyme are available, the potential and the time of application of the potential can be calculated for plugs of known length in order to ensure maximum plug overlap resulting in a high efficiency of the enzymatic conversion of the substrate. However, information on the electrophoretic mobilities of the reactants may not always be available. An easy solution of this problem is the injection of a substrate–enzyme–substrate plug sequence. This way it is ensured that enzyme and substrate get in contact with each other upon application of the mixing voltage. Moreover, the enzyme can be protected from a noncompatible separation buffer.

The EMMA approach has been applied to study enzyme activity and kinetics, the screening and determination of substrates and inhibitors as well as drug metabolism. Unlabeled as well as labeled substrates for increased assay sensitivity have been employed. Selected examples of the EMMA approach for enzyme assays have been summarized in Table 2; further examples have been compiled in the reviews (2–8). EMMA has also been transferred to the chip format.

Table 2
Examples of EMMA CE enzyme assays

Enzyme	Substrate	EMMA mode	Detection	Note	Reference
Uridine diphosphate glucuronosyltransferase	4-Methyl-7-hydroxycoumarin, 4-nitrophenol	Plug–plug mode	UV 310 nm	Enzyme kinetics and inhibition	(46)
Glucose oxidase	1,4-Benzoquinone	Plug–plug mode	UV 245 nm, 290 nm	Enzyme activity	(47)
Penicillinase	Penicillin derivatives	Plug–plug mode	APS, UV 200 nm	Substrate specificity screening	(48)
Acetylcholinesterase	Acetylthiocholine	Plug–plug mode	UV 230 nm	Inhibitor screening	(49)
Trypsin	Cytochrome *c*, apomyoglobin	Plug–plug mode	C^4D	Digestion assay	(50)
Farnesyltransferase	Farnesyl pyrophosphate	Plug–plug mode	LIF 488 nm/520 nm	Inhibitor screening	(51)
β-Glucosidase	*p*-Nitrophenyl-α-D-glucopyranoside	Plug–plug mode	UV 191 nm	Inhibition by Chinese medicine	(52)
L-amino acid oxidase	Dns-Trp	Plug–plug mode	UV 254 nm	Enzyme kinetics	(38)
β-Galactosidase	Fluorescein mono-β-D-galactopyranoside	Plug–plug mode (TDLFP)	LIF 488 nm/520 nm	Enzyme online reaction	(44)
Adenosine kinase	ATP	Partial filling mode	UV 210 nm	Inhibitor and substrate screening	(30)
Hexokinase	ATP	Partial filling mode	UV 254 nm	Inhibitor screening	(53)
Haloalkane dehalogenase, rhodanase	1-Bromobutane, Cyanide	Partial filling mode	UV 200 nm	Enzyme kinetics	(54)
Haloalkane dehalogenase	1,2-Dibromoethane	Partial filling mode	UV 200 nm	Enzyme kinetics, substrate inhibition	(55)

Flavin-containing monooxygenase	Clozapine	Partial filling mode	UV 190 nm	Enzyme kinetics	(56)
Flavin-containing monooxygenase isoforms	Cimetidine	Partial filling mode	UV 220 nm	Enzyme kinetics	(57)
Trypsin	Various proteins	Partial filling mode	UV 200 nm	Digestion assay	(58)
Cytochrome P450	Testosterone, nifedipine	Partial filling mode	UV 260 nm	Enzyme kinetics	(59)
Sirtuin 1	Fmoc-peptides	Partial filling mode	UV 220 nm	Enzyme kinetics, inhibition	(60)
Alkaline phosphatase	Disodium phenyl phosphate	Continuous mode	ECD	Enzyme activity	(61)
Alkaline phosphatase	3-*O*-Methylfluorescein phosphate	Continuous mode	LIF 488 nm/520 nm	Enzyme kinetics	(62)

LIF laser-induced fluorescence, *C^4D* capacitively coupled contactless conductivity detection, *ECD* electrochemical detection, *APS* active pixel sensor

In-capillary assays can also be performed as heterogeneous assays. In this case the enzyme is immobilized either on the surface of the capillary wall or on a suitable carrier such as a monolith that is fixed at the capillary inlet forming a microreactor. After introduction of the substrates the reaction occurs in the microreactor part of the capillary containing the immobilized enzyme and the resulting products are subsequently electrophoretically separated (63, 64).

2. Materials

2.1. Instrumentation and Materials

1. A commercial CE instrument with a capillary cartridge thermostat, a high-voltage source (up to 30 kV) and a photodiode array detector. A P/ACE MDQ CE System or a P/ACE 5510 CE instrument (both from Beckman Coulter, Fullerton, CA, USA) is suitable (see Note 1).
2. Uncoated fused-silica capillaries (e.g., from Polymicro Technologies, Phoenix, AZ, USA) with an internal diameter (i.d.) of 50 μm (see Note 2).
3. A commercial pH meter for pH adjustment of the background electrolytes.
4. An ultrasonic bath for degassing of the solutions.
5. A thermostated water bath for enzyme incubations at 37°C.
6. A commercial centrifuge suitable for centrifugation of Eppendorf tubes (microfuge).
7. Syringe filters containing polyester filter membranes with a pore size of 0.20 μm (e.g., from Macherey-Nagel, Düren, Germany). The use of 0.45 μm filters is also possible.

2.2. Chemicals and Enzymes

All chemicals should be of analytical grade. Water should be deionized, double-distilled water or Milli-Q water (18 MΩ water). Filter all rinsing and buffer solutions for CE analysis using polyester membrane syringe filters and sonicate for 3–5 min.

1. Human recombinant sirtuin1 enzyme (SIRT1) from Biomol (Hamburg, Germany) or Enzo Life Sciences GmbH (Lörrach, Germany) (see Notes 3 and 4).
2. Acetylcholinesterase from *Electrophorus electricus* from Sigma-Aldrich (St. Louis, MO, USA) (see Note 5).
3. Fmoc-Lys-Lys(Ac)-LeuNH_2 (as substrate) and Fmoc-Lys-LysNH_2 (as internal standard) (see Note 6).
4. Amino acid derivatives N_ε-acetyl-N_α-fluorenyloxycarbonyl lysine (*Z*-Lys(Ac)NH_2) and N_ε-(tert.-butyloxycarbonyl)lysine amide hydrochloride (Lys(Boc)NH_2·HCl) from Bachem AG (Bubendorf, Switzerland).

5. Dansyl chloride (e.g., from Sigma-Aldrich, St. Louis. MO, USA).
6. Nicotinamide adenine dinucleotide (NAD^+) (e.g., from Sigma-Aldrich, St. Louis. MO, USA).
7. Acetylthiocholine (e.g., from Sigma-Aldrich, St. Louis, MO, USA).

2.3. Buffers and Solutions

2.3.1. SIRT1 Assay

1. SIRT1 incubation buffer: Prepare a 50 mM tris(hydroxymethyl)aminomethane (Tris) buffer, pH 8.0, containing 137 mM sodium chloride, 2.7 mM potassium chloride, 1 mM magnesium chloride, and 1 mg/mL bovine serum albumin (fatty acid-free BSA, e.g., Sigma-Aldrich catalog # A3803) (see Note 7). For example, dissolve 606 mg Tris, 800 mg sodium chloride, 20.1 mg potassium chloride, 9.5 mg magnesium chloride, and 100 mg BSA in about 90 mL water in a 150 mL beaker. Mix and adjust the pH to 8.0 with 1 M hydrochloric acid (see Note 8). Transfer solution into a 100 mL volumetric flask and make up to 100.0 mL by adding water.
2. SIRT1 enzyme stock solution: Prepare a solution of SIRT1 with an activity of 1 U/μL in incubation buffer immediately before use. For example, thaw the enzyme solution obtained from the supplier with a total activity of 100 U upon receipt and add SIRT1 incubation buffer to give a final volume of 100 μL. Divide into 5 μL aliquots and use immediately or store at −80°C (see Note 4).
3. NAD^+ stock solution (10 mM): Dissolve 3.32 mg NAD^+ in 0.5 mL SIRT1 incubation buffer. Prepare immediately before use.
4. Substrate stock solutions: Prepare 10 mM stock solution of Dns-Lys(Ac)-NH_2 in water containing 30% dimethyl sulfoxide (DMSO). For example, dissolve 2.10 mg Dns-Lys(Ac)-NH_2 in 0.5 mL water/DMSO (7:3, v/v). Prepare 10 mM stock solution of Fmoc-Lys-Lys(Ac)-LeuNH_2 in water. For example, dissolve 3.25 mg Fmoc-Lys-Lys(Ac)-LeuNH_2 in 0.5 mL water. Prepare 10 mM of Fmoc-Lys-LysNH_2 in water as internal standard stock solution. For example, dissolve 2.48 mg Fmoc-Lys-LysNH_2 in 0.5 mL water. All solutions can be stored at −20°C.
5. Offline assay quenching solution: 1% (v/v) trichloroacetic acid in acetonitrile.
6. CE background electrolyte (200 mM phosphate–Tris buffer, pH 2.7): Add about 90 mL water into a 150 mL beaker. Add 1.36 mL phosphoric acid (85%). Mix and adjust pH by addition of 2 M aqueous Tris solution (see Note 8). Transfer into a 100 mL volumetric flask and make up to 100.0 mL with water. The buffer can be stored in a refrigerator for about 1 week. Filter through 0.2 μm membrane filters and degas by sonication before use in the CE analysis.

2.3.2. Acetylcholinesterase Assay

1. Electrolyte: 30 mM borate–phosphate buffer, pH 8.0, prepared by dissolution of the appropriate amount of boric acid and sodium dihydrogen phosphate in water and adjusting the pH with 1 M sodium hydroxide solution. For example, dissolve 185.5 mg boric acid and 468 mg sodium dihydrogen phosphate in about 90 mL in a 150 mL beaker. Mix and adjust pH with 1 M sodium hydroxide solution (see Note 8). Transfer the solution into a 100 mL volumetric flask and make up to 100.0 mL with water. The buffer can be stored at 4°C for about 1 week.
2. Acetylcholinesterase solution: Dissolve the required amount of enzyme in the electrolyte solution to obtain an enzyme activity of about 150 U/mL buffer (see Note 5).
3. Acetylthiocholine solution: 10 mM solution of acetylthiocholine in the electrolyte solution containing 20 mM magnesium sulfate (see Note 9). For example, dissolve 9.89 mg acetylthiocholine chloride and 24.65 mg magnesium sulfate heptahydrate in 5.0 mL electrolyte solution.

3. Methods

3.1. Offline Enzyme Assay

The example of the offline assay is based on the deacetylation of a N_ε-acetyl-lysine derivative by SIRT1 (29).

3.1.1. Synthesis of SIRT 1 Substrate Dns-Lys(Ac)-NH_2

All synthetic reactions have to be performed under a well-ventilated hood using proper protection. Special care has to be taken when handling hydrogen gas.

1. Dissolve 350 mg Z-Lys(Ac)NH_2 in 10 mL methanol in a 50 mL two-necked round-bottom flask equipped with a magnetic stir bar and two valves. Add 1 mL 1 M aqueous hydrochloric acid and 15 mg Pd/C (palladium on carbon) catalyst.
2. Connect inlet valve with a nitrogen tank and flush the flask repeatedly with nitrogen keeping the other valve open to remove the air (see Note 10).
3. Flush the flask with hydrogen and stir the solution under hydrogen atmosphere overnight (see Note 11).
4. Filter the solution and remove the solvents under reduced pressure. Dry under vacuum to obtain N_ε-acetyl lysine amide hydrochloride (Lys(Ac)NH_2·HCl) as an amorphous whitish solid.
5. Dissolve 95 mg (0.5 mmol) Lys(Ac)NH_2·HCl and 200 mg (2 mmol) of freshly distilled triethylamine (see Note 12) in 10 mL dry dichloromethane (see Note 13) in a 50 mL two-necked round-bottom flask equipped with a magnetic stir bar, a gas inlet valve, and a calcium chloride drying tube (see Note 14).

6. Flush the flask with nitrogen and cool to 0°C in an ice bath.
7. Add 140 mg (0.5 mmol) dansyl chloride in 5 mL dry dichloromethane while stirring at 0°C.
8. Remove the ice bath and stir at room temperature overnight.
9. Extract the dichloromethane layer with 10 mL ice-cold 0.1 M hydrochloric acid and three times with 10 mL water. Dry over anhydrous sodium sulfate and evaporate to dryness under reduced pressure.
10. Dry the resulting N_{ε}-acetyl-N_{α}-dansyl lysine amide (Dns-Lys(Ac)NH_2) under vacuum to obtain a greenish fluorescent solid.

3.1.2. Synthesis of the Deacetylation Product Dns-Lys-NH_2·HCl

The synthesis of the deacetylation product is not necessarily required for the assay unless unequivocal peak identification in the electropherogram is desired.

1. Dissolve 140 mg Lys(Boc)NH_2 hydrochloride and 200 mg (2 mmol) freshly distilled triethylamine (see Note 12) in 10 mL dry dichloromethane (see Note 13) in a 50 mL two-necked round-bottom flask equipped with a magnetic stir bar, a gas inlet valve, and a calcium chloride drying tube (see Note 14).
2. Flush the flask with nitrogen and cool to 0°C in an ice bath.
3. While stirring add a solution of 140 mg (0.5 mmol) dansyl chloride in 5 mL dichloromethane at 0°C.
4. Remove the ice bath and stir at room temperature overnight.
5. Extract the dichloromethane layer with 10 mL ice-cold 0.1 M hydrochloric acid and three times with 10 mL water. Dry over anhydrous sodium sulfate ad evaporate to dryness under reduced pressure.
6. Dissolve the resulting solid in 5–10 mL 4 M hydrochloric acid in dioxane (see Note 15) and stir at room temperature for 3 h.
7. Evaporate the solvent under reduced pressure. The product is obtained as greenish, fluorescent solid.

3.1.3. Offline Enzyme Incubation

1. Adjust the temperature of the water bath to 37°C.
2. Dilute 20 μL Dns-Lys(Ac)-NH_2 stock solution with 980 μL SIRT1 incubation buffer and vortex (see Note 16).
3. Add 17.5 μL incubation buffer and 2.5 μL NAD^+ stock solution to a 0.2 mL Eppendorf tube containing 5 μL SIRT1 enzyme solution with an activity of 1 U/μL and vortex. Keep on ice if not immediately used.
4. Preincubate the enzyme containing Eppendorf tube (step 2) and 50 μL of the substrate incubation solution containing 200 μM Dns-Lys(Ac)-NH_2 (step 3) separately for 10 min in the water bath at 37°C.

5. Pipette 25 μL of the substrate solution into the enzyme solution tube, mix by gently vortexing and incubate at 37°C.
6. At selected time intervals gently vortex the incubation Eppendorf tube (see Note 17) and withdraw 5 μL of the incubation mixture into 20 μL ice-cold offline assay quenching solution (1% trichloroacetic acid in acetonitrile) in a 0.5 mL Eppendorf tube. Vortex and keep on ice for 10–15 min.
7. Centrifuge sample for 10 min at 10,000 × *g*. Use the supernatant for CE analysis (see Note 18).

3.1.4. CE Analysis

1. Install a 50 μm i.d. fused-silica capillary with an effective length of 30 cm in the CE instrument according to the instructions of the manufacturer.
2. Rinse a new capillary subsequently with 1 M sodium hydroxide solution for 30 min, water for 5 min, 1 M hydrochloric acid for 30 min, water for 5 min, and the background electrolyte (200 mM phosphate–Tris buffer) for 10 min using a pressure of 34.5 mbar (0.5 psi) (see Note 19).
3. Set UV detection wavelength to 230 nm and capillary temperature to 25°C and observe the baseline applying a voltage of 20 kV until it is stable.
4. Inject the sample electrokinetically for 60 s using a voltage of 5 kV (see Note 18).
5. Apply a separation voltage of 20 kV and record the electropherogram. An electropherogram of the incubation of 100 μM Dns-Lys(Ac)-NH_2 with 0.1 U/μL SIRT1 for 20 min is shown in Fig. 2.
6. Rinse the capillary between analyses with the background electrolyte for 3 min at a pressure of 34.4 mbar (0.5 psi) (see Note 20).

3.2. Plug–Plug EMMA Enzyme Assay

The example of the plug–plug EMMA mode is based on the hydrolysis of acetylthiocholine to thiocholine and acetic acid by acetylcholinesterase (49).

1. Install a 50 μm i.d. capillary with an effective length of 26 cm in the CE instrument according to the instructions of the manufacturer.
2. Rinse a new capillary for 30 min with 0.1 M sodium hydroxide solution, for 2 min with water, and for 3 min with the background electrolyte at a pressure of 34.5 mbar (0.5 psi) (see Note 19).
3. Set the capillary temperature to 37°C and the detection wavelength to 230 nm. Observe the baseline until it is stable.
4. Inject a plug of acetylcholinesterase solution for 4 s at 20 mbar (0.3 psi) followed by a plug of acetylthiocholine solution for 4 s at 20 mbar (0.3 psi) (see Note 21).

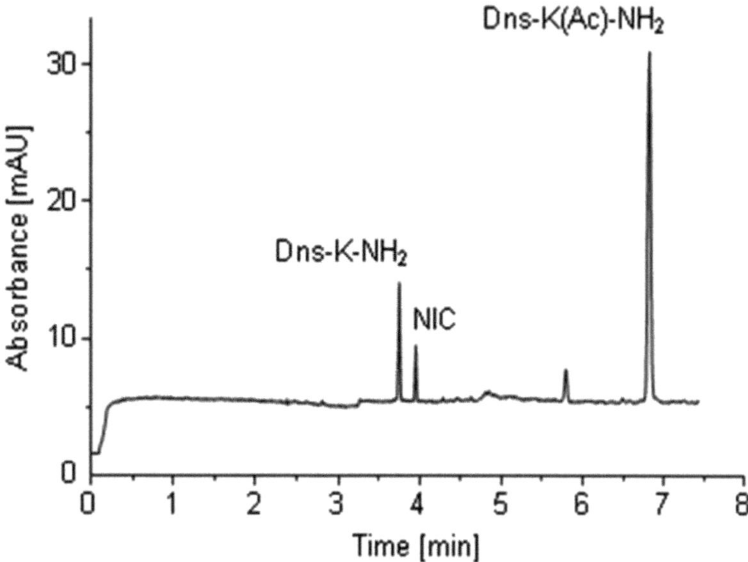

Fig. 2. Electropherogram of offline SIRT1 assay using the non-peptide substrate Dns-Lys(Ac)NH_2.

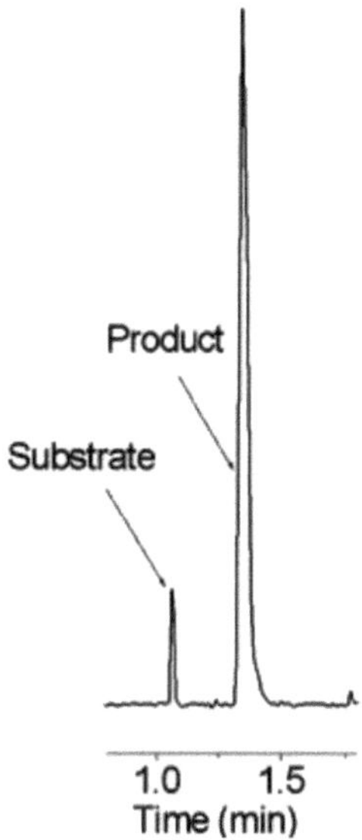

Fig. 3. Electropherogram of plug–plug EMMA monitoring the hydrolysis of acetylthiocholine by acetylcholinesterase (reprinted with permission from ref. 49, copyright by Wiley-VCH).

5. Apply a voltage of 1.0 kV for 15 s to mix substrate and enzyme solutions (see Note 22).
6. Immediately afterwards, apply a voltage of 15 kV to separate the product (thiocholine) and the substrate (acetylthiocholine). Record the electropherogram (Fig. 3). The activity of the enzyme is a function of the peak area of the product thiocholine.

7. Rinse the capillary between analyses with the background electrolyte for 3 min at 34.5 mbar (0.5 psi) (see Note 20).

3.3. Partial Filling EMMA Enzyme Assay

The example is based on the partial filling EMMA of the deacetylation of the Fmoc-labeled tripeptide derivative Fmoc-Lys-Lys(Ac)-LeuNH$_2$ by SIRT1 (60).

1. Install a 50 μm i.d. capillary with an effective length of 30 cm in the CE instrument according to the instructions of the manufacturer.
2. Rinse a new capillary subsequently with 1 M sodium hydroxide solution for 30 min, water for 5 min, 1 M hydrochloric acid for 30 min, water for 5 min, and the background electrolyte (200 mM phosphate–Tris buffer, pH 2.7) for 10 min using a pressure of 0.5 psi (see Note 19).
3. Set the capillary temperature to 37°C and the detection wavelength to 220 nm. Observe the baseline until it is stable.
4. Mix 100 μL of NAD$^+$ stock solution and 1900 μL SIRT1 incubation buffer.
5. Mix 10 μL of the substrate Fmoc-Lys-Lys(Ac)-LeuNH$_2$ stock solution and 5 μL of the internal standard Fmoc-Lys-LysNH$_2$ stock solution with 985 μL SIRT1 incubation buffer containing NAD$^+$ (step 4, see Note 23).
6. Add 45 μL of SIRT1 incubation buffer containing NAD$^+$ (step 4, see Note 23) to a 0.2 mL Eppendorf tube containing 5 μL SIRT1 enzyme solution with an activity of 1 U/μL to obtain a solution with an enzyme activity of 0.1 U/μL (see Note 24). Mix by gently vortexing and keep on ice if not immediately used.
7. Inject hydrodynamically subsequent plugs of SIRT1 incubation buffer (step 4), SIRT1 enzyme solution (step 6), Fmoc-Lys-Lys(Ac)-LeuNH$_2$ substrate solution containing the internal standard Fmoc-Lys-LysNH$_2$ (step 5), SIRT1 enzyme solution (step 6), and again SIRT1 incubation buffer (step 4). Use a pressure of 34.5 mbar (0.5 psi) for 5 s for each injection (see Notes 21 and 25–27). Between injections, dip the capillary into a vial containing SIRT1 incubation buffer in order to avoid cross-contamination of the solutions.
8. Apply a voltage of 1.0 kV for 90 s to mix substrate and enzyme solutions to initiate the enzyme reaction (see Note 28).
9. Turn off the voltage for 1–20 min to allow the in-capillary enzyme incubation (zero voltage amplification time).
10. Immediately afterwards, apply a voltage of 20 kV to separate the product (Fmoc-Lys-Lys-LeuNH$_2$), coproduct nicotinamide, the substrate (Fmoc-Lys-Lys(Ac)-LeuNH$_2$) and the internal standard (Fmoc-Lys-LysNH$_2$). Record the electropherograms. Figure 4 shows the electropherogram of an

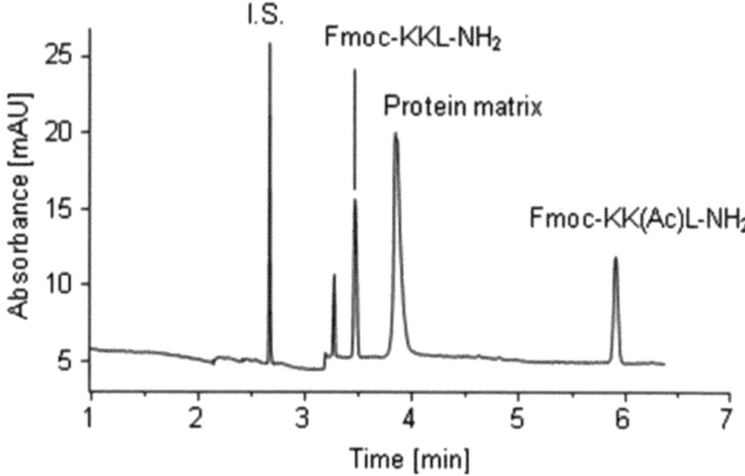

Fig. 4. Electropherogram of partial filling EMMA SIRT1 assay using the tripeptide as Fmoc-Lys-Lys(Ac)-LeuNH_2 substrate. Fmoc-Lys-Lys-LeuNH_2, deacetylation product; I.S., internal standard Fmoc-Lys-LysNH_2.

in-capillary assay using 100 μM Fmoc-Lys-Lys(Ac)-LeuNH_2 as substrate and an enzyme activity of 0.1 U/μL SIRT1 applying a zero voltage amplification time of 15 min. The concentration of the internal standard is 50 μM. The activity of the enzyme is a function of the peak area ratio of the product Fmoc-Lys-Lys-LeuNH_2 to the internal standard Fmoc-Lys-LysNH_2.

11. Rinse the capillary between analyses with the background electrolyte for 3 min at 34.5 mbar (0.5 psi) (see Notes 20 and 29).

4. Notes

1. CE instruments from different companies as well as different instruments from the same supplier may yield slightly different results even when using identical experimental conditions. Thus, the variables may require slight changes when transferring a certain analytical method from one instrument to another so that fine-tuning of the parameters of a published method can be necessary.
2. Capillaries from different suppliers may lead to slightly different separation efficiencies. Even capillaries from the same supplier may vary to a certain extent. Thus, the purchase of larger quantities of capillaries is recommended especially if a method is intended for validated routine analysis in an industrial environment.

3. Sirtuins are NAD^+-dependent class III histone deacetylases, which catalyze the deacetylation of acetyl-lysine residues of histones and other protein substrates yielding the deacetylated protein, nicotinamide, and 2′-*O*-acetyl-ADP-ribose. Seven sirtuin enzymes have been identified in humans with diverse cellular locations and molecular targets regulating a wide range of biological processes such as proliferation, differentiation, metabolism, apoptosis, and senescence (65–67). Recombinant human SIRT1 enzyme is expressed in E. coli and contains a N-terminal His-tag. 1 U is defined as the amount of enzyme that hydrolyzes 1 pmol of the Flour de Lys™ substrate per min in the presence of 500 μM NAD^+ (68). The suppliers sell the enzyme in quantities of 100 U with a varying amount of protein, i.e., different specific activities (U/μg protein).
4. SIRT1 is supplied in 25 mM Tris buffer, pH 7.5, containing 100 mM sodium chloride, 5 mM dithiothreitol (DTT), and 10% glycerol and should be stored at −80°C (68). The enzyme is stable on ice for the period of time of 30–60 min but may lose activity upon prolonged storage. Therefore, the enzyme should be thawed and diluted to the appropriate activity immediately before the assay. The remaining quantities should be immediately snap frozen, for example, in liquid nitrogen or a dry ice–ethanol bath. The manufacturer guarantees stability of the enzyme for 4 freeze–thaw cycles. In order to minimize the freeze–thaw cycles it is recommended to aliquot the enzyme into several tubes when thawing first and store them at −80°C (68). However, the aliquots should not be too small as the enzyme trends to adsorb to the surface of the tubes. In our experience it is best to obtain a lot with a low specific activity (U/μg protein) as the relatively high content of (foreign) protein will stabilize SIRT1. Alternatively, bovine serum albumin (BSA, fatty acid free, e.g., Sigma-Aldrich catalog # A3803) may be added to prevent wall adsorption. In fact, BSA is also added to the incubation buffer to stabilize the enzyme and to maintain its activity during the incubations. We typically use lots with a specific activity below 5 U/μg protein. Upon receipt of the enzyme with a total activity of 100 U, we thaw the enzyme and make up to a total volume of 100 μL with SIRT1 incubation buffer. The required amount of SIRT1 incubation buffer that has to be added to the obtained enzyme batch can be calculated from the amount of enzyme solution stated on the certificate from the supplier. This way an enzyme stock solution with an activity of 1 U/μL is obtained. We divide the solution in 5 μL aliquots in 0.2 mL Eppendorf tubes which are either put on ice for immediate use or snap frozen in a dry/ice ethanol bath or liquid nitrogen and subsequently stored at −80°C.
5. Acetylcholinesterase catalyzes the hydrolysis of acetylcholine to choline and acetic acid. Sigma-Aldrich sells 2 qualities of

acetylcholinesterase with activities of 200–1,000 U/mg and >1,000 U/mg, respectively. 1 U is defined as the amount of enzyme that hydrolyzes 1 μmol of acetylcholine per min at pH 8.0 and 37°C. Both qualities are suitable but the amount required for the experiments may vary depending on the batch obtained. This also applies to enzyme obtained from a different supplier. Thus, the amount of enzyme needed for preparation of a solution with a certain activity may vary from batch to batch and supplier to supplier. The activity stated on the certificate should be used to calculate required μg quantity of enzyme. Enzyme from the same batch and supplier is recommended for a series of experiments.

6. The Fmoc-labeled peptide substrate (Fmoc-Lys-Lys(Ac)-LeuNH$_2$) as well as the internal standard (Fmoc-Lys-Lys-NH$_2$) can by synthesized by classical Fmoc solid-phase peptide synthesis if equipment and experience are available in the laboratory. If required, the product Fmoc-Lys-Lys-LeuNH$_2$ can be synthesized as well. If equipment and/or experience of solid-phase peptide synthesis is not available, the peptides can alternatively be custom synthesized by a company specializing in peptide synthesis. Such companies can be easily identified searching the Internet.
7. BSA in the incubation buffer is required in order to stabilize the SIRT1 enzyme and to maintain its activity during the incubations.
8. Due to the temperature dependence of dissociation equilibria, buffer pH should be adjusted at the temperature that is used during the electrophoretic run. Specifically, the change of the pK_a per Kelvin (or degree Celsius) of organic buffers can be significant.
9. Acetylcholinesterase is a Mg^{2+}-dependent enzyme. Therefore, 20 mM magnesium sulfate is added to the substrate solution to ensure enzymatic activity (49).
10. Flushing the reaction flask with nitrogen is required in order to remove the air (oxygen) from the reaction vessel. Open valve to nitrogen tank and let nitrogen gas stream into the flask with the second valve also open. Close valve to the tank and apply a low vacuum to the outlet valve. This way most of the gas phase is removed. Open both valves again and repeat flushing procedure.
11. The hydrogenation has to be performed in a well-ventilated hood as hydrogen gas is explosive when in contact with oxygen (air). Flushing of the reaction flask with hydrogen prior to hydrogenation is required in order to obtain a hydrogen atmosphere inside the flask. Thus, the inlet valve is connected either to a hydrogen gas tank or to a rubber gas bag (Sigma-Aldrich, No. Z186740) filled with hydrogen gas. Open both valves and let some hydrogen stream through the flask. Close inlet valve and apply low vacuum to the outlet valve. Close outlet valve,

and open valve connected to the hydrogen gas supply. Open outlet valve again and let some hydrogen gas stream through the flask. Close outlet valve and let hydrogenation proceed for the required period of time.

12. Freshly distilled triethylamine (b.p. 89°C) should be used as triethylamine may contain diethylamine which would also react with dansyl chloride.
13. Dichloromethane can be dried over molecular sieves, pore size 4 Å. Alternatively it may be refluxed over phosphorous pentoxide and subsequently distilled (b.p. 39.7°C) or it may be washed with water, dried over anhydrous potassium carbonate and distilled. Do NOT use sodium wire to dry dichloromethane—explosion hazard!
14. The solution will not become clear due to the precipitation of triethylamine hydrochloride.
15. 4 M hydrogen chloride in dioxane can be obtained from Sigma-Aldrich (St. Louis, MO, USA). Alternatively, 30% trifluoroacetic acid in dichloromethane can be used to cleave the *tert.*-butyloxycarbamate (BOC) protecting group. In this case, the reaction is stirred for 2 h at room temperature. The product is obtained as trifluoroacetate salt.
16. The substrate incubation solution can be stored at −20°C for at least 1 week.
17. It is important to vortex the sample as at 37°C some water condenses at the lid of the tube leading to higher concentrations of the analytes in the incubation solutions. This becomes important if multiple sampling from the same vial is intended.
18. For reproducible electrokinetic injection in CE it is necessary to use a constant volume in the injection vials. It is not recommended to transfer more than 22 μL of the supernatant in order to avoid transferring protein precipitate into the injection vial. This may lead to blockage of the CE capillary.
19. Conditioning of the capillary is important in order to obtain reproducible conditions of the inner wall of the capillary. Therefore, careful preconditioning of the capillary is required. Moreover, it is necessary to include all rinsing steps into validation procedures when developing validated CE procedures.
20. Different vials containing the background electrolyte should be used for rinsing of the capillary and for the analytical separation. Buffer should be replaced after a number of injections (typically between 2 and 10 injections) because of buffer depletion.
21. When applying hydrodynamic injection, the actually injected amount of the sample may vary depending on the temperature or the viscosity of the solution. Thus, adjustment of the injection time and/or pressure may be required. In the present examples the samples were injected at ambient temperature.

22. A low voltage of 1 kV is first applied in order to slowly mix the plugs so that sufficient hydrolysis of the substrate by the enzyme is achieved. Depending on the activity of the enzyme and the plug lengths, the mixing voltage may have to be optimized.
23. When using an enzyme activity of 0.1 U/μL online incubations can be performed at least 20 min (zero potential amplification time). Using lower enzyme activities may reduce the available incubation time interval while higher activities may increase the possible incubation times. Moreover, faster substrate conversion is achieved with higher enzyme activities.
24. Sirtuins require NAD^+ as cofactor. In order to ensure the presence of the cofactor after mixing of the plugs, NAD^+ is added to the both enzyme and substrate solutions.
25. If the mobilities of enzyme, substrate, and cofactor(s) are known the injection sequence of the plugs can be estimated. When such data are not available or cannot be determined experimentally, the so-called sandwich mode can be applied by sandwiching the enzyme between the substrate or vice versa, i.e., injecting subsequent plugs of enzyme–substrate–enzyme or substrate–enzyme–substrate. In this case it is advisable to add cofactors to both enzyme and substrate solution. The appropriate mode, i.e., enzyme–substrate–enzyme or substrate–enzyme–substrate, has to be determined experimentally.
26. The partial-filling EMMA assay can also be used to screen enzyme inhibitors. In this case, add the inhibitor to substrate solution at the desired concentration and follow the same procedure as described for the regular partial-filling EMMA enzyme assay.
27. SIRT1 is very temperature sensitive (see also Note 4). Therefore, set the CE sample tray to 4°C to avoid rapid loss of enzyme activity. If the instrument is not equipped with a cooled sample tray, remove the enzyme solution from the CE instrument immediately after completion of the injection sequence and keep on ice.
28. When establishing an EMMA method, take into account that the mixing voltage may affect the enzyme activity. Thus, the appropriate mixing voltage and mixing time have to be optimized in a series of experiments during assay development based on the product yield from the enzymatic reaction.
29. BSA may adsorb to the capillary wall. Thus, reactivation of the capillary wall may be necessary when obvious protein adsorption is observed. Protein adsorption will lead to peak broadening and variation of the migration times. When rinsing with background electrolyte is not adequate to regenerate the capillary, rinse it with 1 M sodium hydroxide solution for 10 min, water for 5 min, and finally with the background electrolyte for 5 min.

References

1. Banke N, Hansen K, Diers I (1991) Detection of enzyme activity in fractions collected from free solution capillary electrophoresis of complex samples. J Chromatogr 559:325–335
2. Van Dyck S, Kaale E, Nováková S, Glatz Z, Hoogmartens J, Van Schepdael A (2003) Advances in capillary electrophoretically mediated microanalysis. Electrophoresis 24:3868–3878
3. Nováková S, Van Dyck S, Van Schepdael A, Hoogmartens J, Glatz Z (2004) Electrophoretically mediated microanalysis. J Chromatogr A 1032:173–184
4. Glatz Z (2006) Determination of enzymatic activity by capillary electrophoresis. J Chromatogr B 841:23–37
5. Zhang J, Hoogmartens J, Van Schepdael A (2008) Advances in CE-mediated microanalysis: an update. Electrophoresis 29:56–65
6. Zhang J, Hoogmartens J, Van Schepdael A (2010) Recent developments and applications of EMMA in enzymatic and derivatization reactions. Electrophoresis 31:65–73
7. Fan Y, Scriba GKE (2010) Advances in capillary electrophoretic enzyme assays. J Pharm Biomed Anal 53:1076–1090
8. Hai X, Yang B-F, Van Schepdael A (2012) Recent developments and applications of EMMA in enzymatic and derivatization reactions. Electrophoresis 33:211–227
9. Horvath J, Dolnik V (2001) Polymer wall coatings for capillary electrophoresis. Electrophoresis 22:644–655
10. Righetti PG, Gelfi C, Verzola B, Castelletti L (2001) The state of the art of dynamic coatings. Electrophoresis 22:603–611
11. Kamande MW, Fletcher KA, Lowry M, Warner IM (2005) Capillary electrochromatography using polyelectrolyte multilayer coatings. J Sep Sci 28:710–718
12. Inagaki S, Esaka Y, Deyashiki Y, Uno B, Hara A, Toyo'oka T (2008) Human liver dihydrodiol dehydrogenase 1-catalyzed reaction generating 9α,11β-prostaglandin F_2 from prostaglandin D_2 followed by micellar electrokinetic chromatography. J Sep Sci 31:735–740
13. Bardelmeijer HA, Lingeman H, De Ruiter C, Underberg WJM (1998) Derivatization in capillary electrophoresis. J Chromatogr A 807:3–26
14. Underberg WJM, Waterval JCM (2002) Derivatization trends in capillary electrophoresis: an update. Electrophoresis 23:3922–3933
15. Mala Z, Krivankova L, Gebauer P, Bocek P (2007) Contemporary sample stacking in CE: a sophisticated tool based on simple principles. Electrophoresis 28:243–253
16. Breadmore MC, Thabano JRE, Dawod M, Kazarian AA, Quirino JP, Guijt RM (2009) Recent advances in enhancing the sensitivity in electrophoresis and electrochromatography in capillaries and microchips (2006–2008). Electrophoresis 30:230–248
17. Mala Z, Gebauer P, Bocek P (2011) Contemporary sample stacking in analytical electrophoresis. Electrophoresis 32:116–126
18. Tzeng H, Hung CH, Wang JY, Chou CH, Hung HP (2006) Simultaneous determination of adenosine and its metabolites by capillary electrophoresis as a rapid monitoring tool for 5′-nucleotidase activity. J Chromatogr A 1129:149–152
19. Friedecky D, Tomkova J, Adam T (2007) Determination of ITPase activity by capillary electrophoresis. Clin Chem 53:1164–1165
20. Iqbal J, Lévesque SA, Sévigny J, Müller CE (2008) A highly sensitive CE-UV method with dynamic coating of silica-fused capillaries for monitoring of nucleotide pyrophosphatase/phosphodiesterase reactions. Electrophoresis 29:3685–3693
21. Pungor E Jr, Hague CM, Chen G, Lemontt JF, Dvorak-Ewell M, Prince WS (2009) Development of a functional bioassay for arylsulfatase B using the natural substrates of the enzyme. Anal Biochem 395:144–150
22. Baršun M, Jajčanin N, Vukelič B, Špoljarić J, Abramić M (2007) Human dipeptidyl peptidase III acts as a post-proline cleaving enzyme on endomorphins. Biol Chem 388:343–348
23. He HL, Chen XL, Wu H, Sun CY, Zhang YZ, Zhou BC (2007) High throughput and rapid screening of marine protein hydrolysates enriched in peptides with angiotensin-I-converting enzyme inhibitory activity by capillary electrophoresis. Bioresource Technol 98:3499–3505
24. Albillos SM, Busto MD, Perez-Mateos M, Ortega N (2007) Analysis by capillary electrophoresis of the proteolytic activity of a *Bacillus subtilis* neutral protease on bovine caseins. Int Dairy J 17:1195–1200
25. Piccard H, Hu JL, Fiten P, Proost P, Martens E, Van den Steen PE, Van Damme J, Opdenakker G (2009) "Reverse degradomics", monitoring of proteolytic trimming by multi-CE and confocal detection of fluorescent substrates and reaction products. Electrophoresis 30:2366–2377
26. Bi H, Qiao L, Busnel J-M, Liu B, Girault HH (2009) Kinetics of proteolytic reactions in nanoporous materials. J Proteome Res 8:4685–4692
27. Hsieh BC, Hsiao HY, Cheng TJ, Chen RLC (2008) Assays for serum cholinesterase activity

by capillary electrophoresis and an amperometric flow injection choline biosensor. Anal Chim Acta 623:157–162

28. Fan Y, Ludewig R, Scriba GKE (2009) 9-Fluorenylmethoxycarbonyl-labeled peptides as substrates in a capillary electrophoresis-based assay for sirtuin enzymes. Anal Biochem 387:243–248
29. Fan Y, Hense M, Ludewig R, Weigerber C, Scriba GKE (2011) Capillary-electrophoresis-based sirtuin assay using non-peptide substrates. J Pharm Biomed Anal 54:772–778
30. Iqbal J, Burbiel JC, Müller CE (2006) Development of off-line and on-line capillary electrophoresis methods for the screening and characterization of adenosine kinase inhibitors and substrates. Electrophoresis 27:2505–2517
31. Lee KJ, Mwongela SM, Kottegoda S, Borland L, Nelson AR, Sims CE, Allbritton NL (2008) Determination of sphingosine kinase activity for cellular signaling studies. Anal Chem 80:1620–1627
32. Qi L, Qiao J, Yang GL, Chen Y (2009) Chiral ligand-exchange CE assays for separation of amino acid enantiomers and determination of enzyme kinetic constant. Electrophoresis 30:2266–2272
33. Konečný J, Juřica J, Tomandl J, Glatz Z (2007) Study of recombinant cytochrome P450 2C9 activity with diclofenac by MEKC. Electrophoresis 28:1229–1234
34. Afshar M, Thormann W (2006) Capillary electrophoretic investigation of the enantioselective metabolism of propafenone by human cytochrome P-450 supersomes: evidence for a typical kinetics by CYP2D6 and CYP3A4. Electrophoresis 27:1526–1536
35. Tan Y, Sun L, Xi Z, Yang GF, Jiang DQ, Yan XP, Yang X, Li HY (2008) A capillary electrophoresis assay for recombinant *Bacillus subtilis* protoporphyrinogen oxidase. Anal Biochem 383:200–204
36. Wang SS, Li Q, Ma PX, Geng LN, Deng YL (2008) Determination of monoamine oxidase activity by capillary electrophoresis. Chin J Anal Chem 36:1711–1715
37. Cunliffe JM, Whorton MR, Sunahara RK, Kennedy RT (2007) A CE assay for the detection of agonist stimulated adenylyl cyclase activity. Electrophoresis 28:1913–1920
38. Qi L, Yang G, Zhang H, Qiao J (2010) A chiral ligand exchange CE essay with zinc(II)-L-valine complex for determining enzyme kinetic constant of L-amino acid oxidase. Talanta 81:1554–1559
39. Li Y, Liu D, Bao JJ (2011) Characterization of tyrosine kinase and screening enzyme inhibitor by capillary electrophoresis with laser-induced fluoresce detector. J Chromatogr B 879:107–112
40. Gratz A, Goetz C, Jose J (2010) A CE-based assay for human protein kinase CK2 activity measurement and inhibitor screening. Electrophoresis 31:634–640
41. Yangyuoru PM, Otieno AC, Mwongela SM (2011) Determination of sphingosine kinase 2 activity using fluorescent sphingosine by capillary electrophoresis. Electrophoresis 32:1742–1749
42. Bao JM, Regnier FE (1992) Ultramicro enzyme assays in a capillary electrophoretic system. J Chromatogr 608:217–224
43. Van Dyck S, Hoogmartens J, Van Schepdael A (2001) Michaelis-Menten analysis of bovine plasma amine oxidase by capillary electrophoresis using electrophoretically mediated microanalysis in a partially filled capillary. Electrophoresis 22:1436–1442
44. Okhonin V, Lin X, Krylov SN (2005) Transverse diffusion of laminar flow profiles to produce capillary nanoreactors. Anal Chem 77:5925–5929
45. Okhonin V, Wong E, Krylov SN (2008) Mathematical model for mixing reactants in a capillary microreactor by transverse diffusion of laminar flow profiles. Anal Chem 80:7482–7486
46. Kim HS, Wainer IW (2006) On-line drug metabolism in capillary electrophoresis. 1. Glucuronidation using rat liver microsomes. Anal Chem 78:7071–7077
47. Urban PL, Goodall DE, Bergström ET, Bruce NC (2006) 1,4-Benzoquinone-based electrophoretic assay for glucose oxidase. Anal Biochem 359:35–39
48. Urban PL, Goodall EM, Bergström ET, Bruce NC (2007) electrophoretic assay of penicillinase: substrate specificity screening by parallel CE with an active pixel sensor. Electrophoresis 28:1926–1936
49. Tang Z-M, Wang Z-Y, Kang J-W (2007) Screening of acetylcholine esterase inhibitors in natural extracts by CE with electrophoretically mediated microanalysis technique. Electrophoresis 28:360–365
50. Schuchert-Shi A, Hauser PC (2009) Peptic and tryptic digestion of peptides and proteins monitored by capillary electrophoresis with contactless conductivity detection. Anal Biochem 387:202–207
51. Wong E, Okhonin V, Berezovski MV, Nozaki T, Waldmann H, Alexandrov K, Krylov SN (2008) "Inject-mix-react-separate-and-quantitate" (IMReSQ) method for screening enzyme inhibitors. J Am Chem Soc 130:11862–11863
52. Guo L-P, Jiang T-F, Lv Z-H, Wang Y-H (2010) Screening α-glucosidase inhibitors from traditional Chinese drugs by capillary

electrophoresis with electrophoretically mediated microanalysis. J Pharm Biomed Anal 53:1250–1253

53. Wang T, Kang J (2009) Hexokinase inhibitor screening based on adenosine 5′-diphosphate determination by electrophoretically mediated microanalysis. Electrophoresis 30:1349–1354
54. Němec T, Glatz Z (2007) Integration of short-end injection mode into electrophoretically mediated microanalysis. J Chromatogr A 1155:206–213
55. Papežová K, Němec T, Chaloupková R, Glatz Z (2007) Study of substrate inhibition by electrophoretically mediated microanalysis in partially filled capillary. J Chromatogr A 1150:327–331
56. Hai X, Konečný J, Zeisbergerová M, Adams E, Hoogmartens J, Van Schepdael A (2008) Development of electrophoretically mediated microanalysis method for the kinetics study of flavin-containing monooxygenase in a partially filled capillary. Electrophoresis 29:3817–3824
57. Hai X, Adams E, Hoogmartens J, Van Schepdael A (2009) Enantioselective in-line and off-line CE methods for the kinetic study on cimetidine and its chiral metabolites with reference to flavin-containing monooxygenase genetic isoforms. Electrophoresis 30:1248–1257
58. Zeisbergerová M, Adámková A, Glatz Z (2009) Integration of on-line protein digestion by trypsin in CZE by means of electrophoretically mediated microanalysis. Electrophoresis 30:2378–2384
59. Zhang J, Hoogmartens J, Van Schepdael A (2008) Kinetic study of cytochrome P450 by capillary electrophoretically mediated microanalysis. Electrophoresis 29:3694–3700
60. Fan Y, Scriba GKE (2010) Electrophoretically mediated microanalysis assay for sirtuin enzymes. Electrophoresis 31:3874–3880
61. Sun X, Gao N, Jin W (2006) Monitoring yoctomole alkaline phosphatase by capillary electrophoresis with on-capillary catalysis-electrochemical detection. Anal Chim Acta 571:30–33
62. Chichester KD, Sebastian M, Ammerman JW, Colyer CL (2008) Enzymatic assay of marine bacterial phosphatases by capillary electrophoresis with laser-induced fluorescence detection. Electrophoresis 29:3810–3816
63. Křenková J, Foret F (2004) Immobilized microfluidic enzymatic reactors. Electrophoresis 25:3550–3563
64. Ma JF, Zhang LH, Liang Z, Zhang WB, Zhang YK (2009) Recent advances in immobilized enzymatic reactors and their applications in proteome analysis. Anal Chim Acta 632:1–8
65. Finkel T, Deng CX, Mostoslavsky R (2009) Recent progress in the biology and physiology of sirtuins. Nature 460:587–591
66. Tanner KG, Landry J, Sternglanz R, Denu JM (2000) Silent information regulator 2 family of NAD-dependent histone/protein deacetylases generates a unique product, 1-*O*-acetyl-ADP-ribose. Proc Natl Acad Sci USA 97:14178–14182
67. Michan S, Sinclair D (2007) Sirtuins in mammals: insights into their biological function. Biochem J 404:1–13
68. SIRT1 product data sheet, Enzo Life Sciences (www.enzolifesciences.com/fileadmin/enzo/BML/se239.pdf)

Chapter 22

A Methodology for Detection and Quantification of Esterase Activity

Ana L. Simplício, Ana S. Coroadinha, John F. Gilmer, and Joana Lamego

Abstract

Carboxylesterases are important enzymes for xenobiotic metabolism and are receiving increasing attention in the context of cancer therapies. Quantification of individual carboxylesterase activity is important since protein levels do not always correlate to activity and significant interorgan, interindividual, and interspecies variations exist. Here we present a methodology enabling the specific quantification of carboxylesterase 2 activity in a pool of other esterases. Method applicability is illustrated for the evaluation of interspecies variation and for activity assessment of transfected cell extracts. The methodology can easily be adapted to the evaluation of other esterases upon careful selection of adequate substrates and/or specific inhibitors.

Key words: Carboxylesterases, Carboxylesterase 2, Esterases, Loperamide, BNPP, Enzyme activity differentiation, Capillary electrophoresis

1. Introduction

Mammalian carboxylesterases (CES and Ces, for human and non-human, respectively) are ubiquitous found enzymes involved in diverse functions ranging from prodrug activities to detoxification of narcotics and metabolism of pyrethroids (1). CES are grouped into five families, from CES1 to CES5, being CES1 and CES2 the main CES found in humans (2).

Assessment of CES is possible through a variety of techniques. Spectrophotometric assays (3), for example, allow the quantification of total CES activity, but identification of the specific CES present in biological samples is problematic. Western blot assays (4) enable CES identification, but no quantitative assessment of their activities is possible (5).

Nicola Volpi and Francesca Maccari (eds.), *Capillary Electrophoresis of Biomolecules: Methods and Protocols*, Methods in Molecular Biology, vol. 984, DOI 10.1007/978-1-62703-296-4_22, © Springer Science+Business Media, LLC 2013

To identify and quantify CES2 activity in biological samples we have developed a methodology based on the combination of substrates and specific inhibitors of CES and a separation technique, such as capillary electrophoresis (CE). Other separation techniques are compatible with the developed methodology. Application to UV/Vis spectrophotometric analysis is, however, not possible due to overlapping wavelengths of absorption of substrates, products, and/or inhibitors (5).

The methodology can be applied for the analysis of different types of samples, such as commercial enzymes, enzyme mixtures, human cell extracts, and mammalian sera. 4-Methylumbelliferyl acetate (4-MUBA) was used as substrate due to its higher stability towards nonspecific hydrolysis in comparison to the classical CES substrate, *p*-nitrophenyl acetate (*p*-NPA). Bis-*p*-nitrophenyl phosphate (BNPP) and loperamide were selected as inhibitors of total CES and CES2, respectively (5). The method was shown to be specific for CES since 4-MUBA is not a cholinesterase substrate at the method concentrations. This methodology was found to be compatible with other esterase substrates such as *p*-NPA, procaine, butyrylthiocholine (BuTCh), and acetylthiocholine (AcTCh), and their corresponding products *p*-nitrophenol (*p*-NP), *p*-aminobenzoic acid (*p*-ABA), and thiocoline (TCh), respectively. Therefore the extension of this method to the detection and quantification of the activity of other enzymes may be easily performed.

The goal of this chapter is to provide a stepwise approach on how to successfully apply the developed methodology for quantification of CES2 activity in different biological samples. Sources of commercially available enzymes are described as well as methods for the preparation of HEK-293T (human embryonic kidney cells) cell extracts and mammalian sera. Operational aspects of the methodology that can critically affect its successful application are described in Subheading 4.

2. Materials

2.1. Buffers and Solutions

1. Aqueous NaOH 0.1 mM (see Note 1).
2. Background electrolyte solution (BES): 30 mM borate-phosphate buffer (30 mM disodium tetraborate, 30 mM sodium dihydrogen phosphate; pH 8.0) (6).
3. Sample dilution buffer: 50 mM tris(hydroxymethyl)aminomethane (Trizma) hydrochloride, 50 mM Trizma base; pH 7.4.
4. Reaction buffer: 90 mM potassium phosphate, 40 mM sodium chloride; pH 7.3 (7) (see Note 2).

2.2. Substrates and Inhibitors

1. Substrate stock solutions: 15 mM 4-MUBA in ethanol; 5 mM procaine, in ethanol; 10 mM *p*-NPA, in methanol; 15 mM BuTCh in ethanol; 15 mM acetylthiocoline (AcTCh), in ethanol (see Notes 3 and 4).
2. Product stock solutions: 15 mM 4-methylumbelliferone (4-MUB) in ethanol; 10 mM *p*-ABA in ethanol; 10 mM *p*-NP, in methanol.
3. Loperamide stock solution: 5 mM, in dimethyl sulfoxide (DMSO).
4. BNPP stock solution: 25 mM, in DMSO.

2.3. Commercial Enzymes

The provided methodology was developed and fully tested for Ces1 and CES2. Nonetheless, it was shown to be suitable for the analysis of other esterases (upon use of adequate substrate and inhibitors) (see Note 5). A list of the tested commercial enzymes and the best way to prepare them is provided below (see Note 6).

1. Human CES2 recombinant enzyme (30,000 pmol (*p*-NPA). $min^{-1}.\mu g^{-1}$, R&D Systems Inc; Minneapolis, USA): Enzyme should be diluted, on ice, to an appropriate stock concentration (i.e., 50 ng/μL), using the same buffer in which the enzyme is provided by the manufacturer (20 mM sodium acetate, 600 mM NaCl, 20% (v/v) glycerol; pH 5.0), under sterile conditions. Aliquot immediately and store at −80°C. Before use, thaw one aliquot on ice, and resuspend to an appropriate final concentration (i.e., 1 ng/μL), using tris(hydroxymethyl)aminomethane (Tris)–HCl buffer. Do not freeze the diluted aliquot.
2. Esterase (mainly Ces1) from porcine liver (≥130 U/mg, Sigma; Buchs, Switzerland): Enzyme should be diluted, on ice, to a final stock concentration of 2.5 μg/μL, according to the manufacturer's instructions (i.e., 3.2 mM ammonium sulfate solution; pH 6.0). Aliquot immediately and store at −80°C. Avoid freezing and thawing. Before use, thaw one aliquot, on ice, and resuspend to an appropriate final concentration (i.e., 1 ng/μL), using Tris–HCl buffer. Do not freeze the diluted aliquot.
3. Butyrylcholinesterase (BuChE) from equine serum (≥500 U/mg, Sigma; St. Louis, USA): Enzyme should be diluted, on ice, to a final concentration of 0.72 μg/μL using 20 mM Tris–HCl; pH 7.5. Aliquot and store at −20°C. Avoid freezing and thawing.
4. Acetylcholinesterase (AcChE) from electric eel (425 U/mg, 62% protein, Sigma; St. Louis, USA): Enzyme should be diluted, on ice, to a final concentration of 0.94 μg/μL using 20 mM Tris–HCl; pH 7.5. Aliquot and store at −20°C.

2.4. Instrumentation

1. Beckman Coulter (Palo Alto, USA) P/ACE MDQ capillary electrophoresis system (or equivalent) equipped with a diode array or LIF detector (see Note 7).

3. Methods

3.1. Sample Preparation

The provided methodology was applied in the analysis of two different biological samples. Preparation of both types of sample is described in this section.

3.1.1. Human Cell Extracts Samples

The transient transfection of human embryonic kidney (HEK-293 T; ATCC CRL11268) cells with CES2 gene has been described previously (5). Extractions were performed using M-PER mammalian extraction reagent (Pierce Biotechnology; Rockford, USA), under sterile conditions, as follows:

1. Cells are harvested through centrifugation at 200 × *g* and 4°C, for 10 min.
2. Supernatant should be removed and the cells lysed using an appropriate volume of M-PER reagent, according to the manufacturer's instructions.
3. Cell debris should be removed by centrifugation at 10,000 × *g* and 4°C for 10 min.
4. Supernatant aliquots should be prepared on ice and may be stored at −20°C.
5. Quantification of total protein amount may be performed using BCA protein assay kit (Pierce Biotechnology) or through other protein quantification method, following manufacturer's instructions.

3.1.2. Mammalian Sera Samples

1. Mammalian sera should be collected without the addition of anticoagulants (8, 9).
2. Store sera for no longer than 24 h at 4°C.
3. Sera centrifugation at 3,000 × *g*, 15 min should be performed.
4. The supernatant obtained may be stored, properly aliquoted, at −20°C.

3.2. Instrument Setup and System Suitability Tests

1. Prepare all required buffers as described in the Subheading 2.
2. Pack the fused-silica capillary (Agilent; Santa Clara, USA), 75 μm i.d., 42 cm total length (32 cm to detector) in a standard cartridge and set temperature to 25°C. Set CE sample tray to 15°C (see Note 8).
3. Pretreat each new silica capillary with NaOH (0.1 mM), water, and BES for 20 min each at 5.0 psi.
4. All injections should be hydrodynamically conducted for 10 s, at 0.2 psi, at the anode. Separations should be performed at a constant current of +70.0 μA for cathodic migration. UV absorption detection should be performed according to Table 1.

Table 1
Maximum absorption wavelengths and migration times of the tested compounds

Compound	Wavelength (nm)	Migration time (min)
4-MUBA	N.d.	
4-MUB	350	4.0
BNPP	280[a]	4.8
Loperamide	N.d.	
p-NPA	N.d.	
p-NP	400	6.0
Procaine	280	2.2
p-ABA	254	6.8
AcTCh	230	2.0
BuTCh	230	2.0
TCh	230	2.5

N.d. Not detected
[a]BNPP is also detected at 350 nm

5. Prepare substrate and product solutions, using reaction buffer. Prepare a calibration curve of the reaction product, each day, with at least three different concentrations up to the substrate concentration (see Note 9).

3.3. Reaction Setup

1. Thaw biological material (e.g., pure enzymes, cell extracts, or mammalian sera) on ice and dilute it with Tris–HCl buffer as described in Materials section (see Notes 10–12).
2. Prepare enzyme reactions in Eppendorf tubes, on ice, to a final volume of 150 μL (see Note 13), using reaction buffer, according to Table 2 (see Notes 14 and 15). Preincubation with BNPP (500 μM) or loperamide (50 μM) (see Note 16) should be performed at 37°C for 8 and 5 min, respectively.
3. Add substrate to each reaction mixture (see Notes 17 and 18), at RT, and incubate again for 8 min at 37°C.
4. Stop each enzymatic reaction through the addition of 150 μL of ethanol and mixing by inversion.
5. Centrifuge at 10,000 × *g* for 10 min and transfer the supernatant to new CE sample tubes (see Note 19).
6. Proceed immediately to the CE analysis (please refer to Fig. 1 for examples of obtained electropherograms).

Table 2
Stepwise order scheme for reaction preparation

	SR	SR with BNPP	SR with Lop	Blank
RB (μL)	150-(z+s)	150-(x+z+s)	150-(y+z+s)	150-s
BNPP (μL)	–	x	–	–
Lop (μL)	–	–	y	–
Test sol (μL)	z			–
Pre-inc (min)	8	8	5	8
Substrate (μL)	s			
RT (min)	8	8	8	8
Ethanol (μL)	150			

Sample reaction (SR), loperamide (Lop), test solution (Test sol), reaction buffer (RB), preincubation time (Pre-inc), reaction time (RT). Appropriate volumes of BNPP (x), loperamide (y), test solution (z), and substrate (s) should be used

3.4. Data Analysis

1. Using P/ACE MDQ version 7.0 (or equivalent), integrate the product peak (see Note 20) in each electropherogram and convert the obtained arbitrary units (AUs) to product concentration through interpolation using the calibration curve (see Note 21).
2. Quantification of CES2-like activity (see Note 22) in a specific sample should be performed according to Fig. 2 and as follows:

   CES2-like activity = sample activity – sample activity in the presence of loperamide.
3. Quantification of other CES activity present in the sample should be performed as follows:

   Other CES activity = sample activity in the presence of loperamide – sample activity in the presence of BNPP.
4. Quantification of other esterases:

   Other esterases = sample activity in the presence of BNPP – spontaneous hydrolysis of the substrate

   Please refer to Figs. 3 and 4 for examples of analyzed data.

4. Notes

1. Milli-Q water should be used in all solutions.
2. Buffers can be kept at 4°C for up to 1 month. Check for any visible contaminants during that period. Before use, each buffer should be filtered with 0.2 μm PVDF filter.

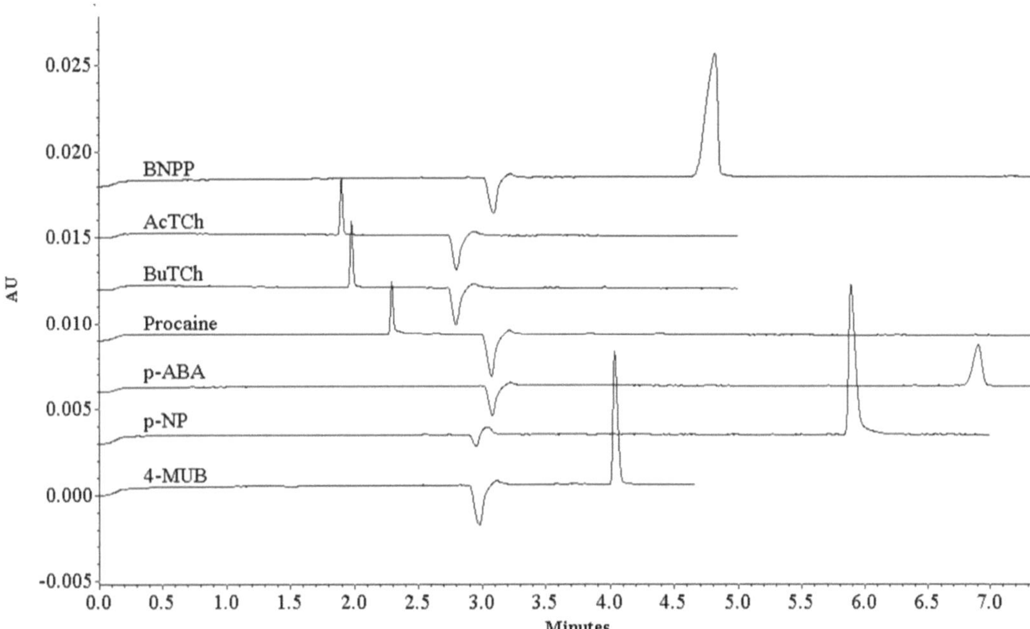

Fig. 1. Electropherograms of substrates, products, and inhibitors detected by CE. Absorption wavelengths and migration times are summarized in Table 1. 0.25 mM 4-MUB, 0.5 mM *p*-NP, 0.1 mM *p*-ABA, 0.1 mM procaine, 0.5 mM BuTCh, 0.5 mM AcTCh, and 0.25 mM BNPP.

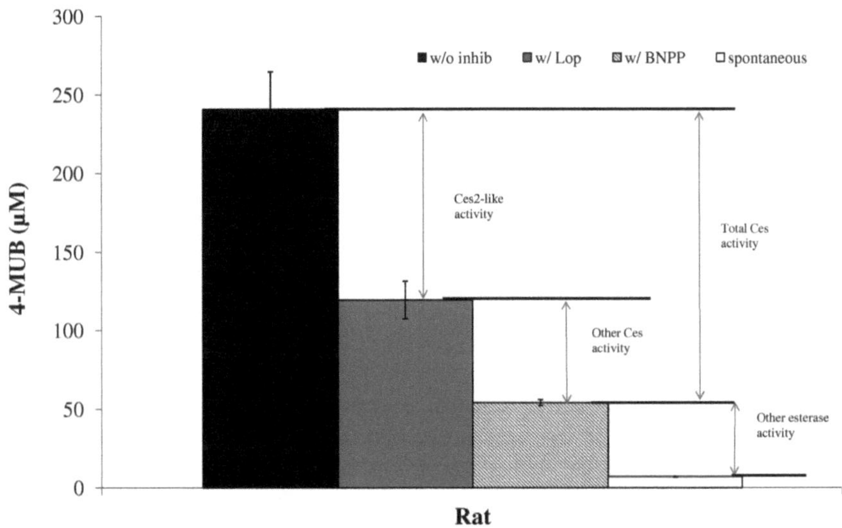

Fig. 2. Quantification of enzyme activity. Rat serum (1%) was tested in the presence (loperamide (Lop) or BNPP) and absence of inhibitors (w/o inhib) as described. Spontaneous hydrolysis of 4-MUBA was also determined (spontaneous). 4-MUB concentration was determined in each reaction. Relative activities can be determined through the difference in the amount of product formed in each case as depicted.

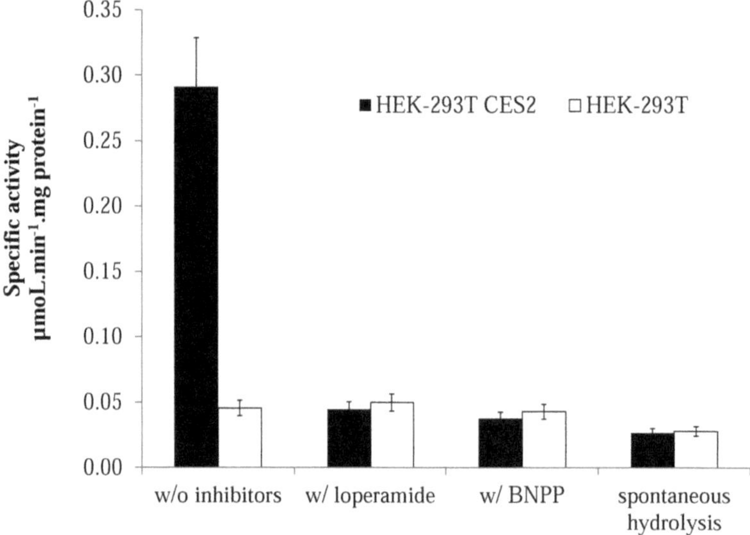

Fig. 3. CES2 activity in cell extracts. HEK-293T cells overexpressing CES2 (HEK-293T CES2) and parental cell line (HEK-293T) were tested as described. Total protein was determined and 6 μg of extract was used per experiment. BNPP (500 μM) and loperamide (250 μM) were used. Error bars represent the standard deviation of intermediate precision.

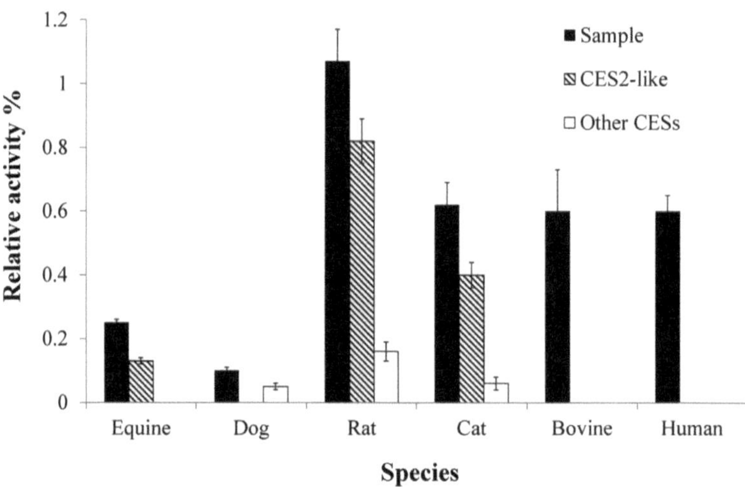

Fig. 4. CES2-like activity in mammalian sera. Appropriate sera dilutions (v/v) were performed to achieve enzymatic linearity range: 10% equine, dog, human, and bovine; 2.5% cat and 1% rat. BNPP (500 μM) and loperamide (250 μM) were used. Error bars represent standard deviation of three different injections.

3. Substrate stock solutions (4-MUBA, *p*-NPA, procaine) should be freshly prepared each day as they are unstable. They should be kept at room temperature (RT) throughout the day since refrigeration will cause substrate precipitation.
4. Substrate and product stock solutions may be prepared with other organic solvent than ethanol, such as DMSO or acetone

(solvents that were previously shown to be suitable for CES activity analysis) (11).

5. The presented methodology was developed for CES2 evaluation in different biological samples, using 4-MUBA as substrate. Nonetheless, the method was also tested for other substrates and products. All the tested compounds are described in Table 1. Moreover, other enzymes were also tested such as Ces1 and cholinesterases. For more details, please read reference (5).
6. Keep all biological material (e.g., pure enzymes, cell extracts, etc.) at 4°C while preparing enzyme reactions. This will help minimize protein degradation (note that CES2 is unstable) and reduce protease activity in the cell extracts (note that no protease inhibitors should be added during cell extracts preparation as this may impair CES activity (10)).
7. As 4-MUB is fluorescent, its detection may be further improved if fluorescence detectors are available.
8. CE sample tray should not be refrigerated to 4°C to avoid substrate precipitation. A temperature of 15°C was found to be suitable to delay solution deterioration.
9. When using a new substrate it is advisable to evaluate some key parameters such as equipment response linearity and the limits of detection and quantification. For 4-MUB, these were found to be:
   - Linear range: up to 1.5 mM ($r^2 = 0.999$), using 7 points
   - Limit of detection: 13.7 μM (calculated from the residual standard deviation of the calibration curve; signal-to-noise ratio of 3.3)
   - Precision: calculated in the range of 23.4 μM to 1.5 mM of 4-MUB through the CV_r which was under 10% in all points
   - Limit of quantification: 23.4 μM (lowest calibration point corresponding to a coefficient of variation of repeatability (CV_r) below 10% ($n = 3$))
10. 4-MUBA spontaneous hydrolysis into 4-MUB at RT is not negligible. A substrate blank is mandatory to properly evaluate enzymatic activity.
11. To account for method repeatability, it is advised to perform 3 replicate reactions for each sample. Results may be presented as the average ± standard deviation. To assess statistical significance between analyzed samples, mean *t* tests may be performed. Using two different concentrations of Ces1, we have assessed the intermediate precision (CVi) of the method in the presence and absence of loperamide, over three different days. A value of 12.8% was obtained. Intermediate precision was calculated by pooled standard deviation.

12. If a large set of enzymatic reactions are prepared at the same time, misleading data will be generated due to the spontaneous hydrolysis of 4-MUBA, even after stopping the enzymatic reaction with ethanol. It is advisable to prepare a smallest set of enzymatic reactions (5 to 6) and have them analyzed before preparing the following set of samples. ON analysis of previously prepared samples is thus not advisable.
13. Total volume of enzyme reaction may be lowered if 0.2 μL sample tubes are used.
14. As rule of thumb, organic solvent (from the substrate and/or inhibitor stock solution) should not exceed 2% (v/v) in the final reaction (11).
15. For a qualitative assessment of CES2 activity in a biological sample, a proper dilution of the biological material is one that does not produce more than 50% hydrolysis of the used substrate. For quantitative assessments, the linearity range of enzyme reaction should be evaluated to choose the best dilution set and the most suited time point to stop the enzymatic reaction. The provided enzymatic reaction parameters should thus be considered as starting points and a proper evaluation of their suitability and adjustments are encouraged to be performed case by case.
16. The provided inhibitor (BNPP and loperamide) concentrations are indicative. Assessment of the suitability of the used concentrations should be performed.
17. 4-MUBA was used at a final concentration of 0.5 mM. Other substrate concentrations may be used keeping in mind that, for specific activity determination, substrate saturation must be assured. Concentrations higher than 1 mM in the final reaction volume could not be obtained due to substrate low solubility.
18. Substrate addition to the reaction mixture should be the last step and performed at RT, as fast as possible to avoid spontaneous hydrolysis and/or the differential start of the reactions.
19. Centrifugation of each enzymatic reaction, prior to sample analysis in CE, is advised to prevent capillary obstruction. If capillary obstruction occurs, it will be detected through dramatic alterations in the normal voltage and/or current and sequence run should be stopped immediately. BES buffer hydrodynamic injection at the capillary outlet, using moderate pressure (i.e., 5.0 psi), will allow the removal of any clog that is obstructing the capillary. If the run could not be stopped immediately, more stringent washing conditions are advised. In this case perform the pretreatment protocol provided in step 3 of Subheading 3.2. Confirm that current/voltage can be properly achieved before resuming your sample analysis. If not, repetition of the steps described in this troubleshooting should be conducted. In case no current and voltage stabilization are achieved it is advisable to change the capillary.

20. If the substrate is detectable with the available detector (such as in the cases of BuTCh and AcTCh), quantifications using a commercially available substrate (instead of the product) are also suitable.
21. After each pretreatment, after each resolved obstruction, and at the end of each day it is advisable to perform an injection of the product in use (as described in Subheading 3.2). The values obtained should be compared with the ones obtained earlier in the day and if any discrepancies are observed, the following acquired data should be analyzed considering this recalibration.
22. Data may be presented as enzyme activity (i.e., μmoL.min^{-1} or μM.min^{-1}) or amount or concentration of product formed (i.e., μmol or μM of 4-MUB). For enzyme-specific activity, data should be presented as enzyme activity per amount of used enzyme. If enzyme amount cannot be determined (such as in the case of biological cell extracts), data may be presented as enzyme activity per total amount of protein determined in the sample.

Acknowledgment

This work was funded by Fundação para a Ciência e Tecnologia, Portugal (SFRH/BD/44025/2008, PTDC/EBB-BIO/111530/2009, PEst-OE/EQB/LA0004/2011).

References

1. Imai T, Ohura K (2010) The role of intestinal carboxylesterase in the oral absorption of prodrugs. Curr Drug Metab 11:793–805
2. Satoh T, Hosokawa M (2006) Structure, function and regulation of carboxylesterases. Chem Biol Interact 162:195–211
3. Crow JA, Borazjani A, Potter PM et al (2007) Hydrolysis of pyrethroids by human and rat tissues: examination of intestinal, liver and serum carboxylesterases. Toxicol Appl Pharmacol 221:1–12
4. Kubo T, Kim SR, Sai K et al (2005) Functional characterization of three naturally occurring single nucleotide polymorphisms in the CES2 gene encoding carboxylesterase 2 (HCE-2). Drug Metab Dispos 33:1482–1487
5. Lamego J, Coroadinha AS, Simplício AL (2011) Detection and quantification of carboxylesterase 2 activity by capillary electrophoresis. Anal Chem 83:881–887
6. Tang ZM, Wang ZY, Kang JW (2007) Screening of acetylcholinesterase inhibitors in natural extracts by CE with electrophoretically mediated microanalysis technique. Electrophoresis 28:360–365
7. Pindel EV, Kedishvili NY, Abraham TL et al (1997) Purification and cloning of a broad substrate specificity human liver carboxylesterase that catalyzes the hydrolysis of cocaine and heroin. J Biol Chem 272:14769–14775
8. Li B, Sedlacek M, Manoharan I et al (2005) Butyrylcholinesterase, paraoxonase, and albumin esterase, but not carboxylesterase, are present in human plasma. Biochem Pharmacol 70:1673–1684
9. Christensen JM, Stalker D (1991) Ibuprofen piconol hydrolysis in vitro in plasma, whole blood, and serum using different anticoagulants. J Pharm Sci 80:29–31
10. Ross MK, Borazjani A (2007) Enzymatic Activity of Human Carboxylesterases. Curr Protoc Toxicol 33:4.24.1–4.24.14
11. Williams ET, Ehsani ME, Wang X et al (2008) Effect of buffer components and carrier solvents on in vitro activity of recombinant human carboxylesterases. J Pharmacol Toxicol Methods 57:138–144

Chapter 23

Enzyme Inhibitor Screening by CE with an On-Column Immobilized Enzyme Microreactor Created by an Ionic Binding Technique

Chao Liu, Qianqian Zhang, and Jingwu Kang

Abstract

Enzymes are an important class of drug targets. In the early stage of the drug discovery, the major task is to find out the inhibitors of a given enzyme target in a compound library. Herein we describe a method for screening the enzyme inhibitors in the complex mixtures (such as the natural extracts) by capillary electrophoresis with an on-column immobilized enzyme microreactor. The enzyme molecules are immobilized on the capillary wall via ionic binding with the positively charged coating, which was created by simply flushing the column with a solution of polyelectrolyte hexadimethrine bromide. The activities of the immobilized enzymes are assayed by performing the electrophoretic separation and thereafter determining the product of the enzyme-mediated reaction. Enzyme inhibition can be read out directly from the reduced peak area of product in comparison with that in the reference electropherogram obtained in the absence of any inhibitor.

Key words: Enzyme inhibitor screening, Capillary electrophoresis, Immobilized enzyme microreactor

1. Introduction

Enzymes are an important class of drug targets (1, 2). In the early stage of the drug discovery, the major task is to find out the inhibitors of a given enzyme target in a compound library (3). Therefore, the development of rapid, cost-effective techniques for enzyme inhibitor screening has received much attention (3). Various enzyme immobilization techniques have been developed for inhibitor screening (4–7). Compared with the free-enzyme-based screening methods, advantages of using the immobilized enzyme combined with separation provided by capillary electrophoresis for inhibitor screening are (a) the enzyme can be reusable, hence a

Nicola Volpi and Francesca Maccari (eds.), *Capillary Electrophoresis of Biomolecules: Methods and Protocols*, Methods in Molecular Biology, vol. 984, DOI 10.1007/978-1-62703-296-4_23, © Springer Science+Business Media, LLC 2013

minimized screening cost; (b) the enzyme stability can be promoted; (c) mixture of compounds can be screened; and (d) the readout data of screening are straightforward (8–13).

Capillary electrophoresis (CE) is not only a separation tool with high separation performance but also a versatile platform for enzyme study and drug discovery. Among the various CE techniques, electrophoretically mediated microanalysis (EMMA) technique is a useful miniature tool for the study of enzymes and for inhibitor screening (14–16). The screening throughput can be dramatically improved by using multiplex capillary electrophoresis with UV absorption detection (17). Here, we describe a new strategy for rapid screening of enzyme inhibitors by CE with the on-column immobilized enzyme microreactor (18). To demonstrate this strategy, angiotensin-converting enzyme (ACE) was employed as a model for the enzyme immobilization, inhibition study, and inhibitor screening. The enzyme molecules are immobilized on the capillary wall via ionic binding with the positively charged coating, which was created by simply flushing the column with a solution of polyelectrolyte hexadimethrine bromide. It is proved that such a prepared enzyme microreactor displays a high enough activity and stability so that the inhibitors can be easily identified. Moreover, the immobilized enzyme microreactor can be easily renewed.

2. Materials

2.1. Instrumentation and Materials

1. A CE system with UV or diode array detector. A P/ACE MDQ CE System (Beckman Coulter, Fullerton, CA, USA) or an Agilent ^{3D}CE system (Agilent, Waldbronn, Germany) are suitable.
2. Fused-silica capillaries with the dimensions of 50 μm I.D. (370 μm O.D.) × 34.5 cm (26 cm to the detection window) (see Note 1) (Sino Sumtech, Handan, Hebei, China or Polymicro Technologies, Phoenix, AZ, USA).
3. A commercial ultrasonic bath for degassing of the buffers and some reaction solutions.
4. A commercial pH meter for pH adjustment.
5. 0.45 μm membrane filters.

2.2. Chemicals and Solutions

Prepare all solutions using ultrapure water (synonyms: Milli-Q water or 18 MΩ water) and analytical grade reagents. Prepare and store all reagents at room temperature (unless indicated otherwise). Diligently follow all waste disposal regulations when disposing waste materials.

1. ACE (3.4.15.1, Sigma-Aldrich, St. Louis, MO, USA) solution: dissolve ACE in 10 mM borate–phosphate buffer (pH 8.0) (see Note 2).
2. Hexadimethrine bromide (HDB) (Sigma-Aldrich, St. Louis, MO, USA): dissolve 0.1 g HDB in 100 mL water.
3. Hippuryl-His-Leu (HHL) (Sigma-Aldrich, St. Louis, MO, USA): 500 nM prepared in 10 mM borate–phosphate buffer (pH 8.0), containing 50 mM NaCl (see Note 3).
4. Cilazaprilat (Roche, Basel, Switzerland): 500 nM prepared in 10 mM borate–phosphate buffer (pH 8.0), containing 50 mM NaCl.
5. Captopril: 500 nM prepared in 10 mM borate–phosphate buffer (pH 8.0), containing 50 mM NaCl.
6. Natural extracts: 0.5 mg/mL prepared in 10 mM borate–phosphate buffer (pH 8.0), containing 50 mM NaCl (see Note 4).
7. Running buffer: 20 mM sodium borate (pH 8.0), containing 0.01% HDB (see Notes 5–7).

3. Methods

3.1. Preparation of a HDB-Coated Capillary

Install capillary into the cassette of the CE system. In the present experiment an Agilent ^{3D}CE system using a 50 μm I.D. fused-silica capillary with an effective length of 26 cm and a total length of 34.5 cm was employed.

1. Flush a new capillary with 0.1 M NaOH solution for 30 min. Then flush with deionized water, HDB solution (0.1%, w/v) (see Note 8), and deionized water for 5 min at pressure of 950 mBar, respectively, to create a positively charged coating on the capillary wall.
2. Set the capillary temperature at 37°C (see Note 9).
3. Set UV or diode array detector at 230 nm.

3.2. Preparation of On-Column Immobilized Enzyme Microreactor

1. Charge the enzyme solution into the capillary by pressure at 50 mbar for 10 s producing a 1.5 cm long plug of the enzyme solution.
2. Leaving the enzyme solution inside the capillary for 5 min allows the enzyme to immobilize on the capillary wall via ionic binding (see Notes 8 and 10).
3. Wash out the un-immobilized enzyme off the capillary by flushing with the running buffer for 3 min from the outlet end of the capillary (Fig. 1).

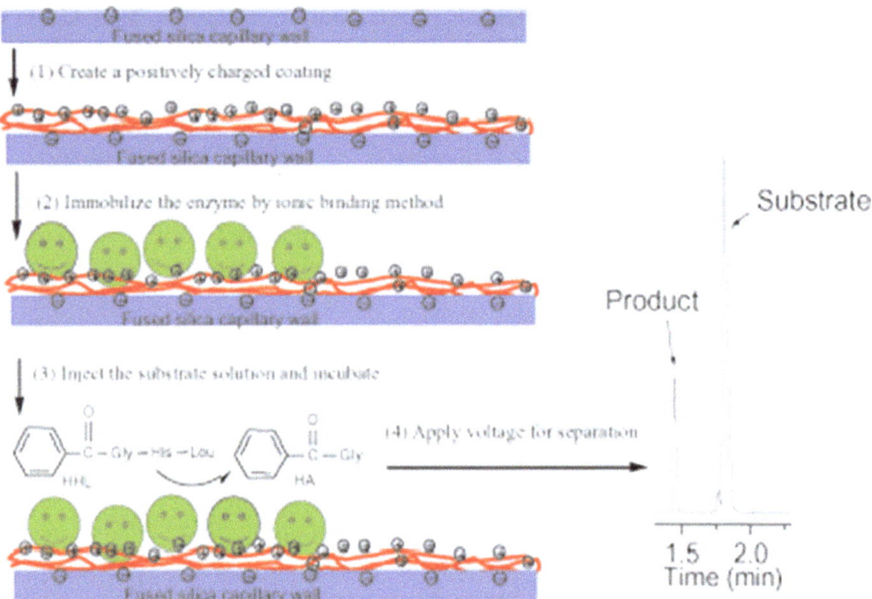

Fig. 1. The schematic representation of the on-column immobilized enzyme microreactor prepared by ionic binding method.

3.3. Enzyme Activity Assay and Inhibition Screening

1. Fill the capillary with the running buffer (pH 8.0), consisting of 20 mM sodium borate and 0.01% HDB.
2. Inject the substrate solutions containing or without containing inhibitor (or candidate compounds) into the enzyme reactor by pressure at 20 mbar for 4 s.
3. Apply voltage of −15 kV to separate the product from the unreacted substrate (see Note 11). Typical electropherograms for inhibitor screening are shown in Fig. 2.

3.4. Regeneration of the Immobilized Enzyme Microreactor

1. Flush the capillary with 1 M NaCl, 0.1 M HCl, and 0.1 M NaOH, consecutively, to elute the immobilized enzyme (desorption) (see Note 12).
2. Flush with deionized water, HDB solution (0.1%, w/v), and deionized water for 5 min, respectively, to create a positively charged coating on the capillary wall.
3. Repeat the preparation protocol for enzyme immobilization.

4. Notes

1. The ends of capillary should be flat.
2. ACE (Sigma-Aldrich, St. Louis, MO, USA) should be stored at −20°C. Therefore, before weighing, it needs to be warmed to room temperature.

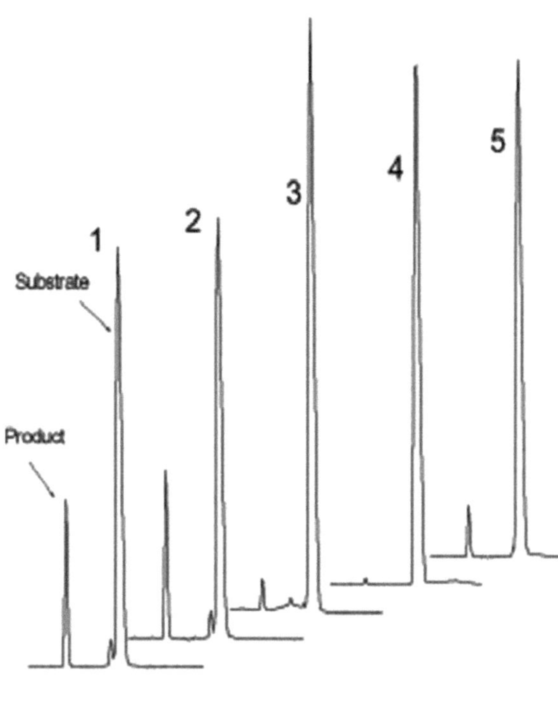

Fig. 2. Typical electropherograms for screening the ACE inhibitors. Conditions: fused-silica capillary, 50 μm I.D., 375 μm O.D. , 34.5 cm (26 cm to detection window); running buffer, 20 mM borate–phosphate buffer (pH 8.0) containing 0.01% HDB; column temperature, 37°C; sample injection, 20 mbar for 4 s; detection wavelength, 230 nm; incubation time, 1 min. Samples: 1 = blank; 2 = extract of Rhizoma coptidis (0.5 mg/mL); 3 = extract of Rhizoma coptidis (0.5 mg/mL) spiked with 38.9 μg of cilazaprilat; 4 = cilazaprilat (500 nM); 5 = captopril (500 nM).

3. ACE is a chloride-ion-dependent enzyme. To achieve maximum enzyme activity, 50 mM NaCl is necessary in the substrate solution.
4. The dissolution can be accelerated by sonication.
5. The running buffer should be prepared daily.
6. 20 mM sodium borate buffer was selected for all experiments because of the sufficient buffer capacity and relatively low Joule heat.
7. The addition of phosphoric acid in borate solution should be performed in small aliquots (e.g., 20 μL) in order to avoid a drop in pH below the intended pH.
8. HDB will adsorb to the capillary wall in a self-assembled way to form a positively charged inner surface to adsorb the studied enzyme. The coating not only acts as the strong anionic

exchanger for enzyme immobilization but also accelerates the migration velocity of the negatively charged analytes. Therefore, very fast separation can be achieved.

9. The thermostat of the cassette should be off after experiment.
10. To achieve the maximum yield of product HA, the incubation time should be over 5 min.
11. The voltage polarity should be reversed because the positively charged coating reversed the EOF.
12. HDB coating is proved to be stable enough to withstand the flush with the running buffer (19–21).

Acknowledgment

The financial support by the National Natural Science Foundations of China (21175146, 20975109, 90713021) and the opening foundations from the National Key Laboratory of Organic Biochemistry are gratefully acknowledged.

References

1. Noble MEM, Endicott JA, Johnson LN (2004) Protein kinase inhibitors: Insights into drug design from structure. Science 303:1800–1805
2. Cushman DW, Ondetti MA (1999) Design of angiotensin converting enzyme inhibitors. Nat Med 5:1110–1112
3. von Ahsen O, Bomer U (2005) High-throughput screening for kinase inhibitors. Chembiochem 6:481–490
4. Wedin R (1999) Taming the monster haystack: the challenge of compound management. Modern Drug Discov 2:61–71
5. Grepin C, Pernelle C (2000) High-throughput screening. Drug Discov Today 5:212–214
6. Nektaria M, Wainer IW (2001) On-line synthesis utilizing immobilized enzyme reactors based upon immobilized dopamine beta-hydroxylase. J Chromatogr B 766:145–151
7. Bartolini M, Andrisano V, Wainer IW (2003) Development and characterization of an immobilized enzyme reactor based on glyceraldehyde-3-phosphate dehydrogenase for on-line enzymatic studies. J Chromatogr A 987:331–340
8. Schriemer DC et al (1998) Micro-scale frontal affinity chromatography with mass spectrometric detection: a new method for the screening of compound libraries. Angew Chem Int Ed 37:3383–3387
9. Kovarik P et al (2005) Capillary-scale frontal affinity chromatography/MALDI tandem mass spectrometry using protein-doped monolithic silica columns. Anal Chem 77:3340–3350
10. Hodgson RJ et al (2005) Inhibitor screening using immobilized enzyme reactor chromatography/mass spectrometry. Anal Chem 77:7512–7519
11. Borch J, Roepstorff P (2004) Screening for enzyme inhibitors by surface plasmon resonance combined with mass spectrometry. Anal Chem 76:5243–5248
12. Cancilla MT et al (2000) Mass spectrometry and immobilized enzymes for the screening of inhibitor libraries. Proc Natl Acad Sci USA 97:12008–12013
13. De Boer AR et al (2004) On-line coupling of high-performance liquid chromatography to a continuous-flow enzyme assay based on electrospray ionization mass spectrometry. Anal Chem 76:3155–3161
14. Lee IH et al. (1998) Picomolar analysis of proteins using electrophoretically mediated microanalysis and capillary electrophoresis with laser-induced fluorescence detection. Anal Chem 70:4546–4548
15. Polakowski R et al (2000) Single molecules of highly purified bacterial alkaline phosphatase

have identical activity. J Am Chem Soc 122:4853–4855

16. Van Dyck S et al (2003) Inhibition study of angiotensin converting enzyme by capillary electrophoresis after enzymatic reaction at capillary inlet. J Chromatogr A 1013: 149–156
17. He Y, Yeung ES (2003) High-throughput screening of kinase inhibitors by multiplex capillary electrophoresis with UV absorption detection. Electrophoresis 24:101–108
18. Tang ZM, Kang JW (2006) Enzyme inhibitor screening by capillary electrophoresis with an on-Column immobilized enzyme microreactor created by an ionic binding technique. Anal Chem 78:2514–2520
19. Kang JW et al (2004) A Mechanistic Study on Enantiomeric Separation with vancomycin and balhimycin as chiral selectors by capillary electrophoresis. Dimerization and enantioselectivity. Anal Chem 76:2387–2392
20. Kang JW et al (1998) Fast chiral separation of amino acid derivatives and acidic drugs by co-electroosmotic flow capillary electrophoresis with vancomycin as chiral selector. J Chromatogr A 825:81–87
21. Kang JW, Wistuba D, Schurig V (2003) Fast enantiomeric separation with vancomycin as chiral additive by co-electroosmotic flow capillary electrophoresis: Increase of the detection sensitivity by the partial filling technique. Electrophoresis 24:2674–2679

Chapter 24

Fluorescent Lipids as Probes for Sphingosine Kinase Activity by Capillary Electrophoresis

Philip M. Yangyuoru, Latanya Hammonds-Odie, and Simon M. Mwongela

Abstract

Capillary electrophoresis (CE) is one among a number of highly sensitive chemical separation techniques used to characterize single or a small number of cells and to develop assays of enzymatic activity. Other commonly used techniques include mass spectrometry and electrochemistry; however, CE using laser-induced fluorescence detection (LIF) is the most sensitive of these techniques. In CE-LIF, fluorescently labeled proteins or lipids are normally separated based on their size to charge ratio in the interior of a small capillary filled with an electrolyte upon the application of an electric field. In this chapter, we describe the application of CE-LIF for the determination of the bioactivity of fluorescently labeled lipids and sphingosine kinase activity.

Key words: Cell signaling, Enzymatic assay, Fluorescent sphingolipids, Sphingosine kinase, Sphingosine-1-phosphate

1. Introduction

Bioactive sphingolipids are under intense investigation due to their involvement in many critical cellular processes. Sphingosine and sphingosine-1-phosphate (S1P) are among sphingolipids that play crucial roles as signal transduction molecules involved in cell survival and migration (1–8). The sphingolipid metabolites, sphingosine and its precursor ceramide, are associated with programmed cell death (apoptosis), while S1P is involved in many critical cellular processes including proliferation, survival and migration, allergic responses, cytoskeletal rearrangements and cell motility, invasion, angiogenesis, vascular maturation, and trafficking of

Nicola Volpi and Francesca Maccari (eds.), *Capillary Electrophoresis of Biomolecules: Methods and Protocols*, Methods in Molecular Biology, vol. 984, DOI 10.1007/978-1-62703-296-4_24, © Springer Science+Business Media, LLC 2013

immune cells (8, 9). S1P, a simple molecule, can play such diverse roles because it functions not only inside cells (9–11) but also as a ligand of cell surface receptors after it is secreted into the extracellular milieu (8, 9). Gene deletion and reverse pharmacological studies have provided evidence that many of the biological effects of S1P are mediated via five specific G protein-coupled receptors, designated as $S1PR_{1-5}$(7).

Cellular levels of S1P are regulated by the concerted action of the enzymes responsible for its formation and degradation (7). There are two mammalian isoforms of the sphingosine kinase (SphK) enzyme (sphingosine kinase 1 (SphK1) and sphingosine kinase 2 (SphK2)). SphK1 is thought to be oncogenic and inhibitors of SphK1 may act as effective chemotherapeutic agents in animal studies (12–15). SphK2 is involved in the immune response and compounds directed at extracellular S1P signaling are showing great promise in clinical trials for treating autoimmune diseases as well as cancer (5, 14, 16, 17). Therefore, sphingosine signaling is emerging as a vital target in biomedical research and clinical medical practice (2, 3, 5–7, 18).

Capillary electrophoresis with laser-induced fluorescence detection (CE-LIF) is a highly sensitive technique that can be used for the study of sphingosine signaling. Newly developed fluorescence-based assays for the determination of sphingosine kinase activity appear promising (19–21). These nonradioactive assays are simpler, faster, and more amenable to high-throughput analyses. In this chapter, we demonstrate that fluorescent lipids can be used as probes for sphingosine kinase enzymatic activity as either an *in vitro* assay or a cell-based assay. To achieve this goal, a CE-based separation of fluorescein-labeled sphingosine (SpFl) and the phosphorylated product fluorescein-labeled sphingosine-1-phosphate (S1PFl) was first developed. The chemical structures of these fluorescent lipids are shown in Fig. 1. This was followed by the application of the CE separation method to measure the activity of purified enzyme. Our results demonstrated the utility of this technique for *in vitro* biochemical determination of kinase and phosphatase activity (22, 23). Furthermore, it was demonstrated that SphFl was readily taken up by mammalian cells without the need for rigorous loading protocols. The technique was then applied to study endogenous SphK activity in erythrocytes using both bulk cell lysates and very small numbers of whole cells. This highly sensitive and quantitative method was fast, simple, and robust for both *in vitro* and cell-based assays. The methodology described should prove to be a valuable analytical tool for quantitative cellular studies of the sphingosine kinases and phosphatases, enzymes of intense interest in both cell biology and clinical medicine.

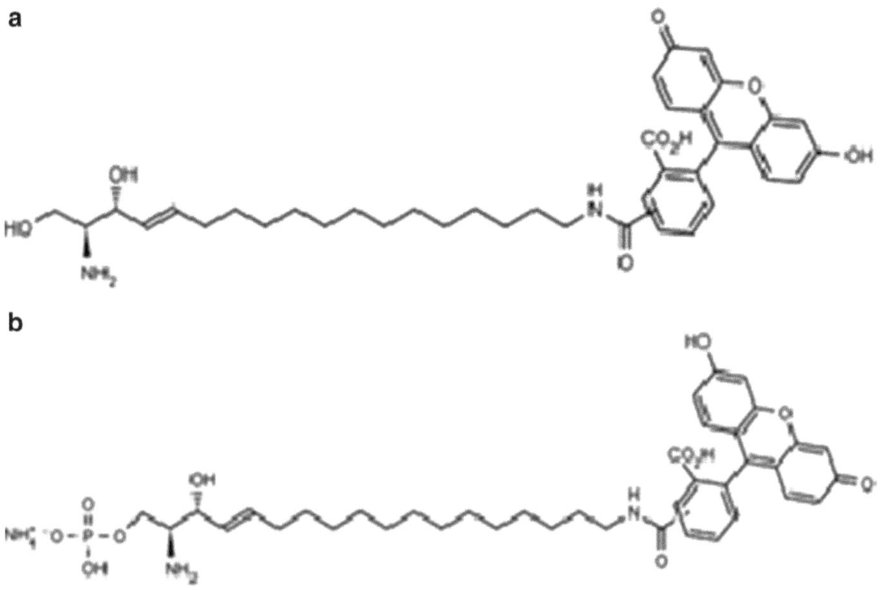

Fig. 1. Chemical structures of (**a**) fluorescein-labeled sphingosine (SpFl) and (**b**) fluorescein-labeled sphingosine 1-phosphate (S1PFl).

2. Materials

1. Sphingosine fluorescein (SpFl) and sphingosine-1-phosphate fluorescein (S1PFl), 5′ and 6′ labeled isomers, and 2-(*p*-hydroxyanilino)-4-(*p*-chlorophenyl) thiazole (2-HCT) (Echelon Biosciences Inc.).
2. Magnesium chloride, sodium chloride, methanol, tri(hydroxymethyl)aminomethane (Tris base), RPMI media, and all other tissue culture reagents (Fisher Scientific).
3. Sodium deoxycholate (SDC), 4-(2-hydroxyethyl) piperazine-1-ethanesulfonic acid (HEPES), 1-propanol, adenosine triphosphate (ATP), and sodium orthovanadate (Na_3VO_4) (Sigma-Aldrich Inc.).
4. *N*, *N*-dimethylsphingosine (DMS) also known as *N*, *N*-dimethyl-D-*erythro*-sphingosine, and phytosphingosine also known as 4-hydroxysphinganine (Avanti Polar Lipids). Dihydrosphingosine or sphinganine (Tocris Bioscience).
5. EOTrol LR (low reverse) polymer solution (Target Discovery). Bodipy-4,4-difluoro-5,7-dimethyl-4-bora-3a,4a-diaza-s-indacene-3-hexadecanoic acid (BODIPY FL C_{16}).

6. The SphK1 and SphK2 used in these studies was a human, recombinant, N-terminal His tagged protein (M.W. = 46.9 kDa) expressed in a baculovirus infected Sf9 cell expression system. The specific activity of the SK1 used in the current study was reported by the supplier as 20 U/mg total protein.
7. Cell line: NIH-3 T3 cells (donation of Dr. Soumitra Basu, Kent State University).
8. Media: DMEM/high glucose medium, 10% fetal bovine serum, 1.0% penicillin/streptomycin.
9. Assay buffer: (50 mM HEPES, 150 mM NaCl, and 5.0 mM $MgCl_2$, 1.0 mM DTT at pH 7.4).
10. The CE separation buffer: 100 mM Tris; 10 mM SDC; 30% 1-propanol; 5% EOTrol™ LR and 2.5 mM $MgCl_2$ at pH 8.5.

3. Methods

3.1. Capillary Electrophoresis

Capillary electrophoresis was performed on a Beckman ProteomeLab™ PA800 CE system (Fullerton, CA). The machine was equipped with LIF detector using an air-cooled 3.5 mW argon laser (Beckman Instruments), with excitation at 488 nm and emission at 520 nm, and was installed with a 32 Karat 8.0 software (Beckman Instruments) for instrument control and data analysis. Separations were performed on 30 cm fused-silica capillaries (Polymicro Technologies, Phoenix, AZ, USA) with 50 μm i.d., 375 μm o.d. (20 cm effective length). New capillaries were preconditioned by successively flushing with 1.0 M NaOH, for 30 min, and rinsed with distilled deionized H_2O, for 20 min at 20 psi. Hydrodynamic injection was carried out at 0.5 psi for 10 s. During the separation the capillary column was thermostated at 25°C and an electric field strength of 833 V/cm was applied in reverse polarity mode (anode at the detector).

3.2. In Vitro Enzymatic Conversion of SpFl to S1PFl (See Note 2)

In vitro study of the conversion of SpFl to S1PFl was carried out by adding 7.20 nM purified SphK2 enzyme to an assay buffer containing 7.43 nM SpFl and 2.10 nM BODIPY FL C_{16} as an internal standard. The reaction mixture was separated under reversed polarity conditions. The more negatively charged S1PFl migrated faster towards the positive end of the capillary (detector), eluting first, followed by BODIPY FL C_{16}, and then SpFl (Fig. 2). The SpFl and S1PFl peaks were well resolved and consistent with the CZE mechanism of separation. The two peaks observed for SpFl were a result of the separation of the 5′ and 6′ sphingosine-labeled isomers. Mg^{2+} probably enhances the separation of the split SpFl peaks to near baseline. However, for the quantitative determination of

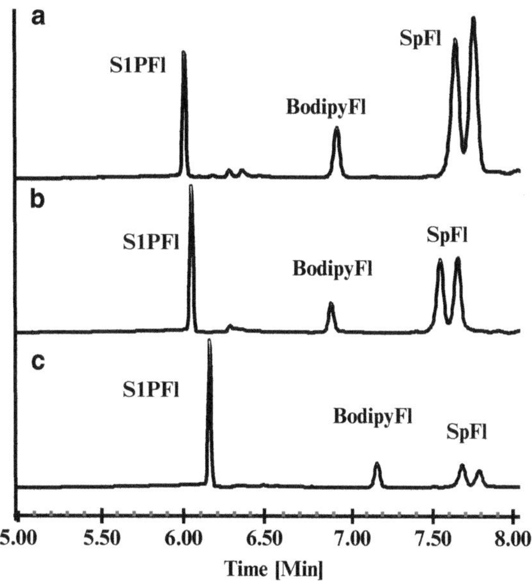

Fig. 2. Electropherogram of separation of SpFl, BODIPY FL, and S1PFl after addition of 7.20 nM SphK2 to 7.43 nM SpFl and 2.10 nM BODIPY FL, 0.25 mM ATP, 0.10 μM Na_3OV_4 in 300 μL assay buffer followed by (**a**) 5 min, (**b**) 30 min, and (**c**) 60 min reaction over a water bath at 37°C. Separation buffer pH 8.50, made of 100 mM Tris, 30% 1-propanol, 5% EOTrol™ LR, and 10.0 mM SDC.

SphK2 activity, experiments were performed using the optimized separation buffer, i.e., 30% 1-propanol, 10 mM SDC which was more reproducible and robust (Fig. 2).

The rate of phosphorylation of SpFl by SphK2 was determined by CE-LIF analysis of aliquots of the enzymatic reaction at selected reaction time intervals. Using BODIPY FL C_{16} (1.75 nM) as internal standard, the ratios of S1PFl, SpFl, and BODIPY FL C_{16} peak areas were calculated. The percentage of phosphorylation was calculated by dividing the average ratio of S1PFl and BODIPY FL C_{16} peak area by the sum of ratios of S1PFl, SpFl, and BODIPY FL C_{16} internal standard peak areas. The activity of SphK2 was determined by the % of SpFl phosphorylated with time. Under these conditions, after one minute of reaction, the SpFl substrate was 50% phosphorylated and gradually increased to 90% after one hour of reaction (Fig. 3). Subsequent experiments were carried out under different substrate, enzyme, and ATP concentrations (Fig. 3).

In addition, *in vitro* studies using purified SphK2 enzyme were carried out to determine the Michaelis–Menten kinetics constants K_M and V_{max} and the turn over number k_{cat} (see Note 3). The initial rates of conversion of SpFl substrate into the product S1PFl were measured at varying substrate concentrations (Fig. 4). Based on the conditions used in this assay, the K_M for SphK2 was determined to be 2.8 ± 0.8 μM with a V_{max} of 2,490 ± 520 μM/min and k_{cat} of

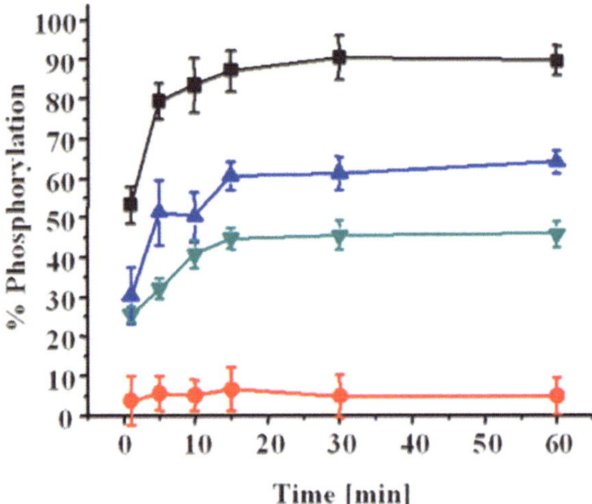

Fig. 3. Percent phosphorylation of SpFl with time in 300 μL assay buffer containing 0.10 μM Na_3OV_4, 1.75 nM BODIPY FL C_{16}, and 12.4 nM SpFl and the following: 14.4 nM SphK2, 0.50 mM ATP (*filled square*); 7.2 nM SphK2, 0.50 mM ATP (*blue-filled triangle*); 14.4 nM SphK2, 0.25 mM ATP (*green inverted filled triangle*); 14.4 nM SphK2, no ATP (*red-filled circle*). The experiment was run nine times and the data collected were averaged. The error bars represent the relative standard deviation.

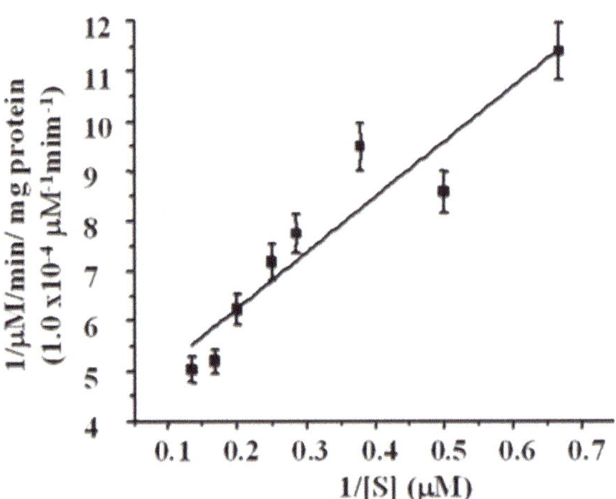

Fig. 4. Lineweaver–Burk plot of initial rate verses SpFl substrate concentration using 21.6 nM of SphK2, 0.10 μM Na_3OV_4, 0.50 mM ATP, and BODIPY FL C_{16} (1.75 nM) used as internal standard in 300 μL assay buffer. The experiment was run nine times and the data averaged. The error bars represent the relative standard deviation.

1,920 ± 402 s^{-1}. The value of K_M observed is slightly higher than the (0.6–1.3 μM) physiological range for sphingosine with SphK (24). This difference is probably due to fluorophore labeling of the substrate. The determined k_{cat} value is also higher than k_{cat} value of 1,265 s^{-1} reported elsewhere for unlabeled sphingosine (25).

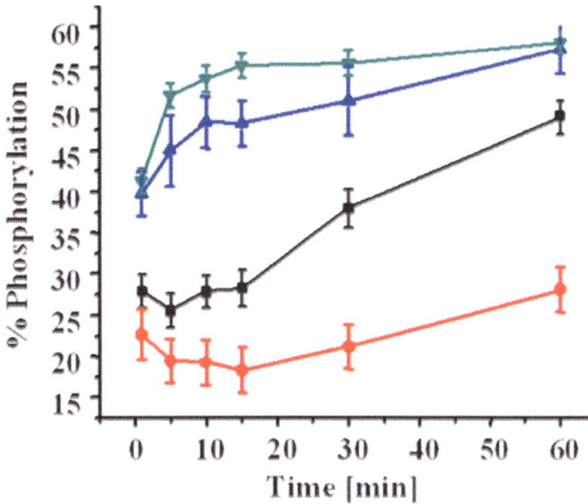

Fig. 5. A plot showing the SphK2 catalyzed phosphorylation of SpFl in the presence of SphK2 (no added inhibitors) (*blue-filled triangle*) and in the presence of inhibitors 0.26 mM phytosphingosine (*filled square*), 0.25 mM *N*, *N*-dimethylsphingosine (DMS) (*green inverted filled triangle*), and 0.23 mM dihydrosphingosine (*red-filled circle*). 1 μL of stock SphK2 (7.2 nM), 0.10 μM Na_3OV_4, and 0.25 mM ATP were added to 300 μL assay buffer. SpFl substrate concentration was 12.4 nM. The experiment was run nine times and the data collected were averaged. The error bars represent the relative standard deviation.

3.3. SphK2 Inhibition In Vitro

The inhibition SphK2 was investigated using phytosphingosine and dihydrosphingosine. Also *N,N*-dimethylsphingosine (DMS) which is known to promote SphK activity was also tested for comparison. These three components were added separately to the *in vitro* reaction mixture just before the addition of the SpFl substrate. From these studies dihydrosphingosine was shown to be a better inhibitor of SphK2 than phytosphingosine (Fig. 5). The addition of dihydrosphingosine resulted in a modest inhibition of the initial rate of phosphorylation of SpFl to (~25%) at the 1 min time point followed by a further drop to ~20% for the 5, 10, and 15 min reaction time experiments, and a slight increase in the rate of phosphorylation to ~ 28% after 1 h reaction. Phytosphingosine resulted in ~ 28% initial rate of phosphorylation of SpFl at the 1 min time point followed by a further drop to ~25% for the 5, 10, and 15 min reaction time experiments, and a reasonable gradual increase in the rate of phosphorylation to ~ 50% after 1 h. In contrast, DMS activated SphK2. As expected, the addition of DMS to the reaction mixture resulted in higher % phosphorylation of SpFl at any given reaction time than is the case of the SphK2 alone (Fig. 5). Our data is in agreement with some of the published studies which monitored the activity of SphK2 using these three inhibitors and D-*erythro*-sphingosine as substrate. The studies by

Liu et al. showed that D-*erythro*-dihydrosphingosine was a better substrate for SphK2 than D-*erythro*-sphingosine (26). Our studies also show that dihydrosphingosine is a better substrate for SphK2 than SpFl and acts as a competitive substrate when added together with SpFl leading to a lower rate of phosphorylation of SpFl. In addition, studies by Liu et al. showed that phytosphingosine is a substrate of SphK2 but it is not as a good a substrate as D-*erythro*-sphingosine.

Based on our observation we can possibly conclude that phytosphingosine is a less competitive inhibitor than dihydrosphingosine; thus, only slightly inhibits the rate of phosphorylation of SpFl. Based on the work by Liu et al., DMS is not a substrate for SphK2 but a substrate for SphK1. Lee et al., concluded that DMS was an inhibitor of SphK1 when SpFl is used as the substrate (22). In this study, we also investigated the effect of DMS on the rate of phosphorylation of SpFl using SphK2. Our studies showed that DMS activates the activity of SphK2 leading to a higher rate of phosphorylation of SpFl. Several other studies have revealed the potency of DMS in inhibiting SphKs activity and in lowering S1P levels (27–30). DMS has been reported to be an inhibitor of both SphK1 and SphK2; however, the data for SphK2 inhibition were obtained with Chinese hamster ovary (CHO-K1) cells expressed enzyme (31). Recently, Vessey et al. reported the rapid and efficient separation of SphK1 and SphK2 from rat and mouse heart cytosol. This study reported that DMS and the immunomodulator, FTY720, inhibit only native SphK1, whereas they activate SphK2 (32).

3.4. Inhibition of SphK(s) in Cells

The determination of fluorescent lipid activity and SphK inhibitors was also investigated using bulk cell lysates (see Notes 4–6). CE analysis of the NIH-3 T3 cell lysate shows an initial 70% phosphorylation of SpFl which gradually decreased to 50% in 30 min (Fig. 6). These results illustrate the interplay of SphK and phosphatase enzymes. The initial phosphorylation of the SpFl substrate is due to the enzyme SphK. However, the phosphatase also present in the cell acts on the S1PFl formed, converting it back to SpFl. Analysis of the media alone collected after incubation with the SpFl substrate did not contain detectible levels of S1PFl but the presence of SpFl, implying that the SpFl diffuses through the cells, is phosphorylated, and remains in the cell. The inhibition studies with the cell lysate indicate that % phosphorylation of SpFl was reduced 15 and 20% with dihydrosphingosine and DMS, respectively. These observations are contrary to the results obtained with purified SphK2 where DMS rather increased the % phosphorylation by probably activating SphK2. DMS has been shown to be a predominant inhibitor of SphK1 (26, 32). Low levels of S1P might be an indication that SphK1 instead of SphK2 is implicated in the cell lysate studies. Similar results were observed when MDA-MB-231 cells were loaded with SpFl (data not shown).

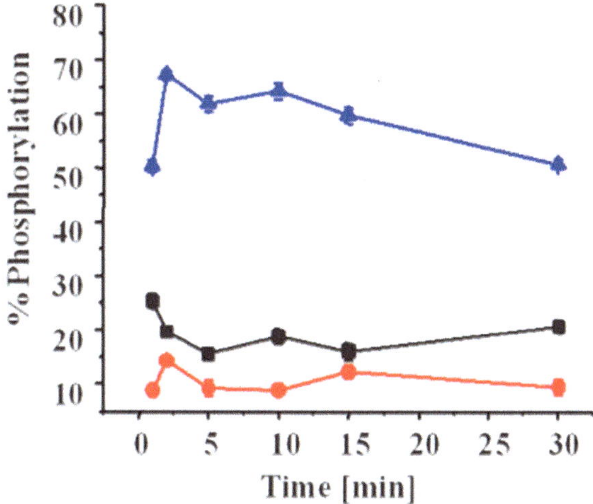

Fig. 6. Time course CE analysis of the inhibition of SpFl phosphorylation in NIH-3 T3 cells incubated with 37.2 nM of SpFl substrate showing (*blue-filled triangle*) no inhibitor added, (*filled square*) 19.0 μM DMS, and (*red-filled circle*) addition of 17.0 μM dihydrosphingosine. The error bars represent the relative standard deviation.

4. Notes

1. *Sample Preparation.* The fluorescent lipids stock solutions were prepared in methanol and stored in –20°C. Due to the lipids poor solubility and ability to adsorb to storage and analysis vials, fresh aliquots of the stock were prepared for each day's experiment prior to analysis or loading into cells with minimal transfers and pipetting to prevent loss of sample (22).
2. *In Vitro Phosphorylation Studies.* The rate of phosphorylation of SpFl by SphK2 was monitored by adding 2.0 μL of stock SphK2 (14.4 nM), unless stated otherwise, to a vial containing 12.4 nM SpFl substrate, 0.50 mM ATP, and 0.10 μM Na_3OV_4 in 300 μL total volume of the assay buffer. BODIPY FL C_{16} (1.75 nM) was used as internal standard. The reaction mixture was placed in a 37°C water bath. At fixed reaction time intervals, the enzymatic reaction was stopped by taking 50.0 μL aliquots of the reaction mixture and mixing with 50.0 μL solution of 5.0 mM Tris and 0.10% triton X-100 followed by CE-LIF analysis.
3. *Michaelis–Menten Kinetics.* The K_M and V_{max} for sphingosine kinase were obtained by plotting a curve of the average of three determinations of SpFl concentration *vs.* the initial reaction velocity. The SpFl concentrations used were 2.5, 5, 10, 15, 20, 30, 40, and 60 μM, using 2% fatty-acid-free bovine serum albumin as a carrier for SpFl, while maintaining the concentration of

SphK2 constant. The K_M and V_{max} were determined assuming Michaelis–Menten kinetics and fitting the data points in Origin (OriginLab Corporation, Northampton, MA).

4. *Cell Culture.* NIH-3T3 and MDA-MB-231 cell lines were grown in a DMEM/high glucose medium (HyClone Laboratories, Inc., UT) to which 10% fetal bovine serum and 1.0% penicillin/streptomycin were added. The cells were maintained at 37°C and 5.0% CO_2 atmosphere. The cells were passed by gently aspirating the media and adding 2.0 mL of trypsin for 2 minutes to dislodge them. Medium was added (20.0 mL) to neutralize the trypsin and 2.0 mL of cells was added to 10 cm plates and incubated overnight.
5. *Cell Loading with SpFl.* After 16–24 h, cells in the above mentioned 10 cm plates were loaded with 1.0 μ L of the stock SpFl solution (74.5 μM in methanol) and incubated for fixed time intervals. The effective concentration of SpFl in each 2.0 mL cell volume was 37.2 nM. To ensure even distribution of SpFl and minimal effect of methanol on the cells, 1 ml of the 2 ml media in the cell plates was pipetted and mixed with the 1 μL of SpFl and then dispensed back to the cell plate.
6. *Cell Loading with Inhibitors.* For some of the experiments, a SphK enzyme inhibitor was added into the cells in the 10 cm plate before the addition of SpFl substrate. Two such inhibitors were investigated in the cell lysate study: DMS and dihydrosphingosine. For the DMS, 5.0 μ L of the stock (7.63 mM in methanol) was loaded into 2 mL cell volume (19.0 μM final) while 10.0 μ L of dihydrosphingosine stock (3.42 mM in methanol) was added into 2.0 mL cell volume to make a final concentration of 17.0 μ M. The inhibitors were introduced into the cells in the same manner as the SpFl to allow for even distribution and minimal effect of methanol on the cells (see Note 5). The cells in the plates were then incubated with or without the inhibitors for different time periods; 0, 5, 10, 15, 30, and 60 min. After incubation, the media was aspirated from the plates and the plates rinsed twice with 1.0 mL PBS and lysed with the Tris-Triton X-100 buffer, centrifuged, and the supernatant frozen in dry ice, before proceeding with the CE-LIF analysis.

Acknowledgment

This research study was supported by the Kent State University research startup fund and the Farris Innovation award.

References

1. Peyruchaud O (2009) Novel implications for lysophospholipids, lysophosphatidic acid and sphingosine 1-phosphate, as drug targets in cancer. Anti-Cancer Agents Med Chem 9:381–391
2. Okada T, Kajimoto T, Jahangeer S et al (2009) Sphingosine kinase/sphingosine 1-phosphate signalling in central nervous system. Cell Signal 21:7–13
3. Gault CR, Obeid LM (2011) Still benched on its way to the bedside: sphingosine kinase 1 as an emerging target in cancer chemotherapy. Crit Rev Biochem Mol Biol 46:342–351
4. Ponnusamy S, Meyers-Needham M, Senkal CE et al (2010) Sphingolipids and cancer: ceramide and sphingosine-1-phosphate in the regulation of cell death and drug resistance. Future Oncol 10:1603–1624
5. Antoon JW, White MD, Slaughter EM, Driver JL, Khalili HS, Elliott S, Smith CD, Burow ME, Beckman BS (2011) Targeting NFκB mediated breast cancer chemoresistance through selective inhibition of sphingosine kinase-2. Cancer Biol Ther 11:678–689
6. Oskouian B et al (2010) Cancer treatment strategies targeting sphingolipid metabolism. Adv Exp Med Biol 688:185–205
7. Takabe K, Paugh SW, Milstien S et al (2008) "Inside-out" signaling of sphingosine-1-phosphate: therapeutic targets. Pharmacol Rev 60:181–195
8. Spiegel S, Milstien S (2003) Sphingosine-1-phosphate: an enigmatic signalling lipid. Nature Rev Mol Cell Biol 4:397–407
9. Alvarez SE, Milstien S, Spiegel S (2007) Autocrine and paracrine roles of sphingosine-1-phosphate. Trends Endocrinol Metab 18:300–307
10. Olivera A, Spiegel S (2001) Sphingosine kinase: a mediator of vital cellular functions. Prostaglandins Other Lipid Mediat 64:123–134
11. Kohno M, Momoi M, Oo ML et al (2006) Intracellular role for sphingosine kinase 1 in intestinal adenoma cell proliferation. Mol Cell Biol 26:7211–7223
12. Xia P, Gamble JR, Wang L et al (2000) An oncogenic role of sphingosine kinase. Curr Biol 10:1527–1530
13. French KJ, Schrecengost RS, Lee BD et al (2003) Discovery and Evaluation of Inhibitors of Human Sphingosine Kinase. Cancer Res 63:5962–5969
14. Maceyka M, Sankala H, Hait NC et al (2005) SphK1 and SphK2, Sphingosine Kinase Isoenzymes with Opposing Functions in Sphingolipid Metabolism. J Biol Chem 280:37118–37129
15. Milstien S, Spiegel S (2006) Targeting sphingosine-1-phosphate: a novel avenue for cancer therapeutics. Cancer Cell 9:148–150
16. Sensken SC (2010) B. C., Nagarajan M, Peest U, Pabst O, Gräler MH. (2010) Redistribution of sphingosine 1-phosphate by sphingosine kinase 2 contributes to lymphopenia. J Immunol 184:4133–4142
17. Lai WQ, Irwan AW, Goh HH, Melendez AJ, McInnes IB, Leung BP (2009) Distinct roles of sphingosine kinase 1 and 2 in murine collagen-induced arthritis. J Immunol 183:2097–2103
18. Ponnusamy S, Meyers-Needham M, Senkal CE, Saddoughi SA, Sentelle D, Selvam SP, Salas A, Ogretmen B (2010) Sphingolipids and cancer: ceramide and sphingosine-1-phosphate in the regulation of cell death and drug resistance. Future Oncol 10:1603–1624
19. Billich A, Ettmayer P (2004) Fluorescence-based assay of sphingosine kinases. Anal Biochem 326:114–119
20. Roberts JL, Moretti PAB, Darrow AL et al (2004) An assay for sphingosine kinase activity using biotinylated sphingosine and streptavidin-coated membranes. Anal Biochem 331:122–129
21. Itatani J, Sportsman R, Boge A (2006) In http://echelon-inc.com/corp/IMAP_TR-FRETSphingosine.pdf.
22. Lee KJ, Mwongela SM, Kottegoda S et al (2008) Determination of Sphingosine Kinase Activity for Cellular Signaling Studies. Anal Chem 80:1620–1627
23. Yangyuoru PM, Otieno AC, Mwongela SM (2011) Determination of sphingosine kinase 2 activity using fluorescent sphingosine by capillary electrophoresis. Electrophoresis 32:1742–1749
24. Vessey DA, Kelley M, Karliner JS (2005) A rapid radioassay for sphingosine kinase. Anal Biochem 337:136–137
25. Pitson SM, Moretti PAB, Zebol JR, Lynn HE, Xia P, Vadas MA, Wattenberg BW (2003) Activation of sphingosine kinase 1 by ERK1/2-mediated phosphorylation. EMBO J 22:5491–5500
26. Liu H, Sugiura M, Nava VE et al (2000) Molecular cloning and functional characterization of a novel mammalian sphingosine kinase type 2 isoform. J Biol Chem 275:19513–19520
27. Cuvillier O (2002) Sphingosine in apoptosis signaling, *Biochim. Biophys. Acta, Mol. Cell Biol.* Lipids 1585:153–162

28. Buehrer BM, Bell RM (1992) Inhibition of sphingosine kinase in vitro and in platelets. Implications for signal transduction pathways. J Biol Chem 267:3154–3159
29. Yatomi Y, Ruan F, Megidish T et al (1996) N, N-dimethylsphingosine inhibition of sphingosine kinase and sphingosine 1-phosphate activity in human platelets. Biochemistry (Mosc) 35:626–633
30. Pitson SM, D'Andrea RJ, Vandeleur L et al (2000) Human sphingosine kinase: purification, molecular cloning and characterization of the native and recombinant enzymes. Biochem J 350:429–441
31. Kim J-W, Kim Y-W, Inagaki Y et al (2005) Synthesis and evaluation of sphingoid analogs as inhibitors of sphingosine kinases. Bioorg Med Chem 13:3475–3485
32. Vessey DA, Kelley M, Zhang J et al (2007) Dimethylsphingosine and FTY720 inhibit the SK1 form but activate the SK2 form of sphingosine kinase from rat heart. J Biochem Mol Toxicol 21:273–279

Chapter 25

Assessment of DNA Damage by Micellar Electrokinetic Chromatography

Paolo Fattorini, Giorgio Marrubini, Pierangela Grignani, Solange Sorçaburu-Cigliero, and Carlo Previderé

Abstract

A simple and inexpensive MEKC method, which is able to assess base damage within DNA samples, is illustrated. After heat-acid hydrolysis of the DNA samples, both the percentage of each canonical DNA base and the relative amount of uncanonical DNA bases can be measured. This method is useful for an evaluation of the integrity of PCR templates used in several fields of investigation.

Key words: DNA damage, Ancient DNA, CE, MEKC, PCR fidelity

1. Introduction

DNA samples recovered from *post mortem* specimens (e.g., ancient, forensic, and formaldehyde-fixed paraffin-embedded samples) usually exhibit molecular damage (1–3) which can be studied by enzymatic, chromatographic, electrophoretic, and hybridization-based methods (4–9). Chemical modification of the DNA bases was originally investigated using MS-based chromatographic techniques (3, 6–9). However, since these methods are laborious and expensive, they are not suitable as routine tests in casework.

In the last decade, CE-based techniques allowed to study the chemistry of DNA in several fields both of fundamental (10–16) and applied research (17–21). Recently, we developed a MEKC (Micellar Electrokinetic Chromatography) (22) protocol to study the integrity of the primary structure of DNA in human samples (23, 24). The protocol described here allows the assessment of percentage of the four canonical bases still present on the backbone

Nicola Volpi and Francesca Maccari (eds.), *Capillary Electrophoresis of Biomolecules: Methods and Protocols*, Methods in Molecular Biology, vol. 984, DOI 10.1007/978-1-62703-296-4_25, © Springer Science+Business Media, LLC 2013

of the DNA molecules. In addition, the high selectivity afforded by MEKC allows also the identification of uncanonical DNA bases resulting from the catabolism of the purinic/pyrimidinic moieties.

As described in details, our protocol optimizes the MEKC analytical conditions which easily accomplish the separation of the basic moieties obtained from the heat-acid hydrolysis of the DNA samples.

The method provides an inexpensive way to assess the basic damage of the genetic materials usually employed in different fields of investigation.

2. Materials

Several chemicals and materials are requested for studying the primary structure of the DNA by CE. Prepare and store all solutions under the indicated conditions. Follow all disposal regulations when disposing waste material.

2.1. Chemicals

Cytosine (C), Guanine (G), Thymine (T), Adenine (A), Uracil (U), 5-methyl-Cytosine (5-met-C), Xanthyne (X), Hypoxanthyne (H), and 5-Bromo-Uracil (5-Br-U) from Sigma (Sigma-Aldrich, MO, USA) are certified with a 98% minimum of purity and are easily commercially available. The other analytical-grade reagents needed are: formic acid (98–100%), HCl (37%, w/v), NaOH pearls, sodium tetraborate anhydrous, sodium phosphate monobasic, sodium dodecyl sulfate (SDS), and ultrapure water (see Note 1).

2.2. Standard Solutions

Three standard solutions are required for accurate experiments. These solutions are: (a) the Master Standard Mix (MSM); (b) the Test Standard Solution (TSS); (c) the Four Modified Bases Standard (4MBS). The first two are employed to calibrate the CE apparatus. The third is helpful for the qualitative analysis of aged/damaged DNA samples. In addition, an internal standard (5-Br-U) is also employed in the CE analysis of DNA samples.

2.2.1. Master Standard Mix

1. MSM is composed of C, G, T, A, U, 5-met-C, X, H, and 5-Br-U and it is used as qualitative standard (see Note 2).
2. Weight each substance separately to obtain 100 mL of 500 μM of each of the nine components.
3. Dissolve the substances in 80 mL of 1% HCl (w/v) by stirring at room temperature in the dark overnight.
4. Add 1% HCl to the final volume of 100 mL.

5. Dispense in aliquots and store them at −80°C in the dark until use.
6. By appropriate dilution with 1% HCl prepare a working solution at final concentration of 100 μM. Store the solution at −20°C (see Note 3).

2.2.2. Test Standard Solution

1. TSS is composed of the four DNA bases (C, G, T, and A) and is employed as standard for quantitative analysis.
2. Accurately weight each substance separately to obtain 100 mL of 500 μM of each of the four components.
3. Dissolve the substances in 80 mL of 1% HCl (w/v) by stirring at room temperature in the dark overnight.
4. Add 1% HCl to the final volume of 100 mL.
5. Dispense in aliquots and store them at −80°C in the dark until use.
6. By appropriate dilution with 1% HCl prepare the following set of TSSs: 100, 50, 25, 12.5, 6.2, 3.1, and 1.5 μM. Store the solutions at −20°C (see Note 4).

2.2.3. Internal Standard (5-Br-Uracil)

1. 5-Br-Uracil (5-Br-U) is employed as internal standard for assessing the consistency of injection of the system in the analysis of the DNA samples.
2. Prepare a 20 mL solution of 1% HCl and 500 μM 5-Br-U.
3. Dissolve the substance by stirring at room temperature in the dark overnight.
4. Dispense in aliquots and store them at −80°C in the dark until use.
5. By appropriate dilution with 1% HCl (w/v) prepare the final working solution (200 μM). Store the solution at −20°C (see Note 4).

2.2.4. Four Modified Bases Standard

1. The 4MBS is composed of U, X, H, and 5-met-C and it is employed for the qualitative analysis of aged/damaged DNA samples (see Note 2).
2. Proceed from item 2 of Subheading 2.2.1 to obtain a final working solution at final concentration of 250 μM. Store the solution at −20°C (see Note 3).

2.3. CE Running Buffer

1. Dissolve in 40 mL of water, 0.503 g of sodium tetraborate anhydrous, and 0.599 g of sodium phosphate monobasic.
2. Stir the solution for about 10 min at room temperature; add 4.326 g of SDS.
3. Put the solution at 37°C without stirring to avoid the formation of air bubbles (see Note 5).

Table 1
The indicated amounts of H_2O and NaOH (5.0 N) are added to 50 mL of the original buffer (pH 7.33) to obtain the buffers A, B, C, and D

Buffer (50 mL)	H_2O (μL)	NaOH (μL)	pH	Current (μA)
A	300	100	7.65	≈213
B	200	200	7.94	≈218
C	100	300	8.18	≈223
D	–	400	8.43	≈230

These buffers, run at 20 kV, generate the currents reported in the last column (data monitored at 2 min of the run). Figure 1 illustrates how the retention times of the analytes are affected by the buffer pH.

4. Add water to the final volume of 50 mL. Therefore, this solution contains 0.05 M sodium tetraborate, 0.1 M sodium phosphate monobasic, and 0.3 M SDS.
5. Filter the buffer through a 0.2 μm nylon GD/X Syringe Filter (Whatman International, UK). This solution has a pH of about 7.33.
6. Adjust the pH by the addition of 200 μL of 5.0 N NaOH (see Note 6 and Table 1).
7. Store the buffer at room temperature until use (see Note 7).

3. Methods

The CE analysis of the DNA samples is performable by a MDQ (Beckman Coulter, CA, USA) apparatus but similar CE systems, produced by different commercial companies, are successfully employed as well.

3.1. Analytical Conditions

To calibrate the CE apparatus, several steps are required. The first is based on the MSM to optimize both the relative and the absolute retention times of the analytes. The second is based on the TSSs and assesses the precision of the measurements and the sensitivity attainable (Limit of Detection, LOD, and Limit of Quantification, LOQ). At least, also the constancy of the injection needs to be checked in the CE analysis of DNA samples. This is done by spiking the samples with a fixed amount of standard 5-Br-U.

3.1.1. Evaluation of the Retention Times

1. Switch on the instrument and put into the "sample in" reservoir plate the vials containing fresh regenerating buffer (0.1 N NaOH), CE running buffer (see Subheading 2.3), and water. A microvial (Beckman Coulter, cat. no. 144709), containing 100 μL of MSM, will be put also into the "sample in" reservoir plate. Fresh running buffer has to be put into the "sample out" reservoir plate (see Note 8).
2. Set the working temperature at 35°C.
3. Set the filters reel at the detection of 254 nm.
4. Insert a suitable cartridge containing an uncoated fused silica capillary (length: 57 cm; external diameter: 375 μm; internal diameter: 50 μm; window at 50 cm) (see Note 9).
5. Wash the capillary with water for 10 min at 20 psi.
6. Wash the capillary with 0.1 N NaOH for 3 min at 20 psi.
7. Equilibrate the capillary with the running buffer for 3 min at 20 psi.
8. Inject the sample at 0.5 psi for 15 s (corresponding to an injection volume of 18 nL).
9. Separate the sample at 20 kV for 35 min (this generates a current of about 218 μA) (see Note 6 and Table 1).
10. After the run, wash immediately the capillary with water for 5 min at 20 psi.
11. Perform a new run starting from step 6.
12. Analyze the raw data by the Beckman Karat (version 5.0) software after 8 min of run (see Note 10).

3.1.2. Assessment of the System Precision, LOD, and LOQ

1. After optimization of the retention times of the analytes, both the precision of their detection and other analytical parameters need to be assessed. To this aim, perform programmed runs of the TSSs (from 100 to 12.5 μM). Put 100 μL of each TSS into a microvial. A minimum number of three replicated runs for each set of dilutions is recommended (see Note 11). These data are collected to assess the system precision in terms of repeatability of the injection and detection system. In order to assess the repeatability of the analytical procedure a minimum number of three TSSs (series of dilution ranging from 100 to 12.5 μM) should be prepared independently one from the other and be processed. These data allow to assess the within-session precision. The intermediate precision (or inter-session precision) is assessed by repeating the analyses of at least three TSSs (series of dilution ranging from 100 to 12.5 μM)

produced independently one from the other over three separate analytical sessions.

2. Wash the capillary with water for 3 min at 20 psi.
3. Proceed from step 6 of Subheading 3.1.1. Since the retention times of the last eluter (Adenine) is <16 min, the run can be shortened at 18 min.
4. Process the data according to established data treatment procedures (see Note 12). In particular, calculate mean ± standard deviation of the areas of each peak obtained from the replicate analyses. For every analyte, report the relative standard deviation % (standard deviation to mean ratio) or CV% which represents the measurement of the precision attained (system repeatability, within-session precision, intermediate precision).

3.1.3. Assessment of the Constancy of the Injection

1. Put 100 μL of 5-Br-U into a microvial.
2. Wash the capillary with water for 3 min at 20 psi and then proceed from step 6 of Subheading 3.1.1 by performing at least three replicate runs. Since the retention times of the 5-Br-U is about 20 min, the run can be shortened to 25 min.
3. Process the data following standard statistical analysis (see Note 13).

3.2. DNA Hydrolysis and CE Analysis of Samples

1. Prepare, in a screw-capped glass vial (Specivial, PBI International, cat. no. 2929), 500 μL of 90% (v/v) formic acid containing 1–3 μg of native DNA (see Note 14). Prepare at least one blank reaction control in each working session (see Note 15).
2. Incubate the samples at 170°C for 30 min in a convective oven (see Note 16). Remove the tubes and leave them cooling at room temperature.
3. Transfer the solutions in Eppendorf tubes.
4. Lyophilize the samples and store them sealed at room temperature in the dark until use (see Note 17).
5. For CE analysis, resuspend the samples in 15–25 μL of the internal standard 5-Br-U by vigorous shaking and incubation at 37°C for 20–30 min. Collect the samples by a short centrifugation at maximum speed. Transfer the samples into the microtubes avoiding the formation of air bubbles.
6. For CE analysis, wash the capillary for 3 min with water at 20 psi and then proceed from step 6 of Subheading 3.1.1 (see Note 18).
7. After CE analysis, the samples can be stored at −20°C for further tests.

4. Notes

1. We observed that purity of the water can influence the retention times of the analytes. For this reason, the chemical–physical properties of the water need to be monitored before buffer preparation.
2. Regrettably, only C, G, T, A, U, 5-met-C, X, H, and 5-Br-U are commercially available. It is clear, however, that other base-analogs can be synthesized in the own Laboratory.
3. MSM and 4MBS, being qualitative standard, can be stored at 4°C for 2–3 weeks.
4. Since repeated freeze/thawing cycles cause degradation of the standard, TSSs and 5-Br-U are recommended to be stored at −20°C in small volume (100 μL) until use.
5. The best way to dissolve SDS is by overnight incubation in a water bath or, alternatively, in a convective oven with occasional and gentle stirring.
6. The retention times of the analytes are strictly related to the pH of the running buffer (see Fig. 1). The optimized working buffer needs the addition of 200 μL of 5.0 N NaOH, which gives a pH of about 7.9. Note also that the current increases with the increment of the pH (see Table 1).
7. The buffer is stable at room temperature for 2–3 weeks. Note that SDS flocculates at temperatures <18°C. If this occurred, warm the buffer at 37°C and filter it again. To prevent flocculation, store, if necessary, the buffer in an incubator at 20–37°C.
8. It is preferable that two vials, each containing running buffer, are placed into the "sample in" reservoir plate. The first will be used only for equilibrating the capillary while the second will be used only to generate current with the running buffer of the "sample out" reservoir plate.
9. The capillary can be constructed by employing a 5 m long capillary roll (Beckman Coulter, cat. no. 338472). In this case, follow the instructions provided taking care that the ends of capillary, being sharp, can be dangerous for your eyes. Therefore, wear always glasses. The window can be constructed by the means of a moderate flame (e.g., lighter) followed by accurate washing with absolute methanol. Check its clearness by a lent. Home-made capillaries, as well as each new capillary, need to be washed with 1% HCl for 10 min and water for 5 min before their routine use. Note that different capillaries, even if from the same roll, can exhibit reduced/increased UV sensitivity. This, rather than to the clearness of the window, is

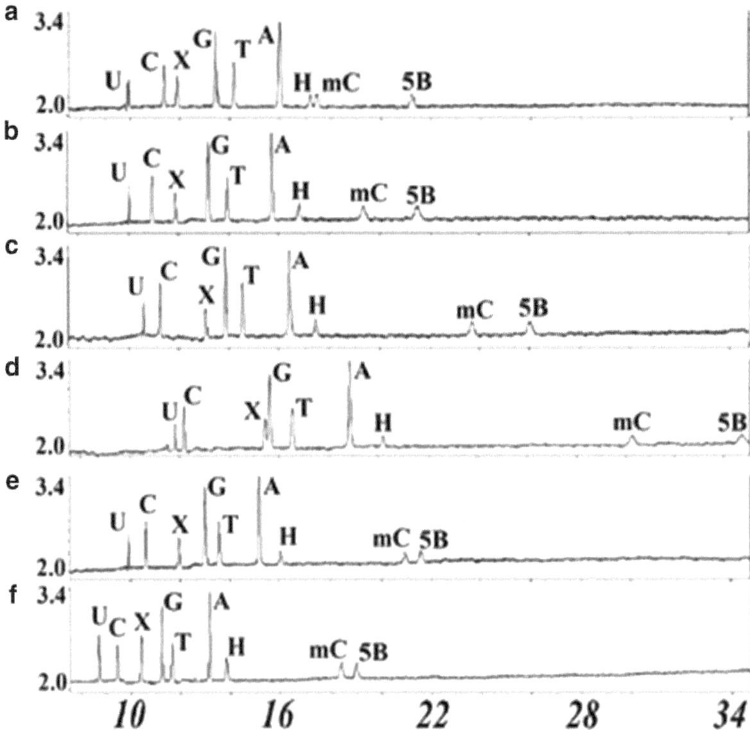

Fig. 1. Electropherograms of the MSM obtained running different buffers at 20 kV as specified in Table 1. **A**: buffer A; **B**: buffer B; **C**: buffer C; **D**: buffer D. Faster elutions of the analytes can be achieved by diluting the buffers with water. **E**: buffer B diluted with water (5:1, v/v) (current at 2 min ≈ 183 μA). This diluted buffer can be also run at 22 kV (current at 2 min ≈ 207 μA) with a further reduction of the retention times (**F**). Traces analyzed after 7.5 min of the run. **5 m**: 5-methyl-Cytosine; **5B**: 5-Br-Uracil. *X* axes: time (min); *Y* axes: mAU (milli-absorbance units) at 254 nm.

usually due to an inaccurate cutting of the capillary. Therefore, carefully check the ends of the capillary by a lent and, if necessary, cut one or both ends again. If quantitative analyses will be performed, a new calibration of the detection system is required after every capillary change.

10. The signals recorded at retention times <8 min are represented by injection artifacts. This noise is easily evidenced by injecting 1% HCl which will be separated and analyzed as usual (see Fig. 2).

11. In the MDQ apparatus (Beckman Coulter, CA, USA) make sure that the reservoir plates are not provided with refrigeration, at least. Therefore, evaporation of the samples occurs after prolonged permanence (e.g., overnight) of the microvials in the apparatus. All this causes an increase in the concentration of the samples leading to inaccurate assessment of their original micromolarity. A well set-up protocol usually shows linearity ($r^2 > 0.998$) between 100 and 12.5 μM. In the routine

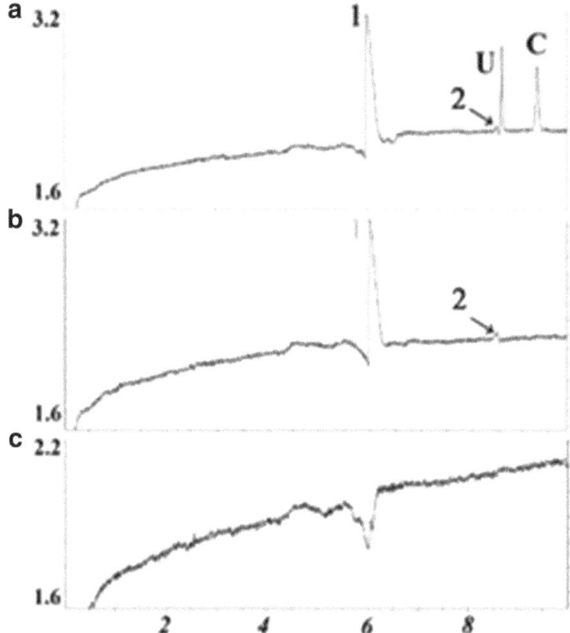

Fig. 2. CE injection artifacts. Two peaks (1 and 2) are always found in the early minutes of the separation (runs at 22 kV with buffer B diluted 5:1 with water). Note that peak 2 is always detectable before U (see also Fig. 1). **A**: MSM; **B**: blank injection control (HCl 1%); **C**: injection of the buffer. *X* axes: time (min); *Y* axes: mAU (milli-absorbance units) at 254 nm.

work, it is usually unnecessary to establish the LOD. However, this can be assessed by the employment of the less concentrated TSSs (6.2, 3.1 and 1.5 μM).

12. Readers interested in all aspects of analytical method validation are addressed to (25) and works cited therein.
13. Calculate the number of injections, minimum volume injected, maximum, range, median, mean, standard error of the mean, confidence interval of the mean ($p=0.05$), variance, standard deviation, relative standard deviation (or CV%).
14. The amount of DNA is intended as assessed by spectrophotometric measurement at 260 nm.
15. We observed that the plastic caps of certain commercial glass tubes release unknown substances which are spiked at 254 nm by the UV-CE detection system. Note also that glycogen (a carrier employed in some DNA precipitation protocols) gives a peak at about 18 min of the run. Therefore, it is recommended to avoid its use.
16. The hydrolysis of the DNA samples is one of the most critical steps of the whole procedure. It is recommended to set up the experimental conditions by hydrolyzing known amounts

(100 μM) of dNTP to assess the mean recovery of each base. Note that hydrolyses under milder conditions than those reported here (e.g., 30 min at 140°C) lead to incomplete releases of the pyrimidinic compounds (C and T) while more drastic hydrolysis conditions afford reduced recoveries of all the basic moieties. Under the conditions suggested here, the mean recovery of the analytes from human DNA is about 0.79 ± 0.07 (23).

17. Stability of the samples was observed after 2 months of storage under these conditions.

18. As reported elsewhere (23), 1 μg of human DNA nominally gives (in a final volume of 25 μL) a solution 24.7 μM of C and G and, respectively, 35.5 μM of T and A. CE analysis provided a ratio between [C]+[G] and [C]+[G]+[T]+[A] of about 0.409 ± 0.014 (23, 24), a result in agreement with the CG % composition of the human genome. In the MEKC analysis of aged/damaged samples it is usual to find uncanonical DNA bases whose qualitative analysis is performable by spiking the sample with 1/10 (v/v) of 4MBS (see also Note 2). The relative amount of unmodified DNA bases (uDNAb) can be calculated by the formula: $(a\mathrm{T} + a\mathrm{C} + a\mathrm{G} + a\mathrm{A})/(a\mathrm{Tot} - a\mathrm{5BrU})$, where a is the peak area of each substance as measured by the Karat software and aTot is the total area obtained by the sum of the areas of all peaks detected at 254 nm after 8 min of run.

Acknowledgments

The authors wish to thank Dr. Sergio G. Tisminetzky and Dr. C. Guarnaccia (ICGEB, Trieste, Italy) for their invaluable support in this work.

References

1. Lindahl T (1993) Instability and decay of the primary structure of DNA. Nature 362:709–715
2. De Bont R, Lorebeke N (2004) Endogenous DNA damage in humans: a review of quantitative data. Mutagenesis 19:169–185
3. Dizdaroglu M (1985) Application of capillary gas-chromatography mass-spectrometry to chemical characterization of radiation-induced base damage of DNA: implications for assessing DNA-repair processes. Anal Biochem 144:593–603
4. Lassen C, Hummel S, Herrmann B (1984) Comparison of DNA extraction and amplification from ancient human bone and mummified soft tissue. Int J Leg Med 107:152–155
5. Pääbo S (1989) Ancient DNA: extraction, characterization, molecular cloning, and enzymatic amplification. Proc Natl Acad Sci USA 86:1939–1943
6. Höss M, Jaruga P, Zastawny TH, Dizdaroglu M, Pääbo S (1996) DNA damage and DNA sequence retrieval from ancient tissues. Nucleic Acids Res 24:1304–1307
7. Fattorini P, Ciofuli R, Cossutta F, Giulianini P, Edomi P, Furlanut M, Previderé C (1999) Fidelity of polymerase chain reaction-direct

sequencing analysis of damaged forensic samples. Electrophoresis 20:3349–3357

8. Previderè C, Micheletti P, Perossa R, Grignani P, Fattorini P (2002) Molecular characterisation of the nucleic acids recovered from aged forensic samples. Int J Legal Med 116:334–339
9. Zimmermann J, Hajibabaei M, Blackburn DC, Hanken J, Cantin E, Posfai J, Evans TC Jr (2008) DNA damage in preserved specimens and tissue samples: a molecular assessment. Front Zool 23:5–18
10. Pizzichini M, Arezzini L, Billarelli C, Carlucci F, Terzuoli L (1998) Determination and separation of allantoin, uric acid, hypoxanthine, and xanthine by capillary zone electrophoresis. Adv Exp Med Biol 431:797–800
11. Alfazema LN, Howells S, Perrett D (1998) Determination of allantoin in biofluids using micellar electrokinetic capillary chromatography. J Chromatogr A 817:345–352
12. Adam T, Friedecký D, Fairbanks LD, Sevcík J, Barták P (1999) Capillary electrophoresis for detection of inherited disorders of purine and pyrimidine metabolism. Clin Chem 45:2086–2093
13. Friedecký D, Adam T, Barták P (2002) Capillary electrophoresis for detection of inherited disorders of purine and pyrimidine metabolism: a selective approach. Electrophoresis 23:565–571
14. Esaka Y, Inagaki S, Goto M (2003) Separation procedures capable of revealing DNA adducts. J Chromatogr B Analyt Technol Biomed Life Sci Chromatogr 797:321–329
15. Ma Y, Liu G, Du M, Stayton I (2004) Recent developments in the determination of urinary cancer biomarkers by capillary electrophoresis. Electrophoresis 25:1473–1484
16. Zhang L, Liu Y, Chen G (2004) Simultaneous determination of allantoin, choline and l-arginine in Rhizoma Dioscoreae by capillary electrophoresis. J Chromatogr A 1043:317–321
17. Cheng ML, Liu TZ, Lu FJ, Chiu DTY (1999) Simultaneous detection of vitamin C and uric acid by capillary electrophoresis in plasma of diabetes and in aqueous humor in acute anterior uveitis. Clin Biochem 32:473–476
18. Terzuoli L, Porcelli B, Setacci C, Giubbolini M, Cinci G, Carlucci F, Pagani R, Marinello E (1999) Comparative determination of purine compounds in carotid plaque by capillary zone electrophoresis and high-performance liquid chromatography. J Chromatogr B 728:185–192
19. Carlucci F, Tabucchi A, Biagioli B, Sani G, Lisi G, Maccherini M, Rosi F, Marinello E (2000) Capillary electrophoresis in the evaluation of ischemic injury: simultaneous determination of purine compounds and glutathione. Electrophoresis 21:1552–1557
20. Zinellu A, Carru C, Sotgia S, Deiana L (2004) Optimization of ascorbic and uric acid separation in human plasma by free zone capillary electrophoresis ultraviolet detection. Anal Biochem 330:298–305
21. Kattygnarath D, Mounier N, Madelaine-Chambrin I, Gourmel B, Le Bricon T, Gisselbrecht C, Faure P, Houze P (2006) Quantification of urinary allantoin by capillary zone electrophoresis during recombinant urate oxydase (rasburicase) therapy. Clin Biochem 39:86–90
22. Silva M (2011) Micellar electrokinetic chromatography: a practical overview of current methodological and instrumental advances. Electrophoresis 32:149–165
23. Fattorini P, Marrubini G, Ricci U, Gerin F, Grignani P, Sorçaburu Cigliero S, Xamin A, Edalucci E, La Marca G, Previderé C (2009) Estimating the integrity of aged DNA samples by CE. Electrophoresis 30:3986–3995
24. Fattorini P, Marrubini G, Sorçaburu Cigliero S, Pitacco P, Grignani P, Previderè C (2011) CE analysis and molecular characterization of depurinated DNA samples. Electrophoresis 32:3042–3052
25. Araujo P (2009) Key aspects of analytical method validation and linearity evaluation. J Chromatogr B 877:2224–2234

Index

A

B

C

Nicola Volpi and Francesca Maccari (eds.), *Capillary Electrophoresis of Biomolecules: Methods and Protocols*, Methods in Molecular Biology, vol. 984, DOI 10.1007/978-1-62703-296-4, © Springer Science+Business Media, LLC 2013

H

I

K

L

M

MIX
Papier aus verantwortungsvollen Quellen
Paper from responsible sources
FSC® C105338

If you have any concerns about our products,
you can contact us on
ProductSafety@springernature.com

In case Publisher is established outside the EU,
the EU authorized representative is:
Springer Nature Customer Service Center GmbH
Europaplatz 3, 69115 Heidelberg, Germany

Printed by Libri Plureos GmbH
in Hamburg, Germany